Software Defined Radio

WILEY SERIES IN SOFTWARE RADIO

Series Editor: Dr Walter Tuttlebee, Mobile VCE, UK

The Wiley Series in Software Radio aims to present an up-to-date and in-depth picture of the technologies, potential implementations and applications of software radio. Books in the series reflect the strong and growing interest in this subject. The series is intended to appeal to a global industrial audience within the mobile and personal telecommunications industry, related industries such as broadcasting, satellite communications and wired telecommunications, researchers in academia and industry, and senior undergraduate and postgraduate students in computer science and electronic engineering.

Mitola: Software Radio Architecture: Object-Oriented Approaches to Wireless Systems Engineering, 0471384925, 568 Pages, October 2000
Mitola and Zvonar (Editors): Software Radio Technologies: Selected Readings: 0780360222, 496 pages, May 2001
Tuttlebee: Software Defined Radio: Origins, Drivers and International Perspectives, 0470844647, £65, 350 pages, 2002
Tuttlebee: Software Defined Radio: Enabling Technologies, 0470843187, £65, 304 pages, 2002
Dillinger, Madani and Alonistioti (Editors): Software Defined Radio: Architectures, Systems and Functions, 0470851643, £85, 456 pages, 2002
Tuttlebee: Software Defined Radio: Baseband Technology for 3G Handsets and Basestations, 0470867701, £75, 275 pages

Software Defined Radio

Baseband Technologies for 3G Handsets and Basestations

Edited by

Walter H.W. Tuttlebee

Virtual Centre of Excellence in Mobile and Personal Communications (Mobile VCE)

John Wiley & Sons, Ltd

Telephone (+44) 1243 779777

Email (for orders and customer service enquiries): cs-books@wiley.co.uk
Visit our Home Page on www.wileyeurope.com or www.wiley.com

Other Wiley Editorial Offices

John Wiley & Sons Inc., 111 River Street, Hoboken, NJ 07030, USA

Jossey-Bass, 989 Market Street, San Francisco, CA 94103-1741, USA

Wiley-VCH Verlag GmbH, Boschstr. 12, D-69469 Weinheim, Germany

John Wiley & Sons Australia Ltd, 33 Park Road, Milton, Queensland 4064, Australia

John Wiley & Sons (Asia) Pte Ltd, 2 Clementi Loop #02-01, Jin Xing Distripark, Singapore 129809

John Wiley & Sons Canada Ltd, 22 Worcester Road, Etobicoke, Ontario, Canada M9W 1L1

Library of Congress Cataloging-in-Publication Data

A catalog record for this book is available from the Library of Congress

British Library Cataloguing in Publication Data

A catalogue record for this book is available from the British Library

ISBN 0-470-86770-1

Typeset in 10/12pt Times by Integra Software Services Pvt. Ltd, India
Printed and bound in Great Britain by TJ International Ltd, Padstow, Cornwall
This book is printed on acid-free paper responsibly manufactured from sustainable forestry
in which at least two trees are planted for each one used for paper production.

Contents

List of Contributors

This Section provides author contact details for all chapters of the book. (For most chapters there are additional contributors, but for brevity, only a single contact person is listed for each).

Rupert Baines
Vice President, Marketing
picoChip Designs Ltd
UK
rubertb@picochip.com

Stephen Blust
Director of Wireless Standards
Cingular Wireless
USA
stephen.blust@cingular.com

Mark Cummings
Senior VP Strategic Development
RFco, Inc.
USA
mark.cummings@envia.com

Chris Dick *et al.*
Xilinx Chief DSP Architect
Director, Signal Processing Engineering
Xilinx
USA
chris.dick@xilinx.com

Paul Ekas
Sr DSP Marketing Manager
Altera Corporation
USA
pekas@altera.com

David Chou
Quicksilver Technology
USA
david.c@qstech.com

Alan Gatherer *et al.*
Texas Instruments Inc
USA
gatherer@ti.com

John Glossner *et al.*
CTO & EVP, Engineering
Sandbridge Technologies, Inc.
USA
jglossner@sandbridgetech.com

Michael Lopez *et al.*
Wireless System Engineer
Embedded DSP Division
Analog Devices
USA
michael.lopez@analog.com

Behzad Mohebbi *et al.*
Director of the Communication Systems Group
Morpho Technologies
USA
behzad@morphotech.com

Jitendra Rayala *et al.*
LSI Logic Inc
USA
jrayala@lsil.com

Eberhard Schüler *et al.*
Product Marketing Manager
PACT XPP Technologies AG
Germany
eberhard.schueler@pactxpp.com

Walter Tuttlebee
Chief Executive
Virtual Centre of Excellence in Mobile & Personal Communications (Mobile VCE)
UK
walter.tuttlebee@mobilevce.com

Zoran Zvonar ***et al.***
RF & Wireless Systems
Analog Devices
USA
zoran.zvonar@analog.com

Foreword

This volume in the Wiley series on Software Defined Radio drills down an additional level beyond the earlier texts into the technology reality required for practical SDR to come into its own in the marketplace. In the Wiley volume, *Enabling Technologies*, Chapter 2, an architecture view that discussed the division of the SDR into a 'front-end' (radio frequency aspects for transmit and receive) and a 'back-end' (the signal processing functionality) was espoused as a simple way of looking at the key areas of development. This present volume investigates the variety of approaches which is becoming available to satisfy the design of an SDR and takes a major step forward – treating the signal processing solutions in some depth. The progress and approaches described in these pages demonstrate that we are not simply seeing simple evolution of DSP, but rather something more profound which will have impact potentially beyond simply radio.

When I envisioned SDR in a commercial wireless mobile communication regime in the BellSouth *Software Defined Radio RFI* (Request For Information) which was released in December 1995, I was thinking how digital signal processing in the classical sense (architectures, performance, power usage, cost, size, etc.) would have to evolve in order to be a practical solution to a wide-ranging SDR universe. In particular, at the time, specific questions were posed to the industry postulating that the evolution of SDR would rely on advances in three important technology segments:

- semiconductor physics and geometry;
- architectures, topologies, and circuits;
- algorithms.

Further, the RFI theorized that the developments in physics and architectures together would enable hardware enhancements leading to hardware platforms (presumably optimized) for SDR. Architecture and algorithms together would enable software operating environments and software applications for SDR. The RFI stipulated that at some level, standards might be required to aid in timely deployment and might include a standard SDR hardware platform with standard hardware interfaces, a standard SDR operating system, or a standard SDR application programming interface (API). Responders were challenged to address research and development in the wireless and DSP industry, focusing on four major areas of enabling technology as it was believed that there was room for significant innovation. They were asked to comment upon the core technologies and, if appropriate, provide roadmaps for evolution including implications on memory, size, power and cost for:

- DSP devices;
- analog to digital converters;

- integrated IF/RF components and subsystems;
- development tools.

Needless to say, at the time, the responses were underwhelming, either due to the newness of the whole of digital signal processing in the field of communications and commercial wireless, or due to the lack of vision of what signal processing and its advances would ultimately allow designers to accomplish in the radio field. Perhaps some of the questions posed in the RFI stimulated some of the research and application development that has evolved. In any event, the intervening years have been fruitful. Standards are emerging in key areas that are meaningful for promoting commercialization without being burdensome to the detailed and individualized approaches being pursued.

It is gratifying to see that not only has traditional DSP emerged in this solution space being optimized for SDR needs, but also that other nontraditional/nonclassical approaches for radio signals, such as reconfigurable hardware and computing structures have come to the fore on an essentially equally footing.

The message is clear from these pages – there is a variety of solutions each having their own merits and drawbacks – not a bad turn of events, as the richness of SDR applicability in specific applications such as basestations, terminal devices (handheld), and mobile (vehicular type) units serving commercial wireless, civil government and defense sectors, can certainly benefit from one or more of these approaches. Whether or not any particular approach should become dominant in any specific application/solution space properly remains a decision for the designer and the marketplace.

My request and hope is that the solutions to the traditional hardware-constrained radio front-end are just as forthcoming, especially to extend the capability/flexibility of the band filtering, amplifying, and digital-to-analog and analog-to-digital conversions to be concomitant with the flexibility being demonstrated in the signal processing in the back-end. This would address some of the lingering questions that I posed back in 1995, but perhaps that is the subject of another text in this series. Walter, are you listening?

Stephen M. Blust
September 2003

Abbreviations

In addition to the glossary below of abbreviations and acronyms used in this book, a comprehensive set of definitions and other useful resources may be found online at *http://www.umts-forum.org/glossary.asp* and at *www.softwaredefinedradio.org*

2G	2nd Generation cellular mobile systems, refers to the digital systems such as GSM, DAMPS and IS-95
3G	3rd Generation cellular mobile systems, refers to the evolving UMTS, cdma2000, and TD-SCDMA standards within the ITU-R family of 3G standards
3GPP	3rd Generation Partnership Project – see www.3GPP.org
4G	Fourth Generation Wireless, often referred to as Beyond 3G, or B3G
A/D	Analog-to-Digital converter
ACELP	Algebraic Code Excited Linear Prediction
ACLR	Adjacent Channel Leakage power Ratio
ACM	Adaptive Computing Machine
ACS	Add-Compare-Select
ACSU	Add-Compare-Select Unit
ADC	Analogue-to-Digital Converter
AES	Advanced Encryption Standard
AFC	Automatic Frequency Control
AGC	Automatic Gain Control
AGU	Address Generation Unit
AHB	Advanced High-performance Bus (AHB), a standardized bus system for SoC, developed by ARM Ltd
AIS	Air Interface Standard
AKA	Authentication and Key Agreement
ALU	Arithmetic Logic Unit
ALU-PAE	XPP Element containing an ALU object
AM	Amplitude Modulation
AML	Application Module Library
AMPP	Altera Megafunction Partners Program
AMPS	American Mobile Phone Standard, a first-generation analog mobile phone standard
AMR	Adaptive Multi Rate speech coder
AMR-NB	Narrow-Band AMR
AN2	Adaptable Node 2

API	Application Programming Interface
ARM	Proprietary microprocessor, made by ARM Ltd, commonly used in cellular handset applications, often as an IP core
ASIC	Application Specific Integrated Circuit
ASP	ACM Simulation Platform
ASR	Automatic Speech Recognition
ASSP	Application Specific Semiconductor Product
Async	Asynchronous data (individual units of data do not have to arrive in order or with limited latency)
AWGN	Additive White Gaussian Noise
AXN	Adaptable X Node
BAP	Baseband Algorithms Processor
Baudot Code	Coding system that transitioned from telegraph (Morse code) to computer code (ASCII)
BDTI	Berkeley Design Technology, Inc.
BER	Bit Error Rate
Bluetooth	A short range wireless PAN AIS
BMU	Branch Metric Unit
BPSK	Binary Phase-Shift Keying
BREG	Object with a PAE
BTS	Base Transceiver Station
C, C++	object oriented programming language
CAN	Controller Area Network
C-API	Application Programming Interface for C
CCK	Complementary-Code Keying
CDMA	Code Division Multiple Access
cdma2000	Qualcomm's IS-95 evolution standard – see www.3GPP2.org
CDR	Clock Data Recovery
CELP	Code Excited Linear Prediction
CIC	Cascaded-Integrated-Comb filter
CISC	Complex Instruction Set Computer
CLB	Configurable Logic Block
CLU	Communications Logic Unit
CMA	Constant Modulus Algorithm
CMAC	Complex Multiply-Accumulate
CMOS	Complementary Metal Oxide Semiconductor
CORBA	Common Object Request Broker Architecture – see http://www.omg.org/gettingstarted/corbafaq.htm
CPICH	Common PIlot CHannel
CPU	Central Processing Unit
CRC	Cyclic Redundancy Check
CVS	Concurrent Versioning System
D/A	Digital-to-Analog converter
DA	Distributed Arithmetic
DAC	Digital-to-Analog Converter
DAN	Domain Arithmetic Node

DBB	Digital Baseband
DBN	Domain Bit Manipulation Node
DCCH	Dedicated Control CHannel
DCT	Discrete Cosine Transform
DDC	Digital Down-Converter
DDR	Double-Data Rate
DFE	Decision Feedback Equalizer
DFN	Domain Filter Node
DIF	Digital IF
DL	DownLink
DLL	Delay Lock Loop
DM	Control Signal for DDR and QDR RAMs
DMA	Direct Memory Access
DMU	Data Memory Unit
DPCCH	Dedicated Physical Control CHannel, a UMTS channel for communicating data formats, parity bits, and other control information
DPDCH	Dedicated Physical Data CHannel, a UMTS channel for communicating data information
DQ	Control signal for DDR and QDR RAMs
DQS	Control signal for DDR and QDR RAMs
DRAM	Dynamic Random Access Memory
DSP	Digital Signal Processing or Processor
DTX	Discontinuous Transmission
DUC	Digital Up-Converter
DVB	Digital Video Broadcasting
EBI	Expansion Bus Interface
EDGE	Enhanced Data for GSM Evolution
EEMBC	Embedded Microprocessor Benchmark Consortium
EFR	Enhanced Full Rate
ETSI	European Telecommunications Standards Institute
EU	Execution Unit
FCC	Federal Communication Commission (US spectrum regulator)
FCRAM	Fast Cycle RAM
FDD	Frequency Division Duplex
FDR	Franklin Delano Roosevelt (President of the USA during World War II)
FEC	Forward Error Correction
FEQ	Frequency EQualizer
FFE	Feed Forward Equalizer
FFT	Fast Fourier Transform
FFT64	Fast Fourier Transform on 64 samples
FIFO	First In, First Out Memory
FIR	Finite Impulse Response Filter
FM	Frequency Modulation
FOMA	NTT DoCoMo's Commercial 3G WCDMA Service, launched October 2001
FPGA	Field Programmable Gate Array
FPPA	Field Programmable Processor Array

FREG	Object with a PAE
FSBM	Full Search Block Matching
FSE	Fractionally Spaced Equalizer
FSK	Frequency-Shift Keying
GFLOPS	GigaFlops (Floating Point Operations Per Second)
GIPS	Giga Instructions Per Second
GMSK	Gaussian Minimum-Shift Keying
GPP	General Purpose Processor
GPRS	General Packet Radio Service
GPS	Global Positioning System
GSM	Global System for Mobile Communication, the predominant 2G digital cellular standard in Europe and many other places as of the writing of this book
GSP	Generic Serial Port
HDL	Hardware Description Language
HDTV	High Definition Television
HLL	High Level Language
HMM	Hidden Markov Model
HSCSD	High Speed Circuit Switched Data
HSDPA	High Speed Downlink Packet Access (feature of the 3GPP WCDMA specification)
HSSP	High Speed Signal Processing
Hz	Hertz (cycles per second)
I/O	Input/Output
IC	Integrated Circuit
ICU	Instruction Cache Unit
IDCT	Inverse Discrete Cosine Transform
IDE	Integrated Development Environment
IF	Intermediate Frequency
IFFT	Inverse Fast Fourier Transform
IIR	Infinite Impulse Response Filter
ILP	Instruction Level Parallelism
IMC	Internal Memory Controller
INT	Integer Unit
IO	Input/Output
IOC	Input/Output Controller
IOE	Input/Output Element
IP	Intellectual Property
IS-95	2G digital Standard for CDMA requiring fall back to analog mode if CDMA is not available
ISA	Instruction Set Architecture
ISI	InterSymbol Interference
Isoc	Isochronous data (individual units of data have to arrive in order with strictly limited latency)
ISR	Ideal Software Radio
ITU	International Telecommunications Union
JD	Joint Detection, a multi-user detection method

JHDL	Java Hardware Description Language
JVM	Java Virtual Machine
LAB	Logic Array Block
LAN	Local Area Network
LBS	Location Based Services
LCC, LCCP	Loosely Coupled Coprocessor
LCD	Liquid Crystal Display
LE	Logic Element
LIU	Load/Integer Unit
LMS	Least Mean Square, an adaptive filter estimation method
LPC	Linear Prediction Coding
LPM	Library of Parameterized Modules
LRU	Least Recently Used replacement algorithm for caches
LSI	Large Scale Integrated Circuit
LSSP	Low Speed Signal Processing
LUT	LookUp Table
LVDS	Low Voltage Differential Signalling
LVPECL	Low Voltage Positive Emitter Coupled Logic
MAC	Multiply-ACcumulate operation found in most DSPs, or in the context of a radio air interface, Medium Access Control
MAP	Maximum *A Posteriori*, or Media Algorithms Processor
MFLOP	MegaFlops (Floating Point Operations Per Second)
MGT	Multi-Gigabit Transceiver
MIMD	Multiple Instruction Multiple Data processing architecture
MIMO	Multiple-In, Multiple-Out, usually referring to wireless communication when both transmitter and receiver have multiple antennas
MIN	Matrix Interconnect Network
MIPS	Mega Instructions Per Second
MLSE	Maximum Likelihood Sequence Estimation
MMACS	MegaMACS (MACs per second)
MMS	Multimedia Message Service
MMSE	Minimum Mean-Square Error
MMSE-BLE	Minimum Mean-Square Error Block Linear Equalizer
MP3	Audio compression scheme
MPEG	Moving Pictures Experts Group
MPEG4	A video compression scheme developed by MPEG
MRC	Maximal Ratio Combiner or Combining
MSK	Minimum-Shift Keying
MUD	Multi-User Detection
MUX	Multiplex
NCO	Numerically Controlled Oscillator
NIST	National Institute for Standards and Technology
NMI	Non Maskable Interrupt
NML	Native Mapping Language for XPP
NRE	Non Recurring Expenditure
NTSC	US analog television standard

OFDM	Orthogonal Frequency Division Multiplexing
OS	Operating System
OVSF	Orthogonal Variable Spreading Factor
PAC	Processing Array Cluster
PAE	Processing Array Element
PAL	UK analog television standard
PAM	Pulse-Amplitude Modulation
PAN	Personal Area Network
PC	Personal Computer
PCI	Peripheral Component Interconnect, a common computer bus structure
PCML	Pseudo Current Mode Logic
PDA	Personal Digital Assistant
PFU	Program Function Unit
PG	Processing Gain
PHY	Physical Layer
PLD	Programmable Logic Device
PLL	Phase Locked Loop
PM	Phase Modulation
PPC	Power PC
ppm	Parts Per Million
PSK	Phase-Shift Keying
PSN	Programmable Scalar Node
QAM	Quadrature Amplitude Modulation
QDR	Quad Data Rate
QoS	Quality of Service
QPSK	Quadrature Phase Shift Keying
RAB	Radio Access Bearer
RACH	Random Access CHannel
RAM	Random Access Memory
RAM-PAE	XPP element containing a RAM object
RC	Reconfigurable Cell
RCF	Reconfigurable Fabric
RF	Radio Frequency
RF FE	Radio Frequency Front-End
RISC	Reduced Instruction Set Computer
RLS	Recursive Least Squares, an adaptive filter estimation method
RRC	Root-Raised Cosine
RTL	Register Transfer Logic
RTOS	Real-time Operating System
SAD	Summing of Absolute Difference
SaDL	Sandbridge Architecture Description Language
SATS	Spatial and Temporal Segmentation
SCA	Software Communications Architecture
SCR	Software Controlled Radio
SDR	Software Defined Radio
SDRAM	Synchronous Dynamic RAM

SDRXPP	System on chip with XPP-core for SDR
SECAM	European analog television standard
SERDES	Serializer–Deserializer
SF	Spreading Factor
SIMD	Single Instruction Multiple Data processing architecture
SIR	Signal to Interference
SISO	Single-Input, Single-Output
SMS	Short Message Service
SMT	Simultaneous Multithreading
SMU	Survivor Memory Unit
SNR	Signal to Noise Ratio
SoC	System on Chip
SoPC	System on Programmable Chip
SOVA	Soft Output Viterbi Algorithm
SRAM	Static Random Access Memory
SRL16	Shift Register Logic 16, a processing element used in the Xilinx Virtex FPGA architecture
SRRC	Square Root Raise Cosine
SS7	Switching System 7 (common telephone switching system interface)
SSoC	Software System on Chip
STTD	Space Time Transmit Diversity
SW	SoftWare
T3	Token Triggered Threading
TCCP	Tightly Coupled CoProcessor
TCI	Texas Instruments' wireless infrastructure modem platform
TDM	Time Division Multiplexed
TDMA	Time Division Multiple Access
TD-SCDMA	Time Division Synchronous Code Division Multiple Access, an evolving 3G cellular standard being developed in China
TEP	Timing and Event Processor
TFCI	Transport Format Combination Indicator
TID	Thread Identifier register
TOA	Time of Arrival
TPC	Transmit Power Control
TTI	Transmission Timing Interval
TTM	Time To Market
UART	Universal Asynchronous Receive and Transmit
UE	User Equipment
UL	Uplink
UMTS	Universal Mobile Telecommunication System, a 3G cellular standard based on CDMA technology that increases both the data rates and flexibility of earlier standards. Variants include UMTS-FDD and UMTS-TDD (for Frequency Division Duplex and Time Division Duplex)
USR	Ultimate Software Radio
VA	Viterbi Algorithm

VHDL	Very High speed integrated circuits hardware Description Language, a language for describing hardware systems
VHSIC	Very High Speed Integrated Circuit
VLIW	Very Long Instruction Word (processor or architecture)
VLSI	Very Large Scale Integrated Circuit
VOIP	Voice Over Internet Protocol
VPU	Vector SIMD Parallel Unit
WAP	Wireless Application Protocols
WCDMA	Wideband Code Division Multiple Access (generally associated with 3G AISs)
WLAN	Wireless Local Area Network
WMSA	Weighted Multi-Slot Averaging
WPAN	Wireless Personal Area Network
WSSUS	Wide-Sense Stationary Uncorrelated Scattering
XMC	External Memory Controller
XPP	Pact's extreme processor technology
ZBT	Zero Bus Turnaround
ZF-BLE	Zero-Forcing Block Linear Equalizer
ZIF	Zero IF
ZigBee	WPAN AIS

Biographies

Series and Book Editor

Walter Tuttlebee
Mobile VCE-Virtual Centre of Excellence in Mobile & Personal Communications

As chief executive of the Virtual Centre of Excellence in Mobile & Personal Communications – Mobile VCE – Walter Tuttlebee heads up a unique, not-for-profit company established by the mobile communications industry and academia to undertake long-term, industry-steered, world-class, collaborative research (www.mobilevce.com). Mobile VCE's activities include software radio research, an area Walter helped to pioneer in Europe in the mid-1990s, with invited presentations at seminal European conferences organized by the European Commission and the SDR Forum. He has subsequently published and spoken widely in the field. Prior to Mobile VCE Walter led R&D teams in Second and Third generation mobile communications. Aside from his technical interests, Walter previously operated in a business development role and at Mobile VCE he is responsible to the Board for the company's strategy and operations.

Walter has also edited books on short range digital wireless, and created on-line industry communities for DECT, Bluetooth and software radio – www.dectweb.org, www.thewireless directory.org, www.softwaredefinedradio.org

He holds an MBA from Cranfield and PhD from Southampton University, is a senior member of the IEEE, a fellow of the IEE and a fellow of the RSA.

Contributors

Rupert Baines
picoChip

Rupert Baines is Vice President of Marketing at picoChip. He spent 6 years with Analog Devices in Boston, working on GSM and wireless infrastructure, which included authoring a paper in the seminal IEEE Communications Magazine issue on Software Defined Radio. He then started their Broadband product line, holding responsibility for the industry-standard ADSL chipset and being directly involved in technology as it moved from research through scepticism to mass-market. He has also worked for operators, most recently as Director of Product Development and Strategy for Atlantic Telecom, where he ran a large pan-European DSL rollout, a MVNO and trialled wireless broadband deployments.

Kathy Brown
Texas Instruments

Kathy Brown is the design manager for the Wireless Infrastructure Business at Texas Instruments, Dallas, responsible for development of IP for 3G wireless infrastructure. Since joining TI in 1998, Kathy has been involved in the design of various CDMA/WCDMA digital baseband modems for terminals and basestations, and has contributed to several patents in this area. Kathy received a BSEE from Stanford University in 1983, and an MSEE from the University of Washington in 1995.

David Chou
QuickSilver

David Chou is a senior member of technical staff in the Architecture Group at Quicksilver Technology and is responsible for methodologies for mapping algorithms onto the ACM. Mr Chou has over 7 years experience in system design and implementation in digital signal processing. Prior to Quicksilver Technology, Mr Chou was a senior engineer at Applied Signal Technology, building all software CDMA radios. His background is in wireless communications (3G, 2G, WLAN) and networking (optimal routing policies). Mr Chou holds a master's degree in electrical engineering and applied mathematics from Columbia University and a BSEE from the University of Iowa.

Mark Cummings
RFCo Inc

Mark Cummings launched MorphICs (SDR baseband processors), SkyCross (SDR antennas) and RFco (SDR RF Front Ends). Mark is the principal inventor on the earliest patent granted on the use of reconfigurable logic for SDR. He chaired the organizing committee of the SDR Forum, was its first steering committee chair, and technical committee chair. He is currently the chairman of the board of the SDR Forum and a special advisor to the IEICE Software Radio Study Group. He has worked in common carriers, large-end users, equipment vendors and in a university as a professor. He helped found IEEE 802, PCMCIA, Smart Card Industry Association, IrDA, WINForum and contributed to X.25. He has over 150 publications in communications and computing.

Chris Dick
Xilinx Inc

Dr Chris Dick is the director of Signal Processing Systems Engineering and the DSP chief architect at Xilinx. Dr Dick joined Xilinx in 1997 from La Trobe University in Melbourne Australia where he was a professor for 13 years and consulted for industry and the military sector. Chris has over 70 journal and conference publications and has been an invited speaker at many international DSP and communications symposiums. Chris's research interests are in the areas of fast algorithms for signal processing, digital communication, software defined radios, hardware architectures for real-time signal processing, parallel computing, inter-connection networks for parallel processors, and the use of Field Programmable Gate Arrays (FPGAs) for custom computing machines and signal processing. Chris is active in the software defined radio area and is on the Software Defined Radio Forum's board of directors. He holds a bachelor's and PhD degrees in the areas of computer science and electronic engineering.

Paul Ekas
Altera

Paul Ekas joined Altera in August 2002 to lead the Altera Code:DSP corporate DSP initiative. Mr Ekas has more than 17 years of business experience in electronic design automation and complex semiconductor systems. Most recently Mr Ekas was the director of product marketing at MorphICs Technology where he was responsible for the 3G WCDMA infrastructure product line. Prior to joining MorphICs, Mr Ekas was with Cadence/Alta/Comdisco Systems where he was responsible for the SPW product line. Mr Ekas has also held sales and engineering positions in Mentor Graphics, Silicon Designs, and Seattle Silicon. Mr Ekas has a BSEE and MSEE from the University of Washington.

Jose Fridman
Analog Devices Inc.

Jose Fridman received a BS in 1987 and an MS in 1991 from Boston University, Boston Massachusetts, and a PhD degree in 1996 from Northeastern University, Boston, Massachusetts, all in electrical engineering. He has been with Analog Devices, Inc., since 1996, and is currently engaged in the development of platforms for wireless handsets. Jose has been an architect for two of the most recent DSP architectures at ADI, Blackfin and TigerSHARC. Jose has served as chair for the IEEE Industry DSP Committee (IDSP), a standing committee of the IEEE Signal Processing Society, and has served as chair for ICASSP Industry and Technology Tracks (ITT) 2001 and 2002. Jose is also serving as a member of the IEEE Design and Implementation of Signal Processing Systems (DISPS) Committee.

Alan Gatherer
Texas Instruments

Alan Gatherer received his BEng degree in electronic and microprocessor engineering from Strathclyde University (Scotland) in 1988 after which he moved to Stanford University, obtaining MS and PhD degrees, both in electrical engineering, in 1989 and 1993. He joined Texas Instruments in 1993, since when he has worked on digital communications research and development in the areas of digital subscriber line, cable modem and wireless. Alan Gatherer is today a distinguished member of technical staff and manager of systems development within wireless infrastructure at Texas Instruments where he leads a team involved in the development of technology for Third-generation cellular telephony. He holds 11 patents in the field of digital communications.

John Glossner
Sandbridge

John Glossner is CTO & EVP of Engineering at Sandbridge Technologies. Prior to co-founding Sandbridge, John managed the Advanced DSP Technology group, Broadband Transmission

Systems group, and was Access Aggregation Business Development Manager at IBM's T.J. Watson Research Center. Prior to IBM, John managed the software effort in Lucent/ Motorola's Starcore DSP design center. John received a PhD in computer architecture from TU Delft in the Netherlands for his work on a multithreaded Java processor with DSP capability. He also received an MS degree in engineering management and an MSEE from NTU. John also holds a BSEE degree from Penn State. John has more than 50 publications and 12 issued patents.

Jun Han
QuickSilver

Jun Han received the PhD degree in Electrical Engineering from the University of California, San Diego (UCSD) in 2002. He is a senior engineer at QuickSilver Technology, working on the mapping of application modules onto ACM, including generic function library and application specific functions in W-CDMA and OFDM systems. He is also a visiting scholar with the NSF Industry/University Cooperative Research Center on Ultra-High Speed Integrated Circuits and Systems (ICAS) at UCSD. His research interests include digital signal processing algorithms in wireless communications and performance analysis of adaptive filters.

fred harris
San Diego State University

fred harris received a BS degree from the Polytechnic Institute of Brooklyn, Brooklyn, NY, in 1961, and a MS degree from San Diego State University, San Diego, CA, in 1967, and the course work required for the PhD degree from the University of California at San Diego, La Jolla, in 1975, all in electrical engineering. He holds the CUBIC Signal Processing Chair of the Communication Systems and Signal Processing Institute, San Diego State University, where he has taught since 1967 in areas related to DSP and communication systems. He holds a number of patents on digital receivers and related DSP technology. He lectures throughout the world on DSP applications to communication systems, as well as consulting for organizations requiring high-performance DSP systems. He has authored and co-authored numerous papers and has contributed to a number of books on DSP and communication systems.

Erdem Hokenek
Sandbridge

Erdem Hokenek received BS and MS degrees from Technical University, Istanbul (Turkey) and PhD from Swiss Federal Institute of Technology (ETH Zurich, Switzerland). After his PhD in 1985, he joined IBM T. J. Watson Research Center where he worked on the advanced development of POWER and PowerPC processors for the RS/6000 Workstations. He also worked in various technical and management positions on the high performance compilable DSP and Cross Architecture Translations. He is co-founder of Sandbridge Technologies Inc.

Jim Hwang
Xilinx Inc.

Jim Hwang received his PhD in electrical engineering from Stanford University, and has been at Xilinx since 1994. He is chief architect of the Xilinx System Generator, and has published papers and holds over 20 patents (issued and pending) in the areas of physical design, module generation, and system level design. His professional interests include system level design tools and methodologies, signal processing in FPGAs and design technology.

Fadi J. Kurdahi
Morpho Technologies and University of California Irvine

Fadi received his PhD from the University of Southern California, in 1987. Since 1987, he has been with the University of California, Irvine, as a professor of EECS and ICS. From 2000 to 2002, he was on leave from UCI working at Morpho Technologies developing reconfigurable computing solutions for communication appliances. He was associate editor of *IEEE Transactions on Circuits and Systems* (1993–95) and is currently editor of IEEE's *Design and Test Magazine* in charge of reconfigurable computing. He was program chair of the 1999 International Symposium on System Synthesis and general chair in 2000. He was a co-recipient of three distinguished paper awards and the best paper award for the 2002 *IEEE Transactions on VLSI.*

Michael J. Lopez
Analog Devices Inc.

Michael J. Lopez received BS and MSE degrees in 1995 from Johns Hopkins University and a PhD in 2002 from the Massachusetts Institute of Technology, all in electrical engineering. From 1995 to 1996 he was with the Interactive Videoserver group at Digital Equipment Corporation, and from 1998 to 2000 held internships at Sanders, a Lockheed Martin Company, Ericsson and Analog Devices. Since 2002, he has been with the DSP Wireless Infrastructure Systems Engineering group at Analog Devices. His research interests include communications signal processing algorithms and DSP architecture.

Filip Moerman
Texas Instruments

Filip Moerman received an MS degree in electrical engineering from the Free University of Brussels in 1991 and has worked for Alcatel Telecom and Alcatel Space Industries in 1995, where he contributed to the definition of the physical layer of SkyBridge, a worldwide, broadband, access network to multimedia services via a LEO satellite constellation. He joined the Wireless Infrastructure Business Unit of Texas Instruments in 2000, where he is working on DSP-based architectures for 3G base stations.

Behzad Mohebbi
Morpho Technologies

Behzad received his PhD in telecommunications engineering from the University of Leeds, UK in 1991. After several years of research as a research associate at the University of Bradford and King's College (London), he worked at companies such as Motorola and Fujitsu as a member of the research team. Behzad joined Morpho Technologies, a fabless semiconductor company, in 2000, where he is in charge of the Wireless Communications Group, studying the application of reconfigurable DSPs to the field of signal processing for wireless communication systems. His interests are software defined radio, digital signal processing and radio propagation science. He is a member of the IEEE.

Mayan Moudgill
Sandbridge

Mayan Moudgill obtained a PhD in computer science from Cornell University in 1994, after which he joined IBM at the Thomas J. Watson Research Center. He worked on a variety of computer architecture and compiler related projects, including the VLIW research compiler, Linux ports for the 40x series embedded processors and simulators for the Power 4. In 2001, he co-founded Sandbridge Technologies, a start-up that is developing digital signal processors targeted at 3G wireless phones.

Jasmin Oz
QuickSilver

Jasmin Oz is a Technical Program Manager at QuickSilver Technology providing support to application developers for the ACM. She has 6 years' experience in the area of algorithm, software and hardware design for signal processing systems. Prior to QuickSilver Technology, she was a DSP software applications engineer at Motorola Semiconductors, working on hardware and software solutions for channel coding algorithms for the Motorola's SC140 DSP. Before Motorola she worked at ELTA, Israel, EW division. She holds a PhD in electrical engineering from Tel Aviv University, a MSc in Physics and a BSc in biology, both from the Technion.

Jitendra Rayala
LSI Logic Corporation

Jitendra Rayala is a staff engineer in the DSP Solutions Engineering Department of LSI Logic Corporation, responsible for the design and development of new algorithms, applications and wireless systems architectures based on ZSP digital signal processors. Jitendra has more than 10 years' experience in digital signal processing and wireless systems. Prior to LSI Logic, he worked at Metricom on system architecture development for next generation Ricochet wireless networks. Jitendra has also held positions with ZSP India and Navigational Electronics, where he was involved in developing signal processing algorithms for millimetric wave radar, transceiver algorithms for ADSL and network echo cancellers for VOIP. Jitendra received his PhD from the Indian Institute of Technology in Kanpur and has published in a number of journals and international conferences.

Rasekh Rifaat
Analog Devices Inc.

Rasekh Rifaat graduated with BSc and MSc degrees in 1994 and 1998 respectively from the University of Manitoba in Winnipeg, Canada. From 1994 to 1996, he worked on signal processing algorithms and software for seismic data signal processing. In 1999, he joined the DSP Wireless Infrastructure Systems Engineering Group at Analog Devices, where his research interests include advanced communications systems as well as DSP algorithms and architectures.

Sharad Sambhwani
QuickSilver

Sharad Sambhwani is technical director of wireless applications at QuickSilver Technology. Mr Sambhwani has over 8 years' experience in CDMA algorithm design and implementation. Before QuickSilver, he worked at National Semiconductors (formerly Algorex, Inc.), where he led a group developing the W-CDMA physical layer for a mobile chipset. At Algorex he helped develop the CDMA blockset used in the Mathworks Simulink Library. Prior to Algorex, he worked at Bell Laboratories developing the physical layer for IS-95 mobiles. He holds a PhD (EE) from Polytechnic University, New York, an MSc (EE) and a BE (EE) from the University of Bombay.

Chaitali Sengupta
Texas Instruments

Chaitali Sengupta is a member of group technical staff with the Wireless Terminals Business Unit at Texas Instruments, Dallas. Her current work focuses on system definition and validation, hardware–software partioning analysis, and architectures and algorithms for 3G wireless systems. She holds a BTech degree from the Indian Institute of Technology, Kharagpur, MS and PhD degrees from Rice University, Houston, and has authored over 20 journal and conference publications in the field of wireless systems.

Eberhard Schueler
PACT

Eberhard Schueler received his Dipl-Ing in electronics and cybernetics from Technische Universität München. He was engaged in hardware and algorithm development for image processing, hardware design for parallel supercomputers and was CTO and director for product management for voice products in an international System House. With PACT he is engaged as product marketing manager and operates as a consultant for technology-driven voice and IT-projects.

Wei-Jei Song
LSI Logic Corporation

Wei-Jei Song is a systems engineer in the DSP Solutions Engineering Department of LSI Logic Corporation, where he is primarily responsible for wireless LAN research. With more than 17 years' experience in digital signal processing, Wei-Jei began his career with Industrial Technology and Research Institute in Taiwan. He has also worked in the AT&T Bell Laboratories, Rockwell Semiconductor and Nortel Networks, and has acquired significant experience in CDMA, GSM, wireless LAN, echo cancellation and VoIP gateways. Wei-Jei holds three patents for wireless technology and has several pending. Wei-Jei received a master's degree in computer and electrical engineering from Rutgers University.

Sundararajan Sriram
Texas Instruments

Sundararajan Sriram holds a BTech degree from the Indian Institute of Technology, Kanpur, and PhD from the University of California, Berkeley. He is a senior member of technical staff in the Wireless Infrastructure Business Unit at Texas Instruments, Dallas, with research interests in algorithm and VLSI architecture design for DSP and communications, an area in which he has more than 12 publications and 10 patent applications. He has also authored a book on embedded multiprocessors.

Cameron Stevens
QuickSilver

Cameron Stevens is a team lead in the Software Tools Group at QuickSilver Technology and is responsible for the delivery of development tools for the ACM platform. Mr Stevens has over 15 years' experience in software development in the fields of computational fluid dynamics, aerospace, operating systems, desktop applications, natural language search and retrieval, multimedia applications, and embedded systems tools, while working for industry leaders such as Microsoft and Amazon.com, as well as smaller companies such as Sunhawk.com and McClelland & Stewart. Mr Stevens holds a bachelor's degree in mathematics from the University of Waterloo, Canada.

Lorna Tan
Pact

Lorna Tan has a BS in electronics and communications engineering and an MS in electrical engineering, specializing in signal processing and communications science, from the University of Southern California. She has worked in several fields ranging from hard disk controllers, audio, modems, reconfigurable devices, wireless communications and power supply ICs. She is currently working in the area of SDR using reconfigurable devices at PACT as a Director.

Qian Zhang
Analog Devices Inc

Qian Zhang received a BE degree in electrical engineering from Tsing Hua University, Beijing, China in 1991, an MS degree in electrical engineering from Union College, Schenectady, New York, in 1995, and a PhD degree in electrical engineering from Syracuse University, Syracuse, New York, in 2000. Since 2000, he has been with Analog Devices, Inc., Norwood, Massachusetts, where he has been developing signal processing systems for

Third-generation wireless communications. His research interests include signal processing theory, software-defined radio, and information theory.

Zoran Zvonar
Analog Devices Inc

Zoran Zvonar is manager of the systems development group within Analog Devices, focusing on the design of algorithms and architectures for wireless communications. He received Dipl-Ing and MS degrees from the University of Belgrade, Yugoslavia, and a PhD degree from the Northeastern University, Boston. He is co-editor of the series on Software & DSP in Radio in the *IEEE Communications Magazine* and, with Joseph Mitola, of the book *Software Radio Technologies: Selected Readings*, published by IEEE Press/John Wiley & Sons, Inc.

Introduction

Walter Tuttlebee

Mobile VCE, Basingstoke, UK

'Software radio is arriving by stealth...'

Software Defined Radio – are you a sceptic or a believer? If you were a total sceptic you would not be wasting your time reading this introduction. However, many are. I am frequently asked for my views on when software radio 'will arrive'. Perhaps the most interesting answer I heard someone give to this question over the past year was 'software radio is arriving by stealth'. As you read the chapters of this book, you may find yourself concluding that perhaps this is, in fact, a very apposite observation.

3G as the Driver for SDR?

Baseband signal processing technology is experiencing a period of radical change, for both handset and basestation applications, with 3G emerging as the major driver. To some, given the downturn that the industry has seen over the past two years, this may seem strange. Ironically, the network roll-out delays and the uncertainties created by the downturn are the very factors that have created fresh opportunity for those developing SDR technologies. The delays have allowed more time for technologies to mature whilst the climate of uncertainty over service take-up has created a demand for flexible solutions able to accommodate varying traffic mixes and unknown future applications.

In the mid to late 1990s, amidst the early hype of software radio, the initial market for SDR was seen by many as being North America, where the diversity of 2G air interfaces was seen to be an obvious area of application for a reconfigurable handset. Unfortunately for some of those early protagonists, particularly those who have recently gone out of business, the time required to develop the technology has, as always, been rather longer than had been hoped. Some of these technologies are, however, now beginning to pull through, with the new flexible baseband approaches described in this book emerging as important potential solutions for 3G kit – not just handsets, but basestations as well.

Who are the Players?

A few years back software radio was largely the domain of West Coast start-ups; today, mainstream players are in the fray, alongside the new boys. This book draws on the wealth of experience from both categories – mainstream DSP market leaders and major players such as

Software Defined Radio: Baseband Technologies for 3G Handsets and Basestations. Edited by W. Tuttlebee
 ISBN: 0-470-86770-1

Texas Instruments, Analog Devices and LSI Logic, FPGA heavyweights Altera and Xilinx, and a selection of newer companies with a range of differing technologies such as Morpho Technologies, PACT, picoChip, QuickSilver, RFCo and Sandbridge.

The chapters from these companies can be seen as representing different approaches to fundamentally common challenges. This diversity of perspectives and opinions arguably reflects the varying technical backgrounds, starting points and objectives of the authors; in this respect, whilst editing this book, I have sought to avoid 'homogenizing' the contributions – each chapter is allowed to convey views with which other contributors might violently disagree. This wide range of views allows the reader to be challenged by perspectives other than his own, offering an opportunity to gain new insights and to perceive alternative approaches, to break out of a 'traditional' view and see the issues from 'outside the box'. With such an approach, each reader will take different things from this book, each synthesizing his own conclusions and solutions, appropriate to the particular problems which he is seeking to address.

Purpose and Audience of the Book

This is the fourth volume in the Wiley series on Software Defined Radio and builds upon what has gone before. The first volume dealt with the origins, drivers and emerging international perspectives on SDR. The second offered a 'helicopter view' of SDR-enabling technologies, addressing the front-end, baseband and software. The third, based on extensive European research, considered system level aspects – architectures, systems and functions.

This new book builds directly on Volume 2, going much deeper into the real-world implementation of baseband signal processing. Principles outlined in Volume 2, such as the digital front-end, parameterization and new signal processing architectures, are exemplified in much greater detail and practicality in this present volume, with the potential, capabilities and limitations of a range of solutions being described. As such, this new book will hopefully resonate with engineering practitioners involved in the development of products for the 3G marketplace and perhaps enlighten them as to alternative approaches to achieving their goal. Equally, however, it will also hopefully find a home on the academic bookshelf as a reference that ties together theory and practice. It will provide a valuable set of perspectives for an emerging generation of silicon designers who, in the coming decade, will be designing architectures that a previous generation would not have considered practical.

Structure of the Book

The book is structured into two primary sections, one describing technologies primarily targeted at handset (or to be strictly correct, terminal) applications, and one describing those targeted at basestation applications. Both sections have contributions from both today's mainstream and from new players. In addition, we initially outline the differing requirements of the target types of end products, from the SDR perspective. We conclude the book by attempting to step back and assess the strategic implications for the industry of the advances that are taking place. Arguably, the technology developments that are occuring today in silicon architectures and design flows could have profound impacts upon the structure of the wireless industry, and beyond.

'When will software radio arrive?'

So, back to the question we started with. In one sense software radio has begun to arrive, is coming and will come – in increasing measure, with increasing impact. It will not all happen overnight – as noted earlier it always takes longer than anticipated to develop and introduce a new technology. However, the profound changes that have begun to impact silicon signal processing technology will enable not only 3G, but also the wider electronics industry.

Part I

Requirements

The requirements and challenges posed by SDR are in some ways common across applications, but vary across implementation platforms. This initial section considers these issues, as a foundation, prior to exploring candidate technologies in later parts of this book.

1

SDR Baseband Requirements and Directions to Solutions

Mark Cummings

RFco, Inc.

This chapter puts the development of SDR baseband technology into functional and historical perspective, then examines the evolving spectrum of requirements from mobile devices such as simple handsets, through complex mobile devices, access points, micro cellular basestations to wide area basestations. Both internal functional requirements and external interface requirements are considered. The requirement challenges are then discussed in the context of currently available (and emerging) technologies for implementing SDR baseband processing subsystems. The chapter concludes with a consideration of likely successful implementation directions.

1.1. Baseband Technology and the Emergence of SDR

The primary intent of SDR is to allow a wireless device to change functionality by switching in different software stacks (or software-like objects) that can be stored locally or downloaded over the air [1].

1.1.1. Wireless Architectures

At the highest level, all wireless systems can be decomposed into four subsystems (see Figure 1.1). The baseband subsystem sits between the RF front-end and the controller. It is responsible for end-user data encoding/decoding and signal modulation/demodulation. The encoding/decoding function can be classified as a low speed signal-processing task and the modulation/demodulation task can be classified as a high speed signal-processing task.

By the end of the Second World War, wireless communication systems had yet to evolve to the functional form shown in Figure 1.1. Rather they consisted of discrete analog electrical

Software Defined Radio: Baseband Technologies for 3G Handsets and Basestations. Edited by W. Tuttlebee
 ISBN: 0-470-86770-1

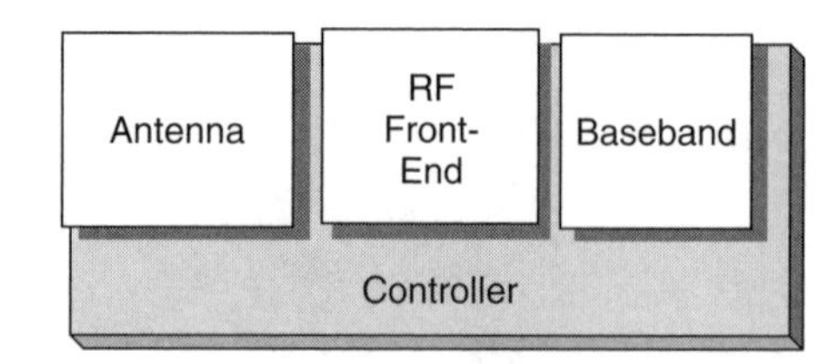

Figure 1.1 A top level view of wireless systems architecture

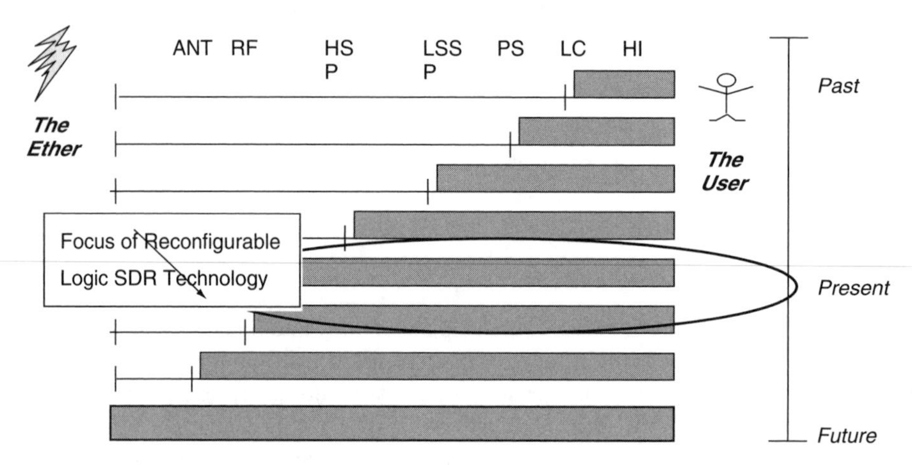

Figure 1.2 Evolution of wireless systems from discrete analog to software-driven hardware

components that performed the following functions (also listed at the top of Figure 1.2, from right to left):

- Human interface,
- Local control,
- Protocol stack,
- Low speed signal processing,
- High speed signal processing,
- RF front-end,
- Antenna.

Examples of the tasks performed by each of these functional units, in a cellular handset include:

- Human interface – controlling the speaker, microphone, keyboard and display;
- Local control – managing the handset hardware and software resources;
- Protocol stack – managing such functions as call set-up and maintenance;
- Low speed signal processing – user data coding, such as interleaving;
- High speed signal processing – modulation such as FM, QPSK, GMSK;
- RF front-end – up/down conversion, amplification and filtering;
- Antenna – interface to the ether.

In these early systems, each function was implemented with discrete analog technology. This resulted in relatively large, expensive, high power-consuming systems, which were difficult to design, manufacture and manage/maintain in the field. The desire to reduce the cost, size and power consumption, and to make devices easier to manage in the field while taking full advantage of technology improvements wherever possible, has driven the technology evolution path we are still on today.

1.1.2. The Impact of Digital Technology

As digital technology arrived and entered the beginning of its period of rapid evolution, a pattern developed. First a function previously performed in discrete analog circuitry was implemented with discrete digital components. Then, as the semiconductor technology developed, these discrete components were integrated into larger and larger units and the specialized, discrete, hard-wired digital logic was replaced with general purpose logic that changed its function with changes in software. This combination of general purpose logic plus software on a single semiconductor came to be called a microprocessor. Several types of microprocessor evolved, each optimized for different functionality. This process is graphically portrayed in Figure 1.2.

1.1.2.1. Early Applications in Radios

The first functions to be aggregated through this process were the human interface, local control, and protocol stack functions. With the appearance of the microprocessor, the discrete logic components that replaced the analog components were, in turn, replaced with a specialized microprocessor, called a microcontroller, and software. Then the low speed signal processing function went from analog discrete components through digital logic to be implemented as another specialized processing engine. Special mathematical functionality (such as multiply accumulate, MAC) was added to microprocessors to create digital signal processors (DSPs) enabling the low speed signal-processing functions to be converted from discrete digital logic to software-programmed DSPs .

1.1.2.2. The Dream of 'Pure' SDR

Next, the high speed signal-processing analog discrete components were replaced with digital logic components. The expectation was that the same process would continue and that high speed signal processing would also soon be implemented by some kind of microprocessor and software. The dream was that this process would continue to the point where a receiver subsystem would consist of an A/D converter at the antenna and everything else would be done by software. However, another factor was by now beginning to come into play – the increasing complexity of the radio systems themselves.

1.1.2.3. The Growing Complexity of Cellular Radio – DSPs and ASICs

Early cellular standards such as AMPS required a baseband bandwidth of 30 kHz. As low speed signal processing was moving into a DSP implementation, second generation cellular standards such as GSM emerged, with a baseband bandwith in the 200 to 300 kHz range.

Other second generation standards, such as IS-95 (CDMA/AMPS dual mode air interface standard, AIS) with an even larger baseband bandwidth of 1.24 MHz, were also coming to the market.

Although monolithic, fixed, single-stream instruction set processor DSPs could theoretically handle the speed of processing required for high speed signal processing of second generation AISs, practical limitations created an impenetrable barrier for the traditional DSP, with its single-stream instruction set processor. This architecture has a single processing unit which must load data (single sample of signal data) and a single instruction, process the data, write the result to a buffer and start over again. The Nyquist theorem requires that a signal be sampled at more than twice its rate of change (in practice 2.5 times its bandwidth in Hertz) in order to preserve the data in the signal. This means that a 1.24-MHz signal must be sampled at approximately 3.25 MHz. In order to avoid quantization error, single samples are typically represented by 13 or more bits in cellular systems. Minimum high speed signal processing for first generation and early second generation AISs require approximately 100 instructions per sample. Assuming that the processor has a 16-bit word size, this means that a DSP attempting to do high speed signal processing for a simple handset would have to operate at a clock speed in excess of 325 MHz. In practice, because of bus delays, the need to write to buffers, etc., the required clock speed turns out to be in the GHz range. Early DSPs could not run at these clock speeds.

As DSP development progressed, it became clear that, since power consumption varied directly with processor speed, it would not be practical to operate a DSP at these clock rates for battery powered applications. It also turned out that, even for systems run on utility-provided power, the nature of their requirements plus issues of size, heat dissipation, etc., would make pure DSP solutions for high speed signal processing impractical.

Discrete digital component implementations avoided the power consumption problem by implementing each of the 100 instructions in 100 discrete digital logic circuits arranged in a bucket brigade. In this way, each circuit could run at the (much lower) clock rate of the sampled and quantized signal (such as 3.25 MHz), dramatically lowering the power consumption. Further power/speed improvements can be made by having each circuit optimized for a single function, thereby avoiding the inefficiencies inherent in a general purpose processor which has to be optimized for a wide range of functions.

So, here, the previous pattern appeared to break. Discrete digital logic components could not be replaced by a microprocessor with software. However, they could be combined into a single chip through the aggregation and integration of the discrete logic components into a single large scale integrated circuit (LSI) and later the very large scale integrated circuit (VLSI). This combination of discrete digital logic into a single chip came to be called an application specific integrated circuit (ASIC). It achieved the cost, size and power consumption advantages inherent in integrated circuits, but it lacked the flexibility inherent in software-driven general purpose processors.

1.1.3. Growth of the Wireless Markets and Multiple Standards

In the late 1980s and early 1990s, a variety of markets for portable two-way wireless communications systems were experiencing rapid growth and rapid evolution of technology for implementing modulation/demodulation and encoding/decoding, resulting in a proliferation of AISs. Each advancement in AISs increased the performance requirements on baseband processing. The pace of requirement growth and the scale of the challenge is illustrated by the early development of IS-95. The first test implementations of IS-95 were completely described in code (software) because of the need to make and test quick refinements. As a result, one early prototype of an IS-95 handset consisted of 64 6800 processors rack mounted in a panel truck.

Demand for universal wireless communications began to appear, see [2]. Users in military, civil government, and commercial telecommunications began to demand 'anywhere' wireless communications and additional wireless services such as geolocation. In each market there was a large and growing number of noncompatible AIS footprints, none of which could by themselves provide the 'anywhere, anytime' service desired. Manufacturers were forced to design and build an inventory of a large number of different models and to manage the logistics of getting the right model to the right footprint at the right time to enable a customer. [6] Network operators were also facing high costs from 'fork-lift upgrades' (complete infrastructure replacements) as they moved from one generation of AIS to another.

1.1.3.1. Parallels with Computing

The situation was very similar to that in computing in the mid-1970s, where each application required a specially built dedicated 'intelligent terminal', which resulted in limited usability, deployment and management problems. The subsequent advent of the PC, with its standard hardware platform, which could be suited to a wide variety of tasks by application of the appropriate software, dramatically changed computing. By the early 1990s, wireless markets began to search for an analogous transformation of wireless equipment.

The lack of flexibility inherent in ASIC processors for high speed signal processing made equipment manufactured to support a given AIS, for the rest of its useful life, limited to that AIS. This made it difficult for commercial users to communicate as they moved in and out of different AIS service footprints. Government users who didn't have the *lingua franca* of the public switched telephone network to fall back on, if they had equipment supporting two different AISs might not be able to communicate at all. ASICs, then, satisfied the desire for lower cost, size and power consumption but didn't satisfy the desire to make devices easier to interoperate or to manage in the field.

1.1.3.2. Software-Defined Radio

In the early 1990s, solutions to the high speed signal-processing requirements that offered software driven flexibility and the ability to change AIS baseband subsystems to support different AISs through software began to appear [3, 4, 5]. The appearance of these solutions, in the company of the market drivers, led to the coining of the term software defined radio (SDR). How SDR fits into the general evolution of RF systems is displayed in Figure 1.2.

These initial solutions can be characterized as being based on reconfigurable logic. There are a variety of approaches within this general area. Generally, they use software or software-like code to configure digital logic to perform the high speed signal processing at relatively low clock rates, thereby achieving the desired power consumption/heat dissipation while they are running, while being capable of being reconfigured between runs to support different AISs [3, 4, 5].

1.2. Evolution of SDR Baseband Requirements

1.2.1. System Optimization

Over time, information processing systems have consistently sought to optimize cost/performance by managing the balance between three elements:

- local processing/local memory,
- communications capacity,
- type and quality of service.

Local processing/memory has to do with the capability to store and process information locally. It includes such factors as the ability to handle different kinds of information (voice, data, image, full motion video, etc.).

Communications capacity has to do with how much raw transmission capability is available. It may be measured in many different ways, depending upon application – number of twisted pair equivalents, Hertz, voice-grade equivalent channels, NTSC/PAL/SECAM/HDTV, number of TV channels, number of simultaneous low speed Async (Asynchronous) data channels, etc.

Type of service includes such factors as audio, image, video, Async, Isoc (Isochronous), etc. Quality of service includes such factors as error rate, availability, latency, fidelity, etc. Cost includes a broad range of factors including equipment cost, cost of ownership, migration cost, opportunity cost of consuming scarce resources, etc.

All those in the communications industry working together over time, seek to optimize systems for the highest possible types and qualities of service at the lowest possible cost. The major players in this optimization process include the equipment designers, system operators, regulators and end users.

Certain threshold phenomena come into play to define the types of service that are 'possible'. In this context, 'possible' is a function of technical feasibility with current technology, reasonable cost, within the range of what can be imagined and desired. If communications resources are considered relatively scarce (expensive) and local processing is considered available and relatively cheap, then communications systems will maximize the utilization of local processing/memory and minimize the utilization of raw communications resources. Full function PCs using low speed dial-up lines are an example of such an optimization. If, on the other hand, communications resources are considered plentiful, but processing/storage is considered scarce, then systems maximize the use of communications resources and minimize the use of local processing/memory. X Windows is an example of such an optimization.

1.2.2. Market Drivers of SDR

SDR is both a product of, and comes to market in the context of, dramatic changes in the world's information technology and usage environments.

The Twentieth Century was shaped by one-way wireless communications systems, namely broadcast radio and television. Prime examples include Franklin D. Roosevelt's radio broadcast of the Fireside Chats, Hitler's radio broadcast of rallies and the US civil rights movement reliance on TV coverage of the response to protest activity. The Twenty-first Century will be shaped by two-way interactive, multimedia, wireless communications. Early examples include the SMS-organized demonstrations that toppled a recent Philippine administration.

1.2.2.1. Usage Trends

The most dramatic consequence of this change process is the movement from the situation where a small fraction of 1% of population uses interactive wireless communications for a small fraction of 1% of the time primarily for voice services, to the situation where 90% of the population of the planet use interactive wireless communications 90% of the time for a wide range of multi media services.[†] This is making the limited quantity of radio frequency spectra a scarce resource (expensive) and is forcing an optimization shift, moving systems to apply additional local processing/memory resources which are seen as more available (less expensive). This optimization move is creating a rapidly increasing rate of technical innovation in air interface standards (AISs).

1.2.2.2. 'Anytime, Anywhere'

Not only is the quantity of usage increasing, but the users are moving to an expectation of 'anytime, anywhere' communications. Because the requirements can change so dramatically from location to location and from time to time, and since once deployed, large communications systems continue to persist in spite of the availability of new, more effective modes (for example the continued operation of the Baudot Telex System), the number of non-compatible AISs available to a particular user is large and growing.

1.2.2.3. User, Manufacturer and Operator Needs

Users don't want to carry multiple devices in order to operate in whichever AIS footprint meets their immediate (time, place, etc.) need. Manufacturers don't want to have to manufacture a wide profusion of different combinations of specific AISs and get the right combination into the right part of the world at the right time in order to make a sale. Operators don't want to have to augment one network completely with a second, and run the two networks in parallel until they can reterminalize their user base.

Hence, the drive to SDR. SDR is seen as being able to allow the user to carry a single device; the manufacturer to reduce product line complexity; and the carrier to offer integrated services across a variety of AIS infrastructures while limiting the cost of technology upgrades. All of these factors come together to increase the demands on the SDR baseband subsystem.

1.2.3. Technology Drivers of SDR

Alongside this fundamental set of market needs, also driving SDR is a series of technology changes that is also increasing the demands on the SDR baseband processor. These technology changes include the following:

- Movement from simple analog modulation, to complex modulation (from FM through QPSK to OFDM, etc.) multimedia, digital modulation, wireless systems (from 1G through 2G, to 2.5G, 2.75G, 3G, 3.XG, 4G...)

[†] In this context, usage may mean connected ready to receive and transmit in real time, not necessarily actually transmiting and receiving 90% of the time.

- Movement from a small number of large geographic area frequency reuse plans (100s of kilometers – metropolitan area radio and TV stations – to 1000s of kilometers – clear channel broadcast radio stations) – to a large number of small geographic area frequency reuse plans (metropolitan cell sites – 5 to 50 kilometers, microcellular cell sites from 50 meters to 5 kilometers; to access points 1 to 100 meters, to PAN access points from 0.1 to 2 meters).
- Introduction of MIMO (multi-in, multi-out) technology, sometimes called smart-antenna technology, which involves multiple antenna/RF/baseband paths that are combined to achieve improved spectral efficiency, signal quality, etc.
- Increased focus on overcoming climbing noise levels resulting from congestion, leading to requirements for adaptive high Q filters and other techniques either to reduce noise levels, improve signal levels or improve performance in low signal-to-noise ratio environments

When all of these factors are combined, the rate of increase in processor demands is greater than the rate of increase in processor capability deriving solely from Moore's law. This is graphically represented in Figure 1.3.

The SDR baseband subsystem must meet these rising requirement levels, seeking better end-user performance (important to the consumer), more cost effective network implementation (important to the carrier), reduced product-line complexity (important to the vendor), and improved spectral efficiency (important to the regulator, carrier, and end-user). These requirements are expressed somewhat differently depending on which of several form factor/ application environments is being targeted.

1.2.4. Requirements by Application Environment

In the environments where SDR is commonly considered, applications can be broken down into two general categories – end-user and infrastructure. End-user systems generally provide

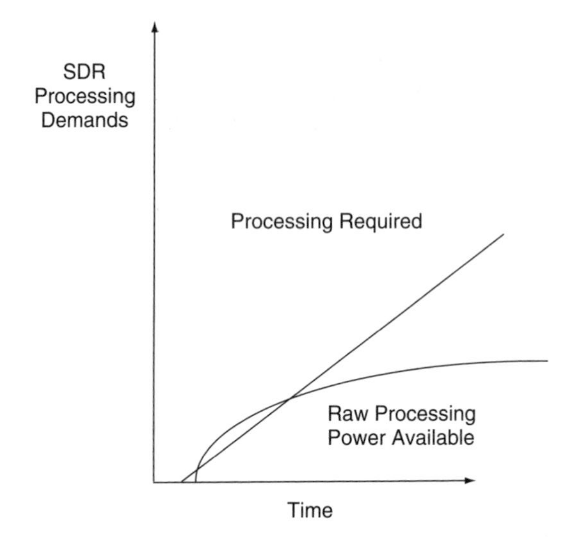

Figure 1.3 Relative increases in available and required processing power

one or a small number of simultaneous channels. Infrastructure systems generally provide many simultaneous channels supporting a large number of end-users at the same time, while also providing access to larger networks in the supported AIS, or gateways into other networks. End-user systems can be broken down into mobile and fixed systems. In the following sections we will consider the differing requirements that emerge in a variety of end-user and infrastructure applications.

1.3. Mobile End-user Systems

Mobile systems can be simple, low functionality devices or more complex systems. They can be pocket sized or relatively large vehicular-mounted systems. New categories are evolving rapidly and so any taxonomy will constantly encounter new categories. For current purposes we will consider the following categories of mobile systems:

- handsets;
- converged devices (e.g. PDA-phones);
- notebook computers and other devices of a similar size;
- vehicular systems.

These four categories can be differentiated by progressive increases in size and power consumption, going from handsets through PDAs and notebooks to vehicular systems. Handsets are designed to be as small as possible and to function without recharging the power supply for several days. They started out providing a single channel of a single AIS. What is emerging is a set of requirements that includes multiple modes and bands of cellular, WLAN (wireless local area network), WPAN (wireless personal area network) and geolocation services such as GPS (Global Positioning System). Services supported include voice, MMS (multimedia message service), data, still image (photographs) and full motion image (video). Cellular services can include first, second and third generation services (currently work is under way on 3.5 G and 4 G networks). WLAN services can include 802.11x (data, VOIP, etc.) and cordless telephony. WPAN can include Bluetooth, ZigBee, etc.

1.3.1. Simple Handsets

1.3.1.1. High Speed Signal-Processing Requirements

Single AIS handsets today require modulation/demodulation of a single RF carrier at a time. In TDMA systems, hand-off between cells is accomplished by fast switching between channels. In CDMA systems today, resources typically applied to improve signal quality (fingers of Rake filters, etc.) are reallocated to produce a system where two code division channels within a single RF carrier are handled simultaneously to assist in 'soft' hand-offs between cells.

More complex 'simple' handsets are now appearing and will soon become dominant. The number of bands supported is increasing. Triple-band handsets are now common; even five-, six-, and seven-band handsets are appearing, and the number of bands will continue to increase. Although different bands of the 'same' modulation technique appear to have the same modulation/demodulation requirements, there are differences. For example, European GSM900 (900 MHz) and GSM1800 (1800 MHz) have somewhat different requirements.

The number of modes supported in 'simple' handsets is also increasing. In the US, all 2 G standards required fall back to 1 G (AMPS). In Europe, ETSI originally required all 3 G handsets to support at least one version of GSM. In the USA, the FCC introduced a requirement for E911 (emergency mobile phone location service); in Asia and Europe the consumer location based services (LBS) are beginning to appear. The LBS requirement is being satisfied today by some vendors through the inclusion of a GPS receiver in the handset, or time of arrival (TOA) software in the handset, or other network based solutions. Support for wireless headsets and other PAN devices require yet another mode.

Some of this growing number of modes and bands can be supported by changing the functionality of hardware with software (SDR) in a sequential manner as the user moves in and out of different footprints or experiences changes in his immediate requirements. Others require simultaneous support. If the handset's software can change fast enough, some of what appear to be simultaneous channel support requirements can be achieved in a sequential fashion.

In handsets, the high speed signal-processing unit must, in all states, minimize power consumption. To this end, it must be able to reduce functionality in the standby mode to the minimum necessary to support receiving a call initiation page and sending 'keep alive' messages in the transmit mode. At the same time it must be able to switch quickly between the wide variety of modulation/demodulation schemes. It must also be able to service multiple, apparently simultaneous, channels in a potentially constantly varying mix.

The initial approach to these 'composite' requirements was with a composite processor, that is, a separate processor for each AIS. This brute-force method seems eminently workable if one relaxes the power, size and cost constraints. However, logistic problems that result from different generations of low speed signal-processing solutions can make this approach difficult. In any case, the pressures of power, cost, and size favor a single chip integrated solution. Also, the TOA analysis for a single AIS is best performed in the high speed signal-processing portion dedicated to that AIS. TOA results can be significantly improved by combining information from multiple AISs and multiple base stations.

In the past, smart antenna, or MIMO (multi-in, multi-out) technology has only been applied to base stations. Because systems requiring multiple independent RF paths and RF front-ends have been relatively large, expensive and power intensive, it has not appeared practical to implement MIMO in handsets. This situation is, however, beginning to change. There are RF front-ends in development that can provide multiple independent RF paths at low cost, size and power. Therefore, in the near future, SDR high speed signal processors will have to also support MIMO algorithm processing.

1.3.1.2. Low Speed Signal-processing Requirements

Early 2G handset low speed signal processors only had to support conversion of voice back and forth from analog to digital, with some fairly simple interleaving and forward error correction techniques added. Now, asynchronous data, music compression, and image compression must also be supported, along with location reporting and multiple simultaneous channels. Additionally, several simultaneous channels will have to appear to be handled simultaneously.

Because of the size of the displays, the resulting image resolution is likely to remain relatively low. Because of small high quality headphones, requirements for audio quality ('resolution') are relatively high. Again, because of size/power limitations, memory size will be limited thus favoring lower 'resolution' compression/decompression with concomitant demands on low speed signal processing. The handset usage model may be structured to limit the aggregate processing demand by downloading multimedia material in compressed form and directly storing it. In this model, decoding (decompression) only occurs when the user is 'off-line', that is, not concomitant with the actual AIS processing. Coding can be handled in a similar fashion.

Asynchronous data will likely consist of short text messages and small screen adapted web pages. The small screen size makes it difficult to handle large documents on a handset, therefore, users will limit the message/e-mail usage to relatively short messages. This will produce a limited demand on low speed signal processing in terms of volume of data. Because of expectations established in the user community through use of high speed broadband wired access to the web, speed of delivery of web pages to handsets is an issue. These demands will foster requirements to speed the flow of web data through the low speed signal-processing system.

The requirements for low speed signal processing in an SDR handset, then, are to be able to code/decode audio streams (voice and music), relatively low resolution image, fast web pages and small size text messages while moving seamlessly between a variety of AISs. The challenge is in accomplishing this processing with small physical size processors consuming limited amounts of power.

1.3.1.3. Packaging

Because of the premium on size and power, there are incentives to minimize the number of separate components. Some of the incentives result from the fact that additional power is required to go off-chip (line drives, etc.) and that packaging can greatly increase the volume occupied by a single die (unpackaged semiconductor). There can also be cost implications. The result is that some vendors are packaging the entire SDR baseband solution (processor or processors, and scratch-pad memory) in a single semiconductor. Others are combining the SDR controller with the low speed signal processor and scratch-pad memory. Both of these approaches use separate nonvolatile memory stores for both user data and program storage. As the number of functions implemented in software increases, and as the number of alternative software stacks for each function increases, the size of nonvolatile storage will have to increase. How many of these alternative software stacks should be stored on the network and how many stored on the handset is a complex optimization involving latency, cost of storage, cost of communications, implications for cost/complexity of network management and business models.

Yet other vendors are focusing on packaging the controller, low speed signal processor and high speed signal processor on a single semiconductor.

As we will see later in this chapter, there are a number of alternative high speed signal-processing architecture alternatives, each with different implications for packaging. As a result, there are likely to be a variety of approaches to minimizing chip count.

1.3.1.4. RF Baseband Interface Requirements

Software efficiency (development, debugging, life-cycle management) is greatly enhanced if it is possible to establish standard APIs (application programming interfaces). This is true in general, but especially true for SDR devices. Because of the range of alternative architectures available and the ongoing technology innovation in the SDR field, it is unlikely that the internal architectures for SDR RF front-ends and SDR baseband processors will become standardized. However, it will greatly assist network operators, equipment manufacturers and software developers if a standard interface between the RF front-end and the baseband processor can be developed. Such a standard will have to accommodate variety in number and type of channels, but should provide a standard meta language for describing and reporting a devices' exact configuration.

1.3.2. Converged Devices

Today, in addition to simple cell phone handsets, there are a wide variety of personal portable information devices including:

- PDA/organizer;
- digital camera;
- portable game console;
- walkman;
- video walkman;
- MP3 player;
- broadcast radio and TV receiver;
- satellite audio;
- navigation system.

This list is constantly growing. Today, there are converged devices that combine cell phones with different subsets of this list. For example, camera phones are popular in Japan. Nokia has launched a game/cell phone combo, and cell phones with some form of audio for music are appearing. Within a short period of time, converged devices will support all of these functions in a single device. Users will want a single device to support the ability to switch between these functions and sometimes to run these functions concurrently. This will place additional requirements on SDR baseband processors.

These devices tend to be larger than handsets, with larger batteries and with usage patterns that generate shorter times between battery recharges. These factors somewhat reduce the power conservation constraints faced by stand-alone handsets.

These devices have larger human interfaces, thus offering richer communications experiences for the user. Medium sized screens lead to medium resolution graphic capture, transmission and display. Medium sized text files and web pages become practical. One consequence of this is that, although WLAN support is not often required on simple handsets, it is becoming more common on converged devices.

Because they deliver a broad range of applications that users come to feel are critical to their everyday functioning, the demand for 'anytime, anywhere' operation is increased. This leads to a demand for support of a wider array of AISs and support for more functions simultaneously.

1.3.2.1. High Speed Signal Processing

As compared with the handset, the SDR converged device, high speed signal processor has more power available and can occupy more volume. However, offsetting this, it must process more modes, more bands (AISs) and support a larger number of apparently simultaneous channels. This means that technology solutions that cannot meet the stringent size/power/cost constraints of the simple handset, but have somewhat more capability, may find an applications niche in the converged device area.

1.3.2.2. Low Speed Signal Processing

As compared with the handset, the SDR converged device, low speed signal processor must provide moderate resolution image and web pages while also coding/decoding a greater (possibly simultaneous) flow of information to and from local nonvolatile storage. Here again, the usage model may be structured to limit the aggregate processing demand by downloading multimedia material in compressed form and directly storing it. In this model, decoding (decompression) only occurs when the user is 'off-line', that is, not concomitant with the actual AIS processing. Coding can be handled in a similar fashion.

1.3.2.3. Packaging

As compared with the handset, the biggest difference in SDR converged device packaging is the larger local nonvolatile memory for storing both program code and application data. This requires large nonvolatile stores that cannot be practically integrated with other functions at this time.

The other major difference is the amount of application processing required. One approach is to combine the application processing and control functions on a single processor. Another approach is to attempt to use a single powerful application processor to perform signal processing as well.

1.3.2.4. RF Baseband Interface Requirements

As compared with the handset, the biggest difference in SDR converged device RF front-end-to-baseband interface is the increase in the number of modes and bands that must be supported. There is also an increase in the area available to mount antennas, and so there is likely to be a greater MIMO requirement. These factors may yield a requirement for a greater number of apparently discrete RF baseband interfaces. However, the same API meta language used for the SDR handset should be able to describe and report these interfaces thus achieving economies of scale.

1.3.3. Notebooks

Notebook computers are increasingly incorporating wireless communications as a standard feature, and some portable DVD players are also taking this form factor. This is driven by a higher bandwidth interface between the user and the device (large screen, full size keyboard, etc.). These devices also have larger batteries, shorter recharge intervals and more volume available for processors and memory. However, because of their larger size, they are not subject to as great an 'anytime, anywhere' requirement. Similarly, they are not well suited to simple voice services. These devices have larger internal nonvolatile memory stores, but they also have large removable local stores that can be used to store both SDR program data and user data.

Notebooks, are typically only required to operate for a few hours on a single battery charge and are often operated when plugged into external power sources.

1.3.3.1. High Speed Signal Processing

As compared with SDR handsets and converged devices, the SDR notebook requires fewer simultaneous channels, and a smaller palette of locally stored AISs from which to draw.

1.3.3.2. Low Speed Signal Processing

As compared with SDR handsets and converged devices, the SDR notebook requires coding/ decoding of large text files, high resolution graphic data and large web pages. The user is less likely to accept the 'off-line' download usage model and is more likely to require access to graphic and audio information as it is downloaded over the AIS. The notebook user is also likely to be a source of such information for such applications as multimedia conferencing and other peer-to-peer applications.

1.3.3.3. Packaging

As compared with SDR handsets and converged devices, the SDR notebook has less of a necessity to limit component count. The high speed signal-processing requirements may be somewhat reduced, while the power/size/cost constraints are relaxed. Therefore, one approach being pursued to meet SDR processing requirements is the use of large high speed applications processors combined with 'accelerators' where necessary. Large nonvolatile memories and removable memory media are common in this form factor.

1.3.3.4. RF Baseband Interface

Because of the increased size and power available, there are SDR baseband architectures that are more practical in the notebook than in the handset or converged device. There are also differences in the noise environment in notebooks that result from the type of processors and the type of computer buses used. This may lead to RF front-end-to-baseband interfaces that are somewhat different than for the handset and converged device. However the handset meta language should be able to describe and report these interfaces.

1.3.4. Vehicular Systems

Vehicular SDR systems begin to take on some properties more common to SDR infrastructure, whilst still retaining some of the properties of individual end-user systems. Vehicular systems are evolving into integrated highly functional systems that can take advantage of the larger antennas, larger volume and larger power sources available to deliver a variety of services to a variety of users over one or more in-vehicle wired or wireless networks. In some cases, they also provide services to users outside the vehicle within a short distance.

In addition to having larger antennas, vehicular systems are allowed by regulatory agencies to use more powerful transmitters. The power of transmitters in personal devices is limited due to health concerns about their close proximity to people. Vehicular systems still have power limitations as compared with some infrastructure systems, again because of health concerns.

Compared with all of the types of services described previously, vehicular SDR systems provide additional unique services. These systems can be grouped into transportation information systems, telemetry systems, and guidance and control systems. The boundaries between these three types of unique service are not exact and firm – one type of service may bleed into another type.

Examples of vehicular information services include wireless toll payment, traffic alerts, etc. Telemetry may be gathered and delivered to the integrated SDR hub either wirelessly or with wires. Examples of telemetry include reporting airbag deployment to remote locations or reporting a low air pressure condition of a tire on a vehicle to the driver. Examples of guidance and control systems are the use of radar-like systems to determine the distance to the next car on the highway. Because information systems and guidance/control systems have such different requirements for reliability and latency, they are sometimes implemented in physically separate systems.

1.3.4.1. High Speed Signal Processing

As compared with personal SDRs, vehicular systems require more modes and bands (AISs) and more simultaneous channels. However, there is more power and more volume available to accommodate system components.

1.3.4.2. Low Speed Signal Processing

As compared with notebooks, the low speed requirements are quite similar. The major difference is in the number and volatility of simultaneous channels.

1.3.4.3. Packaging

SDR is a particularly attractive approach for vehicular communications systems. Although the total space and power available for communications is large, if each function is implemented in a separate radio, the space and power required would quickly outstrip what is available. The ability of SDRs to support a wide range of wireless applications with a single relatively small package is extremely valuable to the vehicle designer.

The other major characteristic is that vehicular systems provide a somewhat hostile environment characterized by high temperature ranges, high G-force loading, high vibration levels and the possible presence of caustic substances. These characteristics tend to favor small ruggedized packages. This puts a premium on reducing part count and heat dissipation.

1.3.4.4. RF Baseband Interface

Because of the increased size and power available, there are SDR baseband architectures that are more practical in vehicular systems than in personal devices. There are also differences in the noise environment in vehicles that result from engines, spark plugs, etc. This may lead to RF front-end-to-baseband interfaces that are somewhat different. However, because these services are still primarily end-user focused, the handset meta language should be able to describe and report these interfaces.

1.4. Fixed End-User Systems

Fixed end-user systems have many of the same requirement profiles as mobile systems. However, there is a relaxation of requirements in the areas of low power, small size, and multifunctionality. Fixed devices can be connected to utility power. Since they are not being carried around, size and weight is not such an issue, nor is accommodating many different types of functionality in a single package.

At the same time, the demand for mobility by itself in our society, as well as expressed through economies of scale, is driving many fixed wireless systems to become mobile. Given these forces at work, the proportion of fixed wireless systems deployed is likely to decline. That is, although there may be increases in the absolute number of fixed systems, the rate of growth of mobile systems will be much greater.

1.4.1. Infrastructure System Requirements

Infrastructure systems can consist of one or many different types of system. For example, large cellular network infrastructures consist of basestations which are connected via an internal point-to-point network link to a basestation controller, which is in turn connected to other controllers, such as separate addressing, billing and maintenance systems. On the other hand, simple WLANs often consist of an access point in a computer connected to a wired network. In this section, we will be primarily concerned only with that portion of the infrastructure that services personal or vehicular devices.

There are two major factors that differentiate SDR infrastructure systems from mobile end-user systems. The first involves SDR usage models and the second involves inherent functionality.

Mobile end-user systems are constantly moving in and out of noncompatible AIS footprints. This requires these types of systems quickly to change from one to another AIS. They also support simultaneous concurrent noncompatible AISs, so these types of system may be required to maintain some AIS channels while reconfiguring others. In this respect, mobile personal systems can be characterized as having a relatively high volatility of AIS change/reconfiguration.

Infrastructure systems are stationary by nature. The predominant usage model is that mobile devices will change to configure to the AIS supported by the relevant infrastructure as they move in and out of infrastructure footprints. Therefore, except in unique specialized cases, infrastructure systems are not supporting a variety of constantly changing AISs. Rather, the SDR usage model focuses on graceful technology migration from one AIS to another. These changes, rather than being measured in fractions of a second, are measured in days, months or years. However, as new AISs are developed and deployed, change requests (network operators requests for features and capabilities different from what was described in the original purchase contract) in the order of 100 per month per infrastructure product are appearing.

The inherent functionality difference is that infrastructure systems are required to support multiple mobile devices. Therefore, infrastructure devices typically process blocks of channels rather than individual channels.

1.4.2. Basestation Requirements

Radio systems have sought from the earliest implementations to make efficient use of the RF spectrum. This has been accomplished by frequency/geographic segmentation to allow frequency reuse. As RF systems have evolved from predominantly point to point to broadcast to interactive, the geographic space assigned to given frequencies has contracted. With the appearance of cellular systems, the concept of the cellular basestation was defined. Initial basestations were implemented as concrete buildings containing rack mounted equipment with utility provided power, adjacent to dedicated antenna towers. Each basestation serviced a relatively large area. For example, typical macro-basestation cell sites ranged from 50 to 100 kilometers.

As basestations evolved, they began to be packaged in smaller form factors and serve smaller cells – microcells. Today there are still macro basestations. In addition, however, there are smaller packages that can be mounted in a utility room on top of a building. Finally, there are small weatherproof containers with heat dissipation capacity that can be mounted on a telephone pole with attached antennas. These are commonly referred to as 'shoe box basestations'.

During this same period, WLAN access points began to appear. As such access points proliferated from being a relatively rare, end-user-deployed device, to provide carpets of footprints covering campuses, and filling hotspots for carriers, they began to take on the characteristics of 'picocellular' basestations.

One of the key aspects relative to SDR basestations is the extremely high requirement for reliable fast changeover. In the case of a single mobile device, a failure in a software download and install that requires a reload is not catastrophic. In the case of a network consisting of 1000 basestations serving 100,000 users, such a failure could be considered extremely serious. There have been examples of networks (both in the case of wired Internet access and wireless networks) where software changes (not associated with SDR) have caused major disruptions. In these cases, the network operators suffered significant immediate financial losses and longer term damage to their reputations for quality and reliability. When an SDR basestation is migrated to a new or modified AIS, the interruption in service if any, must be minimized.

1.4.2.1. High Speed Signal Processing

Requirements for SDR basestation high speed signal processing are more demanding than mobile systems in that basestations must modulate/demodulate relatively large blocks of channels. At the same time, as compared with the handset, the SDR basestation high speed signal processor has more power available and can occupy more volume. The processing load can vary dramatically from one AIS to another.

The introduction of smart antenna technology is creating another vector requiring SDR support. For example, one usage model is as follows: A network operator installs a base station in a suburban area that has recently been connected to high capacity transportation systems. Population and subscriber penetration is very low, but is expected to grow. As the area becomes more intensively used, population density and subscriber penetration grows. The operator wishes to increase the traffic handling capability of the basestation by adding smart antenna capability. Again, this is a graceful migration requirement.

These requirements combine with equipment and logistics costs. Providing more processing power than is needed for a particular AIS/smart antenna combination in order to provide a future migration path can have significant cost implications. On the other hand, savings associated with not having a wide variety of physical configurations deployed in a single network operator are also significant.

1.4.2.2. Low Speed Signal Processing

Here again the major differences in requirements for SDR basestation low speed signal processing and mobile system processing is that basestations must decode/code relatively large blocks of channels. These channels interface with a variety of formats including circuit switched voice, SS7, TCP/IP, etc.

In many cases, the basestation can avoid having to decode fully, individual streams of end-user data. Instead, the basestation decodes only the coding (if any) added to overcome unique wireless noise environments, and merely passes the rest of the data to another network component until it arrives at the destination end-user where it is finally fully decoded.

Over time, the back-end network to which the basestation hands off traffic may change, thus changing the coding/decoding requirements. Therefore, SDR requirements include the ability of basestations to migrate gracefully between different formats and mixes of formats as the network operator migrates their network back-end structure.

1.4.2.3. Packaging

Larger SDR basestation baseband processors are generally implemented on printed-circuit boards that can be plugged into backplanes. Smaller shoe-box basestation baseband processors are migrating to ‘single board’ packaging. This process is analogous to the movement from mini-computer backplane architectures to PC single-board architectures.

1.4.2.4. RF Baseband Interface

Different AISs may lend themselves to different architectures for high speed signal processing. For example, 2G basestations typically interface to the RF front-end through a single transmitter and a single receiver channel that delivers the complete band being serviced to the high speed signal processor. As WCDMA moves to 5 MHz (and beyond) baseband bandwidth blocks, some are considering separate interfaces to each block. This allows the processing load to be parsed across three, dedicated, high speed signal-processing subsystems. Similar approaches are being considered for multiple RF interfaces supporting multiple smart antenna RF baseband channels. This can lead to a requirement for reconfigurable RF/baseband interfaces to support migration to different interface architectures as AIS migration occurs.

1.5. Today’s SDR Baseband Technology Approaches

1.5.1. Technologies and Criteria

There are (at least) nine technology conceptual approaches to implementing SDR baseband subsystems that can be seen today. These are:

- general purpose processors with specialized software;
- high speed DSPs;
- multiple ASICs;
- parameterized hardware;
- switchable microcode;
- multiprocessor array (including VLIW);
- reconfigurable logic;
- analog signal processing;
- combinations of the above.

Each of these technology approaches has strengths and weaknesses. Each can be characterized by a profile of the following parameters [7]:

- power consumption;
- size;
- cost;
- field upgradeability;
- silicon evolution;
- tools maturity;
- technology risk;
- time to market (TTM);
- staff availability.

Each of the technology alternatives can be characterized as having strengths and weaknesses in each of these different areas. Breaking down the analysis into these separate areas can be helpful in determining the suitability of a particular technology to a particular application. Such an analysis has been performed [7]. In the following, the most significant aspects of the alternatives being given the most attention are examined in an effort to predict the likely direction of a generalized solution.

The most challenging portion of the baseband processor continues to be the high speed signal-processing subsystem. Although the requirements in the low speed signal-processing portion have become more demanding through the addition of simultaneous channels and new modes, it is the high speed modulation and demodulation functions that remain the most demanding.

When we look at SDR baseband technology alternatives in the context of the spectrum of requirements sets, the fundamental challenge is power, size, heat dissipation and cost. Because of the way that the requirements sets are evolving, as the size/power/heat/cost constraints are relaxed, the processing demands increase.

1.5.2. Capabilities and Constraints

1.5.2.1. Conventional DSPs

The oldest, most mature, easiest to staff and shortest time-to-market solution is to use a monolithic single stream instruction set DSP. The problem is that in order to meet the system requirements, the clock speed requirements are either unattainable or produce results which don't fit the size/power/heat/cost constraints.

Power consumption in CMOS (the lowest cost/power consuming semiconductor process generally available) is directly proportional to the combination of the number of transistors and the speed with which they move through different states (clock speed). Heat dissipation is a function of power consumption. Cost is a function of die size (the size of the raw semiconductor when it is cut out of the wafer on which it was produced and before it is packaged).

To illustrate the situation, let's go back to the handset example from the beginning of the chapter of 100 instructions per sample (number of instructions per sample are increasing due to the evolution of AISs) for the modem subsystem resulting in a clock speed in excess of 1 GHz. Performing these high speed signal-processing functions in a DSP gives us a solution that has many advantages, including being fully software defined and thus fully software flexible, but doesn't meet the power/heat/cost constraints. Now add the increasing processing demands and the problem becomes more severe.

1.5.2.2. Evolved Submicron DSPs

It has been postulated that, as semiconductor technology advances and feature sizes and resulting voltages are reduced on chips, DSPs will be able to meet the constraints. However, limitations in voltages, bias currents, etc., are creating an asymptote in power consumption (and related heat dissipation). The result is that this architecture can't meet the constraints. This defines one limit case.

1.5.2.3. Optimized Hardwired Logic

Another end of the spectrum is to implement each of the 100 modem instructions in a separate piece of fully optimized hard wired logic, then string them together in the order of processing (ASIC). The theoretical result is a modestly sized die running at 4 MHz. This architecture is characterized as highly scalar and deeply pipelined. It clearly meets the size/power/heat/cost constraints. However, it doesn't meet the SDR flexibility requirement. This defines another boundary case.

1.5.2.4. Software Configured FPGAs

The first breakthrough in SDR came with the use of FPGAs. Under software control, FPGAs could be configured in a similar highly scalar deeply pipelined fashion. Unfortunately, in order to achieve software reconfigurable flexibility, FPGAs require more transistors to implement a given logic function than does hardwired logic. There are also challenges in stringing the 'instructions' together in the correct order. These problems are characterized as place and route difficulties. This architecture does achieve the desired software programmability; however, it pays for it in bigger dies that cost more and burn more power. Some maintain that the power differential can be as much as 50 times that of ASICs. Early SDR commercial basestations and military devices were implemented this way because they could relax the size/power/heat/cost constraints. FPGA technology is not as mature as DSP technology so there are also costs in development parameters (cost, TTM, risk, etc.). This defines another limit case.

1.5.2.5. 'On-the-Fly' FPGA Reconfiguration

Some maintain that innovative FPGA routing structures plus reconfiguration times can be decreased to the point where one part of the chip can be reconfigured fast enough for only a limited number of 'hard wired' logic instruction capable hardware elements to be needed. In this architecture, while one element is processing, the next element is being configured for the next instruction, and then the one after is ping-ponged back to the first element. If this were done, the silicon area would be reduced but the clock speed would be increased. The final result in size/power/heat/cost for such an architecture is not clear because, to date, none has been publicly demonstrated.

1.5.2.6. Multiple Processor Arrays

Another approach is to have multiple instruction set processors. At the limit, there would be one for each of the 100 instructions strung together. Now, the resulting system could lower the clock speed 100-fold, but if the processors where the same size as the single stream instruction set processor, the die size would increase 100-fold (or more because of the routing problem). This is another limit case.

1.5.2.7. Optimized Processors

Another approach is to optimize the processors in this array. One way to do this is to use reconfigurable microcode. In this architecture, special machine instructions (CISC on steroids) are created that are optimized for functions required to support a particular AIS. Now time/power/space resources must be used to store, determine the correct instantiation of, retrieve, and load the correct microcode at the correct time. This also consumes silicon space, time power, etc. (instruction set resource management). One way of reducing the resource management problem is to have all processors load an instruction (or microcode) at the same time. This is often referred to as a single instruction multiple data (SIMD) architecture. If the algorithm being implemented lends itself to this kind of processing (for example a series of FIR, finite impulse response, filters that are the same and connected in series), it may increase the efficiency of the resulting processor. This is only practical if the AIS can be represented by an algorithm which fits the constraints. If so, in this way, it may be possible to reduce the number of instructions tenfold and thereby reduce the clock speed. This is another limit case.

1.5.2.8. Parameterized ASICs

Finally, it is possible to create ASIC blocks that are tailored to one very particular function, but within that function to allow for software control of a particular parameter. For example, a FIR filter that can be programmed for one or another limited number of different taps, or a rake filter which can be programmed for a limited number of different forks.

1.5.3. A Way Forward?

For several years now, proponents of each of these different architectures have been competing in the SDR space. Each has sought to optimize their own particular solution to the

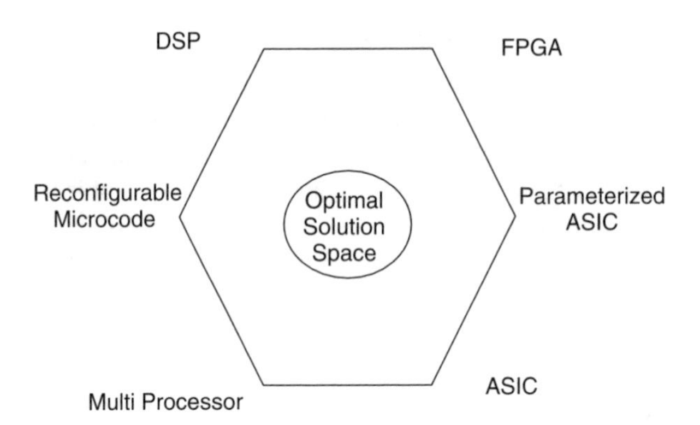

Figure 1.4 SDR baseband processor solution space

point where it can meet the full set of SDR requirements. Many observers now believe that the truly successful solution path is not the optimization of one particular limit case, but rather some combination of the architectures that lies within the interior of the space defined by the limit cases. The relationship of the limit cases is graphically represented in Figure 1.4.

For example, a limited number of small single stream instruction set processors each with reconfigurable microcode (multiple instructions – multiple data architecture), connected with a small set of parameterized ASIC blocks by a reconfigurable interconnect matrix, might produce the optimal processor architecture. Such an SDR baseband architecture might have several different instantiations, one optimized for each of the different AIS/application requirement sets. This optimal solution space is represented in Figure 1.4.

One of the key challenges in creating innovative hardware architectures like these lies in the programming models employed and the resulting availability of tool sets and qualified programmers. The best imaginable solution would be one that could make full use of existing (without requiring any new ones) programming models and tool sets. If this is not fully achievable, the closer it can be approached the better.

1.6. Conclusions

In this chapter we have traced the evolution of SDR baseband processing, discussed the range of requirements for SDR baseband processing systems in different form factors and applications, and considered some of the technologies that are available to implement SDR baseband processing subsystems. We have seen that evolution of AISs and usage models is creating processing demands that are increasing faster than Moore's law increases in semiconductor performance. This has led us to the conclusion that SDR solutions require innovations in processor architecture. Although some are strong proponents of a single architectural approach, given the evolving requirements and the technologies available to meet them, it is possible that the optimal solution will be found in a combination of available processing architectures.

Acknowledgments

The author would like to acknowledge assistance in building the author's understanding of the subject matter from Professor Emeritus Michel J. Flynn, Professor Tadao Nakamura, Professor Martin Morph, Ravi Subramanian and John Watson.

References

[1] M. Cummings and S. Heath, 'Mode Switching and Software Download for Software Defined Radio – The MMITS Forum Approach', *IEEE Communications*, February, 1999.
[2] *Software Defined Radio Forum Technical Report (TR) 2.1*, Edited by SDR Forum Technical Committee (Chair M. Cummings), November 1999, SDR Forum, NY, USA.
[3] M. Cummings *et al.*, Method and Apparatus For Communicating Information, USA Patent No. 5 907 580, 25 May, 1999.
[4] M. Cummings *et al.*, Method and Apparatus For Communicating Information, USA Patent No. 6 175 589 B1, 16 January, 2001.
[5] M. Cummings and S. Haruyama, 'FPGA in the Software Radio', *IEEE Communications*, March, 1999.
[6] M. Cummings, 'A Multimode SDR radio frequency implementation', Chapter 3 in *Software Defined Radio: Enabling Technologies*, W.H.W. Tuttlebee (Ed.), John Wiley & Sons, Ltd, Chichester, 2002.
[7] 'Puzzle of software defined radio has multiple solutions', *Electronic Engineering Times*, 16 October, 2000.

Part II

Handset Technologies

An unrelenting demand of handsets is low power consumption – the early slow take up of 3G has in part arisen from consumers' battery life expectations being conditioned by that of mature 2G products. Yet the processing demands of 3G are seriously greater than 2G. SDR offers the promise of baseband technologies that deliver Reconfigurability *and* low power consumption.

2

Open Mobile Handset Architectures Based on the ZSP500 Embedded DSP Core

Jitendra Rayala and Wei-Jei Song

LSI Logic Corporation

2.1. Introduction

Over the last decade, wireless communications has grown rapidly to provide high quality voice services to hundreds of millions of users worldwide. One of the main reasons for this has been the availability of increasingly affordable and small mobile handsets. These mobile handsets have been evolving over the years, adding newer features and lower power consumption, thereby increasing the battery life and reducing the overall size of the handset. Programmable DSPs have been instrumental in making these handsets available with the right features at the right moment in time. This ability of being able to implement complex wireless systems on programmable devices is the first step towards the future goal of realizing complete software radio systems.

With the advent of commercially available and reliable Internet services over a fiber-optic backbone, a demand for high data rate Internet services over wireless devices has emerged, resulting in the need for wireless Internet connectivity over a metropolitan area alongside the voice services that the present networks provide. In order to address this need for data services, newer standards have been developed, collectively called the 2.5G/3G technologies. 2.5G/3G mobile handsets are expected to provide high-rate data services for users in addition to today's voice services. It is expected that with the availability of high data rates, networks will be able to provide reliable multimedia services such as audio/video streaming and messaging, telemedicine, etc., and access to banking services and corporate networks. In order to realize these services, mobile handsets need to be designed with new required features.

In recent times, the design cycles for mobile handsets have reduced dramatically, from 2 to 3 years to around 1 year, or even less. This has put enormous pressure on engineering design

Software Defined Radio: Baseband Technologies for 3G Handsets and Basestations. Edited by W. Tuttlebee
 ISBN: 0-470-86770-1

teams to bring new products to market in a very short time. In the early days of mobile handset design, most of the functionality was implemented in fixed hardware. The DSP was used mainly for the voice coders as the baseband MIPS requirement for earlier wireless standards such as IS-95 and GSM was beyond the reach of programmable DSPs at that time. With the tighter time-to-market constraint, a complete fixed hardware implementation is no longer feasible and, with the demand for multimedia services and the availability of much more powerful and lower power DSPs, it is no longer economical. Also, since future mobile handsets will have to support multiple standards, the underlying hardware cannot be changed as standards evolve. These two reasons make it imperative for designers to use software programmable solutions and hence drive the requirement for software defined radios.

> According to the Software Defined Radio Forum [1] (quoted *verbatim* here), '*the term software defined radios (SDRs) is used to describe radios that provide software control of a variety of modulation techniques, wide-band or narrow-band operation, communications security functions (such as hopping), and waveform requirements of current and evolving standards over a broad frequency range.*'

SDR represents a major design change in communications technology because it brings solutions that are flexible and programmable.

In the mobile communications space, the use of SDR is currently limited to its use in base stations. However with the advent of low power, high performance, programmable DSPs, SDR is quickly becoming the key technology enabler in mobile handset design. In order to take advantage of SDR technology, there is a need for open architectures in both the hardware and software. Open hardware architectures will allow designers to target different applications with a common underlying hardware and open software architectures will allow portability and reuse of software components for different applications.† In this chapter, we consider various issues in defining open hardware architectures for mobile handsets and present the foundations of an open architecture based on a high performance programmable DSP core.

2.2. Handset Requirements

In order to realize the dream of software based radios, the handset architecture has to be open, flexible, scalable, power efficient, and able to support complex algorithms. These aspects will drive the evolution of future handset architectures.

2.2.1. Flexibility/Multifunctionality

A flexible architecture would allow the mobile handset to provide support for multiple wireless standards. With the proliferation of different wireless standards in different geographical areas, as well as the different service levels within the same standard, the need for supporting at least a few of them in a handset is growing rapidly. This would allow the handset to roam in multiple networks with different air interfaces. This means the handset must be able to support hugely varying requirements of 2G/2.5G/3G standards. Also, with a flexible architecture, as standards continue to evolve, the functionality of the handset can be made to follow the standards without changing the underlying hardware.

† This philosophy is evidenced in the establishment within the SDR Forum of a Hardware Abstraction Layer Working Group tasked with addressing such issues.

2.2.2. *Multimode/Adaptive Modulation*

In order to support roaming in different networks, the handset should be able to switch in and out different modulation schemes as and when it enters or leaves different networks. This can be achieved more easily with a software programmable architecture than with a completely fixed hardware architecture. Also, when multiple modulation schemes are supported within a given network, depending on the network conditions, the mobile should be able to adapt to different modulations to guarantee quality of service (QoS). Reprogrammability allows the mobile handsets to be upgraded to support enhanced services as and when they are introduced into the network without replacing the unit.

2.2.3. *Scalability/Modularity*

A scalable architecture would allow introduction of newer standards without complete redesign of the hardware platform. If the hardware and software can be partitioned in a *modular* way, then newer hardware modules could be added to the system incrementally without changing the underlying architecture, thereby achieving scalability. Modularity also allows reuse of most of the components of the system, so greatly easing manufacturing, and would allow the designer to characterize and optimize the power consumption incrementally.

2.2.4. *Power Efficiency*

A fixed hardware designed solution may provide better power efficiency than a programmable solution in a single-functionality device at the expense of flexibility. However, when mobiles must provide multifunctionality with multimode operation, a programmable solution would offer better power efficiency. In the case of fixed hardware, entire individual systems have to be glued together whereas a programmable solution can reuse some of the common components. For example, a Turbo/Viterbi decoder that is common to multiple standards could be made to be flexible and reused.

2.2.5. *Algorithm Support*

In order to realize the kind of services that future mobile handsets are expected to provide, various algorithms and technologies have to be incorporated and integrated. The mobile handset must have adequate resources to support all the required tasks and at the same time be sufficiently flexible, as discussed earlier. As an example of a multiple standard, multimode mobile handset, we consider in this chapter an open architecture for a dual mode GSM/EDGE and WCDMA mobile that can handle both the baseband algorithms and multimedia applications. The following discusses some of the algorithms that can be expected in such a dual-mode mobile handset. This is only meant to provide a glimpse of the complexity of tasks that multi-mode mobile handsets have to handle. We do not discuss all the design and implementation aspects of the above standards.

2.2.5.1. Equalization

In GSM/EDGE mobile handsets, equalization has to be performed using maximum likelihood sequence estimation (MLSE) [2] to mitigate the multipath effects of the channel. This is implemented efficiently using the Viterbi algorithm. The computational complexity of MLSE is directly related to the number of states of the underlying trellis structure of the Viterbi algorithm. The number of states in turn is related to the maximum length of the channel impulse response that the EDGE receiver is expected to compensate. For example, for a channel impulse response length of $(L+1)$ and a signal constellation with size 2^n, the number of states required would be $N=2^{nL}$. As can be seen, as L increases, the number of states increases exponentially. The GSM/EDGE mobile handsets have to be designed such that they have enough performance to handle this computationally demanding task.

2.2.5.2. Rake Receiver

CDMA is the predominant multiaccess/modulation technology among all the air interfaces that are incorporated into the 3G standards. WCDMA is a variant of the direct sequence spread spectrum technique that typically uses larger bandwidth than the coherence bandwidth of the channel. This results in the channel appearing to be frequency selective, so a rake receiver can be used very effectively for demodulation [3].

A rake receiver consists of number of fingers, each of which de-spreads the received signal with a correlator; the outputs of these fingers are phase rotated and are combined with an appropriate combining strategy. The number of rake fingers depends on the channel profile and the chip rate. A separate channel estimation algorithm measures the channel profile. The signal after the rake combining is de-interleaved and channel decoded. Since WCDMA uses a high chip rate, it would be extremely difficult to implement this complexity of processing on traditional programmable DSPs, so hardware acceleration is needed.

A WCDMA base station may use multiuser detection after the rake search to improve the uplink performance in the presence of co-channel interference. This process needs the knowledge of the code sequences of all the users. Only base stations have this knowledge; also these algorithms are computationally intensive. However, if short codes are used, mobiles can also incorporate multiuser detection, making use of the cyclostationary properties of the signals. Future mobile handsets may employ multiuser detection to improve the downlink performance; consequently the architecture should be able to support this without a requirement for extensive changes. This necessitates the need for modular and reconfigurable architecture.

2.2.5.3. Adaptive Antennas

As the number of users in a given network and the demand for higher data rates increases, there is need to improve the usage of frequency bandwidth and reduce the co-channel interference among the users. One method of doing so is to introduce diversity at the transmitter and receiver using multiple antennas. By using multiple antennas, we can improve both the signal-to-noise and the signal-to-interference ratios substantially [4].

Multi-antennas are classified into two categories:

- switched beam arrays, and
- fully adaptive arrays.

Switched beam arrays require less computational complexity than do fully adaptive arrays. However, fully adaptive arrays provide higher performance and are more flexible [5]. The type of array to use is determined by factors such as cost, performance required, etc. Various algorithms have been developed for both adaptive and switched beam arrays, most of which are iterative in nature, perform many decision-oriented computations and require flexibility. The choice and performance of the algorithms depend on the network conditions, modulation scheme, etc. Software-based solutions can facilitate the choice of algorithm dynamically depending on the operating conditions. Further, the algorithms have to work alongside other baseband algorithms in the handset; however some of the adaptive algorithms may need high computational throughput and extra hardware support. Many of the adaptive algorithms have similar computational structures; by identifying possible commonality in the algorithms, extra hardware support can be provided for the blocks with high computational complexity, whilst other parts of the algorithms can be implemented in software, to retain flexibility.

2.2.5.4. Channel Decoding

In both GSM/EDGE and WCDMA, convolutional channel coding is used to increase the performance of the communications link under noise and multipath effects. In WCDMA, apart from convolutional codes, turbo codes can be used to improve the link performance. Soft output Viterbi algorithm (SOVA) can be used to decode the convolutionally coded sequences in order to obtain better performance. Turbo-coded sequences can be decoded using an iterative algorithm based on SOVA [6]. The computational complexity of these algorithms depends on the data rate. At lower data rates, they can be implemented on programmable DSPs but for higher data rates they may need extra hardware support. A good hardware–software partitioning can be obtained since SOVA is common for both convolutional and turbo codes.

2.2.5.5. Speech Codecs

Speech coding is an essential part of any mobile handset and has been instrumental in making wireless telephony ubiquitous. Speech coding is used to remove the redundancy in a speech signal; because of this, the speech signal can be coded at a lower bit rate so that the coded signal requires a smaller bandwidth.

The speech codecs used in GSM/EDGE are based on narrowband adaptive multirate (AMR) coding. AMR has been shown to provide good performance across a wide range of network conditions. Narrowband AMR coding has recently been extended for wideband speech signals for use in WCDMA and has been adopted as an ITU-T standard [7]. Speech coding algorithms are very complex in nature and they perform intensive signal processing operations. They also involve large decision-oriented computation and, for these reasons, fixed hardware implementation of speech coding algorithms is extremely difficult, making programmable DSPs essential for their implementation on mobile handsets.

2.2.5.6. Multimedia

WCDMA networks provide higher data rates and as a result, network service providers now can offer new services like streaming of video and audio content to consumers. There are many applications such as multimedia streaming, multimedia messaging, telemedicine, etc., that can take advantage of the high data rates. However, both video and audio data needs to be compressed even for these high data rates. MPEG-4 [8] is one of the multimedia coding standards that future mobile handsets will need to support.

Video coding refers to compression of video signals and is a computationally intensive process. Motion Estimation and Transform Coding are the two main algorithms that consume the majority of the computation. MPEG-4 video coding also needs considerable decision-oriented computation. The actual amount of computation depends on the frame size and the frame rate. At low frame rates and small frame sizes, video coding can be implemented on programmable devices. However large frame sizes and faster frame rates would need hardware acceleration.

Audio coding refers to compression of audio signals and, like speech coding, the coding algorithms are very complex in nature and perform intensive signal processing operations. They also involve large decision-oriented computation and, for these reasons, fixed hardware implementation of audio coding algorithms is extremely difficult, making programmable DSPs essential for their implementation on mobile handsets. As we see from audio and video coding, we need both programmability and hardware acceleration, so these are amenable for good software–hardware partitioning.

2.2.5.7. Speech Recognition

Present generation mobile handsets use a very simple user interface in the form of numeric keypads. Users navigate using menus and scroll bars. Since the handsets are intended for use when the user is moving, the current user interfaces are inconvenient to use. Automatic speech recognition (ASR) is one of the technologies that can significantly enhance the user interface of a mobile handset. Various algorithms have been developed for ASR [9], all of which are computationally and memory intensive.

Since mobile handsets have limited memory and power resources, different speech recognition algorithms have to be developed. A speech recognition system using a combined DSP and general-purpose processor (GPP) has been described [10]. In that system, the computationally intensive parts of the algorithms such as the hidden markov model (HMM) are executed on the DSP and the memory intensive parts, such as grammar, dictionary, etc., are run on the GPP. Since the recognition algorithms are decision-oriented and computationally intensive, a fixed hardware implementation is extremely difficult and a programmable solution is needed. It is also easier to modify and update the algorithms with a programmable solution.

2.2.5.8. Biometrics

Future mobile services will require a reliable knowledge of the identity of the user. For example, identification could be used to verify that the user of the mobile handset is the actual owner. Other application examples would include access to banking services, access to information on corporate networks, access to multimedia services, etc. Biometric identification is one of the ways to obtain reliable user identity recognition by using a person's physiological or behavioral characteristics such as retina, speech, fingerprint, iris, face, etc. [11], or a combination of these, to perform automatic personal identification.

Since mobile handsets have limited power and memory resources, the number of characteristics that can be used for identification is limited. Identification based on fingerprint is one good candidate for implementing on mobile handsets. Finger-print identification involves the following steps: capturing the image, pre-processing and filtering, extraction of features and finally pattern matching. All these steps require complex algorithms and decision-oriented computations. Implementing them in fixed hardware would be extremely difficult and programmable devices are needed. However, some of these algorithms are computationally intensive, so the programmable device requires enough processing power to implement them. If the performance required by these algorithms is beyond the capability of the programmable device, then the architecture should be able to support the partitioning of the algorithms onto both hardware and software.

2.2.5.9. RF Control

A multistandard mobile handset has to operate in multiple frequency bands. The RF transceiver in the mobile should be able to change its frequency band depending upon the network with which it is operating and also when it moves across networks. In order to identify the network, the baseband has to make measurements and determine the modulation scheme to use. That means the baseband processor should be able to tune the transceiver to a particular frequency band adaptively. Currently, multistandard RF transceivers [12] are being developed that can be tuned by a baseband processor. Since the baseband processor has to perform measurements and take decisions adaptively, this can be done more easily by using a software solution than a fixed hardware solution.

2.3. An Open Mobile Handset Architecture

The dual requirement of high computational throughput and good flexilibility with programmability necessitate good hardware–software partitioning of the architecture. In Figure 2.1 an open multimode mobile handset architecture is shown. There are two main subsystems in this architecture, apart from the analog front-end. The first subsystem consists of a baseband algorithms processor (BAP) and a MAC/link layer. The second subsystem consists of a media algorithms processor (MAP) and an OS/user interface layer. In this chapter we mainly discuss the baseband and media processors. The MAC/link and OS/user interface layers may

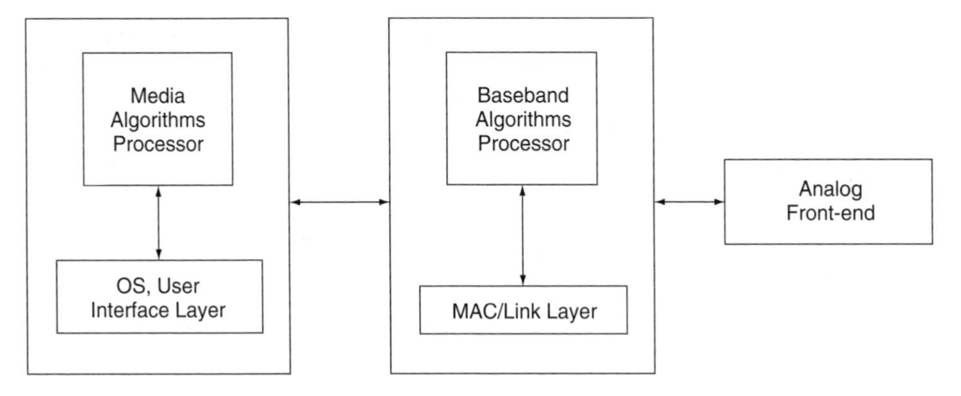

Figure 2.1 Open mobile handset architecture

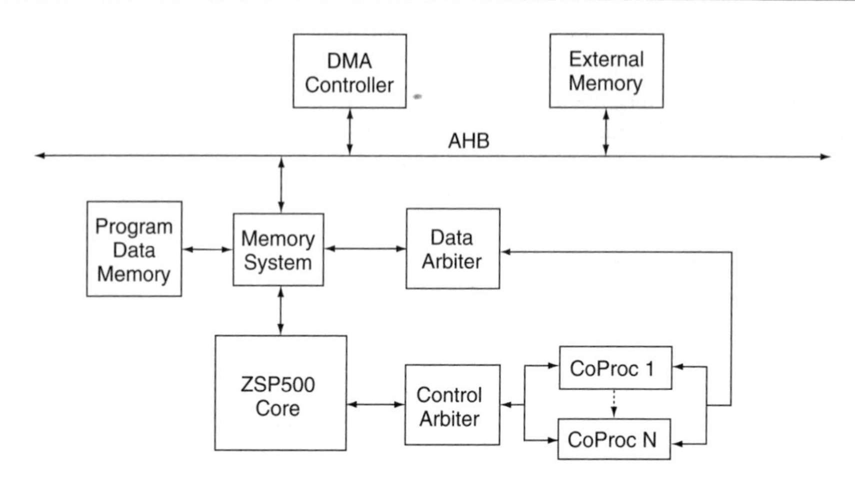

Figure 2.2 Open baseband/media processor architecture

be implemented on general-purpose processors (GPP), or they may be combined and implemented on one GPP with appropriate interfaces to the baseband and media processors.

The baseband processor mainly deals with the demodulation and channel decoding algorithms and the media processor deals with media applications such as audio and video coding, etc. In the previous section, some of the complex algorithms that mobile handsets have to support were discussed. A good partition and mapping of these algorithms would be:

- The baseband algorithms processor may handle algorithms like equalization, rake receiver, channel decoding, adaptive antennas and RF control.
- The media algorithms processor may handle algorithms like speech coding, video and audio coding, speech recognition and biometrics.

The baseband and the media processors use a similar architecture for implementing the complex algorithms. Their configurations may differ in a particular implementation depending on the target algorithms. An open architecture for both baseband and media processors based on ZSP500 DSP core is shown in Figure 2.2.

From the figure we note that there are two crucial pieces in the architecture that are required to achieve the twin objectives of high performance and flexibility/programmability. These are the ZSP500 high performance programmable DSP core and the hardware coprocessor support. The ZSP500 programmable DSP provides the flexibility, programmability and also at the same time high performance. Any additional performance required and modularity is obtained through the coprocessor support. In the following, we discuss the components of this architecture in more detail.

2.3.1. The ZSP500 DSP Core

As seen earlier, for multimode mobile handsets, we need a programmable DSP with high performance and low power dissipation. The ZSP500 is a programmable DSP based on multi-MAC, multi-ALU, four-way superscalar RISC architecture. The ZSP500 has been designed to accommodate both the baseband processing algorithms and media processing algorithms.

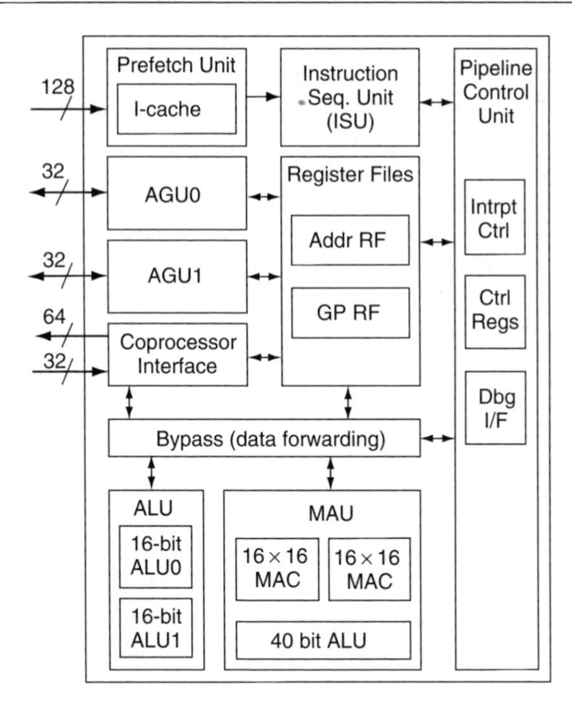

Figure 2.3 ZSP500 block diagram

It has a rich instruction set architecture (ISA) that has been tailored for the kinds of algorithm discussed above. The core architectural elements and the instruction set have been designed after careful study of various standards and applications. A simplified block diagram of the core architecture is shown in Figure 2.3 and a brief overview is given in the following.

2.3.1.1. Architecture Overview

The ZSP500 is a multi-multiply-accumulate (MAC), multi-arithmetic-logic-unit (ALU) and superscalar architecture. The architecture is scalable to satisfy the performance, power and cost trade-offs of current and future mobile handsets.

The key features of the core architecture are:

- **Superscalar Architecture**: It is a four-way superscalar, meaning that up to four instructions can be executed simultaneously if the instructions do not have any dependencies. When running at 250–300 MHz, it can reach a peak performance of 1000–1200 MIPS.
- **Hardware Scheduling**: It uses static grouping rules to determine the number of instructions to be issued to each clock cycle at run time. The actual number of instructions executed depends on the data, and on the required hardware resources being available to execute the instructions.
- **Arithmetic Features**: There are two MAC units and two ALUs. Two multiply-accumulate operations can be performed in one cycle. One complex multiplication can be completed in two cycles and one 32×32 bit multiply-accumulate can be performed in a single cycle.

The 32×32 multiplication lends itself to efficient implementation of audio codecs. The MAC units also have an extra ALU, so there are a total of three ALUs. With the ALU units, three 16-bit arithmetic operations or two 16-bit and one 32-bit arithmetic operation can be performed in single cycle. All the ALU operations can be performed in parallel with the MAC operations, so high throughput can be achieved.

- **Register Files**: There are sixteen 16-bit general purpose registers, eight 32-bit address registers and eight 16-bit index registers. Two general-purpose registers can be combined to provide a 32-bit register. There are no restrictions on the usage of the general-purpose registers. All pairs of general-purpose registers can be used as accumulators. The address registers and the associated index registers provide very efficient pointer and table lookup support.
- **Load/Store Architecture**: A variety of load and store instructions optimizes memory operations. Data that is required for an operation is moved from memory to a register to perform the operation, and the result is moved back to memory. All arithmetic instructions operate only on registers, which conserves processor-to-memory bandwidth. There are two dedicated address generation units (AGU) for the purpose of loading and storing data. The AGU units use the address and the index registers and all the address calculations are performed by the AGUs, so the ALU/MAC units are left free for performing mathematical computations. During each clock cycle, the AGUs can load/store up to four 16-bit words per cycle. At 250 MHz, ZSP500 can sustain 2 Gbytes/s of data bandwidth.
- **Other Features**:
  - The ZSP500 uses static branch prediction using a set of static rules to minimize penalties due to branching.
  - Packets of up to eight instructions can be marked for conditional execution. Their results are written, dependent upon a selected condition flag being true or false.
  - Bit fields can be extracted from a general purpose register, and zero or sign extended into another register. This feature greatly enhances the ability of the ZSP500 to process bitstream data found in audio and video codecs and bits-to-symbol mapping in communications efficiently.
  - The ZSP500 has a low latency interrupt service mechanism with programmable priority for all the interrupts, and
  - has one-cycle add–compare–select (ACS) for efficient implementation of the Viterbi decoding algorithm.

As mentioned earlier, many of the algorithms to be supported involve not only pure data-oriented signal processing computations but also decision- or control-oriented computations, both of which will influence the architectural efficiency of a DSP. In order to achieve high performance, a DSP must therefore provide support for both kinds of computation, and the ZSP500 provides a well-balanced support for both types of computation. For example, the ALU/MAC/ACS instructions provide support for signal processing computations whilst the static branch prediction/bit manipulation/ALU instructions provide support for control-oriented tasks.

The above are only some of the key features of the ZSP500 architecture; details of many other architectural features may be found elsewhere [13]. The architecture provides the high performance needed for baseband and media applications, as is evident from the high benchmark scores that it has received in all the industry standard benchmarks.

2.3.1.2. Performance Benchmarks

According to BDTI [14], the ZSP500 achieves a benchmark score of 2570 when running at 325 MHz. BDTI benchmarks include not only computationally intensive signal processing algorithms such as fast fourier transform (FFT), Viterbi decoding, etc., but also control-oriented applications. So a good BDTI score represents the suitability of a DSP, not only for data-oriented signal processing algorithms but also for control-oriented applications. The high BDTI score reflects the architectural efficiency of the ZSP500 and shows the level of performance that it can provide for mobile handset applications.

2.3.1.3. Low Energy Consumption

Today's wireless applications need increasingly lower energy consumption. Lower energy consumption will mean that the mobile handset can provide longer talk time, data transmit time and standby time. As the number of features that a mobile handset must support grows, so the need for the DSP to be more energy efficient becomes increasingly important. The ZSP500 has been designed for low energy consumption and there are three factors that contribute to this capability.

(i) **Hardware Power Reduction**: The target technology for the ZSP500 implementation is based on CMOS logic. Devices based on CMOS logic consume current only during switching. A distributed clock on the device controls the switching activity – controlling the clock activity to the various processor subsystems can reduce the energy consumption.

The ZSP500 has been implemented with extensive clock control or clock gating. Clock gating reduces the switching activity by temporarily disabling the clock to flip-flops within the device. The intelligence built into the pipeline control unit will determine what logic will be switched during a particular cycle. Clock gating will then disable the clock signal to the logic that is not utilized during that particular cycle. Clock gating has the ability to reduce the overall switching activity within the device on a cycle-per-cycle basis. Control of clock gating is not available to the programmer but is controlled by the intelligence built into the pipeline control unit, thus simplifying the programming requirement.

At the system level, the programmer can control the energy consumption using a set of modes. Three modes are available to the programmer by which the clock to complete subsystems can be stopped – these modes are idle, sleep and halt. In the *idle* mode, the ZSP500 core remains inactive but any on-chip peripherals present remain active; an external interrupt or non-maskable interrupt (NMI) or reset can wake up the device. In the *sleep* mode, both the core and peripherals remain inactive and any external interrupt or NMI can wake up the device. In the *halt* mode everything becomes inactive and only a NMI or reset can wake up the device.

(ii) **High Performance**: Instantaneous power is often used as a metric for characterizing the power consumption of a DSP; however, energy is a more appropriate metric because it tells us how the power is consumed over time. A high performance processor may have slightly higher instantaneous power consumption but would take less time to complete an algorithm than would a processor with lower performance and slightly lower instantaneous power consumption. Since energy is power consumed over time, the processor with higher performance could consume less energy, because it can complete a given task in shorter time and for the rest of the time could be put into an idle mode. So in power-critical applications, energy consumption should be considered rather than instantaneous power consumption as a metric. According to BDTI [15], the ZSP400, which is a predecessor of the ZSP500, is rated as the most energy efficient processor in the market today. The ZSP500 typically provides 40–50% better performance than the ZSP400.

(iii) **High Code Density**: Code density is one of the most important metrics for DSP because it is directly related to the amount of memory that is needed to implement a given application. A DSP that needs larger memory would consume more energy for two reasons:

(a) The larger memory will itself consume more energy because chip die area increases with the memory size.

(b) Since more instructions have to be fetched from memory, their transfer to-and-from the processor will consume more energy.

Code density is the number of bits required to perform a given task. High code density would mean lower energy consumption because die size is reduced and the number of instructions to fetch is also reduced. According to EEMBC [16], the ZSP400, predecessor to the ZSP500, has best-in-class code density among all processors. The ZSP500 typically provides 10–15% higher density than the ZSP400, so it has one of the best code densities. Consequently it will consume less energy at the system level.

2.3.1.4. Programming

The increasingly short product design cycles and the time-to-market issue have been explained earlier. For mobile handsets, time-to-market is influenced by the programming model and the ease of programming. An easier programming model and a rich but orthogonal instruction set allows faster code development. Apart from writing the code, optimizing and debugging the code becomes easier. When trying to obtain the absolute best performance from the processor, ease of programming becomes essential, especially at the assembly language level. To obtain good performance from the processor, assembly level programming is imperative and an assembly language with simple and efficient syntax and semantics is essential.

The instruction set and the programming model of the ZSP500 has been designed with the above objective in mind. Along with the key features mentioned earlier, there are other features of the ZSP500 that make it possible to generate efficient assembly language code in a very short span of time. We briefly discuss them below:

- **Instruction Set**: The ZSP500 is a RISC based superscalar architecture with a simple RISC-like orthogonal instruction set. Most instructions execute in a single cycle. Because of this, the ZSP C-compiler is very efficient and generates compact program code. As mentioned earlier, according to EEMBC [16], the ZSP processors have best-in-class code density. This is the result of the efficiency of the ZSP C-compiler and the simple RISC-like orthogonal instruction set.
- **Memory Structure**: The ZSP500 has a simple, contiguous data space supporting memory mapped I/O and can support both Von Neumann and Harvard memory schemes. The core has a stallable memory interface in which all data and addresses are registered at the core

boundaries. The core makes very few assumptions regarding memory sizes or whether the memory is internal or external in a system implementation. The core has a small instruction cache and an external memory controller supports a data cache. These caches enhance the performance and reduce the energy consumption.

- **Hardware Scheduling**: At run time, the processor determines how many instructions it can execute by using a set of rules. This grouping of instructions is performed dynamically at each cycle. The number of instructions that are grouped is dependent upon the resources required and the availability of data for execution of those instructions. This releases the programmer from the burden of grouping the instructions during compilation. Programmers can write logically correct and sequential code and be confident that the code will execute correctly when multiple instructions execute in parallel. Programmers can also use the static grouping rules to optimize the assembly code in order to improve the performance of the code. The majority of the grouping rules are simple and intuitive. The software development tools provide all the information about resource usage and what grouping rule would be used by the hardware on a cycle-by-cycle basis. The programmer can use this information to reorder the code very easily and force the hardware to group the instructions in an optimum way so that as many resources as possible are used efficiently.
- **Static Branch Prediction**: A set of static rules is used to minimize penalties due to branches. As mentioned earlier, many of the modern algorithms involve not only data-oriented signal processing computations, but decision- or control-oriented tasks. By using static branch prediction, branch penalties can be reduced substantially by profiling the way that branches occur in a given algorithm and performing branching accordingly. This removes the need for programmers to write nonportable code that uses dead time or delay slots following branches.
- **Interrupt Structure**: The core has a low latency interrupt service mechanism with programmable priority for all interrupts. Flexible interrupt control gives the programmer the ability to dynamically alter the priorities of the interrupts depending on response to certain events or the results of computation. There are no restrictions on which sections of the code are interruptible. Hence the programmer need not worry about some noninterruptible section increasing the latency and hence affecting the real-time behavior of the program.

2.3.2. The Role and Use of Coprocessors

As mentioned earlier, an efficient approach for achieving performance and functionality with flexibility is to combine a programmable device like a DSP with hardware acceleration. There are two ways in which hardware acceleration can be provided.

2.3.2.1. Hardware Acceleration through Extended Instructions

The simplest way to achieve hardware acceleration is to identify the most basic operations of an algorithm and provide instructions that can handle those operations. For example, sum-of-products is a basic operation in an algorithm like an FIR filter. So if a MAC instruction is available, then FIR filters can be implemented efficiently. Similarly, various other instructions can be defined and tailored to reflect algorithmic structures. The ZSP500 provides various special-purpose instructions such as ACS (add–compare–select) and complex multiplications, etc. An instruction like ACS is very useful for implementing a convolutional channel decoder.

An efficient implementation of a convolutional channel decoder using the ACS instructions has been described [17]. However, when the computational performance requirement is large, it may still be beyond the capability of these dedicated instructions or may represent inefficient use of a programmable DSP.

2.3.2.2. Hardware Acceleration through Coprocessors

The second type of hardware acceleration is through the use of coprocessors. Coprocessors are computational engines that are external to the core, communicate with the DSP through an interface, and extend the arithmetic capabilities of the DSP. Coprocessors can be categorized based on functionality into two categories, the tightly coupled coprocessor (TCC) and loosely coupled coprocessor (LCC) [18].

2.3.2.3. The Tightly Coupled Coprocessor

In a TCC, the coprocessor can be viewed as an extension of the data path of the DSP. A TCC is tightly coupled to the pipeline of the DSP and may have access to the registers and other internal data paths of the DSP; similarly the DSP has complete access to the coprocessor. The DSP will initiate a task on the coprocessor, the task will be completed a few cycles later and the results will be communicated to the DSP. This approach makes sense when the performance requirements are moderately higher than those provided by the DSP stand alone and the coprocessor is expected to work on small amounts of data. Since the coprocessor is tightly coupled to the pipeline, large data transfers would become a bottleneck. As the DSP speed and pipeline length increases, a TCC may be absorbed into the DSP and may become another dedicated instruction.

As described earlier, in video coding, motion estimation is one of most computationally expensive algorithms. In one example implementation, a software based implementation for a full search algorithm takes about 83% of the total computation [19]. A TCC implementation of the motion estimation algorithm, which is well suited to hardware acceleration, has been described by Wichman *et al.* [19], wherein the coprocessor approach yielded a 63% reduction in the overall computation.

2.3.2.4. The Loosely Coupled Coprocessor

In the LCC approach, the coprocessor is decoupled from the pipeline of the DSP and runs independently of the data path of the DSP. This approach makes sense when the amount of data to be processed and/or the performance requirements are large. TCCs are generally small compared with the DSP. LCCs can be large and may even be as large as a DSP. The size depends on the processing requirements, amount of local memory needed, etc. In this case, the DSP would typically send the address of a block of data to be processed and will continue to work on other tasks, whilst the LCC completes the task and then interrupts the DSP to communicate the results.

LCCs are ideal for computationally intensive tasks like a rake receiver, Viterbi and turbo decoder, etc. Recall the open architecture for both baseband and media processors as depicted in Figure 2.2. This architecture has been designed to support various combinations of multiple LCCs and to provide modularity in a plug-and-play fashion.

2.3.2.5. Architectural Elements to Support Coprocessing

There are four elements that are important to support high performance operation of the DSP in a coprocessor architecture:

(i) **Memory System**: The memory system is responsible for providing master/slave interfaces to ensure appropriate data movement. It provides a slave interface to the DSP core memories and master interfaces for the coprocessors. It also provides multiple parallel interfaces, to ensure that adequate memory bandwidth is available for the core and coprocessors.

(ii) **Data Arbiter**: The data arbiter performs the arbitration among the coprocessors when there are contentions in the memory accesses. It also ensures that the coprocessors can access data across multiple interfaces. The coprocessors may have their own local memory, or may use the DSP core on-chip memory, or may even access external memory through DMA engines. For example, a rake receiver may receive data directly from the analog front-end and after processing would write the data into the DSP core on-chip memory. The DSP would then perform symbol-rate computations. That means the coprocessor should be able to access the data across multiple interfaces. The data arbiter would resolve the contentions and ensure that coprocessors can access data across multiple interfaces.

(iii) **Control Arbiter**: With multiple coprocessors, there could be contentions when they try to interrupt the DSP core after completing their tasks. Even though the DSP core can handle multiple interrupts by assigning priorities, an external arbiter allows smoother handling of the interrupts. However, the main advantage of a control arbiter is energy conservation. With a control arbiter, unused coprocessors can be shutdown easily, minimizing power consumption. For example, when the mobile handset switches from WCDMA to GSM/EDGE, the rake receiver and turbo decoders could be shutdown thereby reducing power consumption.

(iv) **DMA Controller**: As mentioned earlier, a rake receiver coprocessor may receive the data directly from the analog front-end. In order to transfer high-speed data into the coprocessor without the intervention of the DSP core, high-speed DMA engines are necessary. This reduces or eliminates the overhead on the DSP and also reduces overall power consumption by eliminating secondary data transfers.

The DSP core, along with the coprocessors and the other elements, can provide the necessary flexibility and programmability with high performance. This architecture provides a foundation on which various configurations targeted for different applications can be defined and realized.

2.3.3. Development TOOLS for SoC design

In the previous sections, we have discussed various algorithms and presented an open architecture for mobile handsets. However, architecture and algorithm design is only one part of the story. The other part is the implementation of the algorithms, either in software or hardware, and integration of the various subsystems to work together. The subsystems may come from different sources and these have to be integrated and debugged.

2.3.3.1. DSP Core Availability Options

In the above architecture, the DSP core is the primary component for obtaining flexibility and software programmability. The other components such as coprocessors have to be custom designed with a specific task in mind and, once designed, they remain the same. However the DSP core has to be chosen carefully because flexibility and programmability depend solely upon it. So the path to integration requires access to a DSP core based on an open architecture and should be available in different hardware forms. The ZSP500 is such a DSP core based on an open architecture and is available in three forms:

(i) **Soft IP**: Soft IP refers to that form of intellectual property (IP) where the DSP core is basically in the form of synthesizable RTL that the user can incorporate and integrate into their own RTL design flows. This provides enormous flexibility to the designer because it allows him to be independent of manufacturing vendor.
(ii) **Standard Products**: ZSP500 cores will also be available as a standard product with complete development kits. This enables rapid prototyping of the system and also faster algorithm development. The algorithms can be developed and tested in parallel with the SoC design using the standard parts, thereby considerably reducing development time.
(iii) **ASIC Hard-Macro**: If mobile handset makers do not want to develop the SoCs themselves, they can utilize design services from ASIC design companies such as LSI Logic. The ZSP500 is available as an ASIC hard-macro IP from LSI Logic so that the DSP can be incorporated into a design very easily for the user.

2.3.3.2. Multi-Source Tools

In software-defined radios, software development times will be a critical consideration. The ZSP500 not only provides a simple instruction set and programming model, but choices of development tools are also available. Just as multiform hardware availability is important for SoC designs, availability of development tools from multiple sources is also an important consideration. Multisource availability reduces the risk to the designers because they are not dependent on a single source for the development tools. LSI Logic and Green Hills provide independent development tools for the ZSP500 core. They include all the necessary tools for developing, testing and debugging software on the ZSP500 core.

2.3.3.3. Hardware Test and Debug

For hardware test and debug in a SoC environment, ZSP500 cores have configurable off-core debug/emulation and embedded real-time trace units that provide exceptional debug capabilities. These units support various features such as combinational program, data value, data address and external source break points, trace triggering capabilities with counter and masking support. This gives programmers true real-time in-system debugging which is essential for hardware–software integration in SoC designs.

2.4. Summary

In this chapter, we have presented the foundations of an open architecture for multimode mobile handsets. The architecture is flexible, programmable, and modular and can handle the complex algorithms required by new generations of mobile handsets. It will evolve along with the mobile communications standards in future. Good hardware–software partitioning can be achieved using the architecture. To provide flexibility and programmability, the architecture is based on programmable DSP with coprocessor extensions to provide better performance and modularity in hardware.

The ZSP500 is a high performance programmable DSP that can be used in the above architecture. We have discussed various aspects such as performance, energy consumption, etc., that are important for mobile handsets, and have shown how the ZSP500 DSP core addresses those aspects. We have also discussed various aspects of coprocessors and development tools needed for realizing the SDR based mobile handsets.

Acknowledgments

The authors wish to thank Brendon Slade, LSI Logic Corp. for his invaluable suggestions and modifications, which have enhanced the quality of the chapter considerably.

References

[1] Software Defined Radio Primer: *http://www.sdrforum.org/sdr_primer.html.*
[2] R. Steele, *Mobile Radio Communications*, IEEE Press, New York, 1992.
[3] L.B. Milstein, 'Wideband Code Division Multiple Access', *IEEE Journal, Selected Areas in Communications*, 1998, **18**, 1345–54.
[4] J.H. Winters, 'Smart Antennas for Wireless Systems', *IEEE Personal Communications*, 1998 (February) 23–27.
[5] Jian Lu *et al.*, 'Research on smart antenna technology for terminals for the TD-SCDMA system', *IEEE Communications Magazine*, 2003, **41** (June), 116–19.
[6] Z. Wang and K.K. Parhi, 'High performance, high throughput Turbo/SOVA decoder design', *IEEE Trans. Communications*, 2003, **51**, 570–79.
[7] ITU-T G.722.2, *Wideband Coding of Speech at Around 16 kbits/s Using Adaptive Multi-Rate Wideband (AMR-WB)*, ITU, January 2002.
[8] *Coding of Moving Pictures and Audio–Overview of the MPEG-4 Standard*, ISO/IEC JTC1/SC29/WG11 N4668, March 2002.
[9] L.R. Rabiner and B.H. Juang, *Fundamentals of Automatic Speech Recognition*, Prentice Hall, Englewood Cliffs, NJ, 1993.
[10] Y.H. Kao and P.K. Rajasekaran, 'Designing a low cost dynamic vocabulary speech recognizer on a GPP-DSP system', *Proceeding of International Conference on Acoustics, Speech and Signal Processing*, IEEE, Istanbul, June 2000.
[11] A. Jain, L. Hong and S. Pankanti, 'Biometric identification', *Communications of the ACM*, 2000, **43**(2), 90–98.
[12] Xiaopeng Li and Mohammed Ismail, *Multistandard CMOS Wireless Receivers: Analysis and Design*, Kluwer Academic Publishing, Boston, 2002.
[13] *ZSP500 Digital Signal Processor Core Technical Manual*, LSI Logic, December 2002.
[14] The BDTI benchmark scores: *http://www.bdti.com/bdtimark/BDTImark2000.htm.*
[15] DSP Insider: *http://www.bdti.com/dspinsider/archives/dspinsider_030201.html*
[16] Embedded Microprocessor Benchmark Consortium (EMBC) Benchmarks for ZSP400 and ZSP500: *http://www.eembc.org/*
[17] D. Wilson, 'An efficient Viterbi decoder implementation for the ZSP500 core', *Proceedings of GSPx and International Signal Processing Conference*, Dallas, March 2003.
[18] Alan Gatherer *et al.*, 'DSP-Based Architectures for Mobile Communications: Past, Present and Future', *IEEE Communications Magazine*, January 2000, pp. 84–90.
[19] S. Wichman, R. Trombetta and P. Chiang, 'Motion estimation using the ZSP500 DSP core', *Proceedings of GSPx and International Signal Processing Conference*, IEEE, Dallas, March 2003.

3

DSP For Handsets: The Blackfin Processor

Jose Fridman and Zoran Zvonar

Analog Devices, Inc.

The Blackfin processor is a recent technology development targeted at cellular handset applications. In this chapter we introduce some of the main architectural features of this processor and discuss the way that they relate to a wireless Digital Baseband (DBB) platform. We also introduce in general terms the SoftFone architecture of a complete DBB, and the role that Blackfin plays in the overall DBB. The Blackfin is a 16-bit fixed-point core that combines some of the best features of traditional DSPs and microcontrollers. It compares favorably with dual-MAC DSPs on DSP specific benchmarks, and with microcontrollers on microcontroller-specific benchmarks. In addition, the core supports a rich set of instructions to enable an efficient implementation of multimedia algorithms found in 2.5 and third-generation (3G) wireless applications. The Blackfin core has been integrated into a number of systems, including the MSP500 DBB family member AD6532, as well as into the general-purpose DSP systems ADSP-21535 and ADSP-21531/2/3, which range in clock speed from 300 to 600 MHz. This chapter presents an overview of the Blackfin architecture, describes key engineering issues, their solutions and details associated with the first implementation of the core in a wireless system. The utility of the Blackfin architecture for practical 2.5 and 3G wireless applications is illustrated with application examples and performance benchmarks for typical DSP and image/video kernels.[†]

3.1. Handsets and the Progress of SDR

3.1.1. The Appeal of SDR

As a concept, a fully programmable Software Defined Radio (SDR) is very appealing to a designer of mobile wireless handsets. Taking the term wireless handset to encompass a wide

[†] SoftFone and Blackfin are trademarks of Analog Devices, Inc.

Software Defined Radio: Baseband Technologies for 3G Handsets and Basestations. Edited by W. Tuttlebee
 ISBN: 0-470-86770-1

range of hand-held devices (e.g. phones, personal digital assistants, smart phones with advanced audio and video services, and many incarnations not yet conceived), changing radio personality, possibly in real time, whilst guaranteeing Quality-of-Service (QoS), is the ultimate goal. One can argue that all the factors that are expected to push wider acceptance of software radio [1] are indeed critical for the handsets: multifunctionality (leveraging other services in the same device), global mobility, compactness and power efficiency, ease of manufacture and ease of upgrades.

This is even more evident for the emerging generation of handsets, which explicitly require support for multiple modes of operation, where, for instance, a GSM/GPRS modem must coexist with an FDD WCDMA modem. This is what we refer to as a *multimode handset*. In addition to cellular standards, we are also experiencing a convergence with multiple wireless connectivity standards such as IEEE802.11 Wireless LAN and Bluetooth, which also are increasingly being incorporated into handheld devices. The multiplicity of wireless standards places a tremendous burden on what is already a highly complex hardware and software system. Further, in a variety of scenarios, several modes of operation of these standards must be active at the *same* time. If there ever was such a thing as a true SDR, the cellular terminal would be one of its most useful applications.

3.1.2. Multistandard Terminals

This ideal multimode handset terminal would be a single hardware device capable of operating in any given wireless standard, given the appropriate software program. Again, in theory, such a device would be a very cost effective solution, primarily due to the fact that the same hardware would be used for all operating standards. In addition, if such a fully programmable wireless terminal were to exist, it would not only be possible to change the elements associated with the air-interface standard, but it would also be possible to implement arbitrary application-level functionality by downloading executable programs. Examples of this kind of application are security functions, complex video as well as audio codecs, gaming engines, Internet web browsers, etc.

3.1.2.1. Pragmatic Constraints

The reality of cellular handsets today, both from a technical as well as an economic standpoint, greatly limit the extent to which we can, as yet, approach a software defined radio. The first multimode solutions that are economically feasible rely on *parallel* (also refered to as *selectable*) single-mode approaches, which consist essentially of two, largely independent, terminals that reside in the same form factor. The hardware cost of more complex software programmable approaches tends to be high, since, generally speaking, the cost of a hardware device increases significantly in proportion to its ability to perform more than one function. In an extreme case of a largely programmable device, the cost can become prohibitive. This, in turn, has great implications in the cost-sensitive and cost-driven wireless terminal market.

3.1.2.2. The Trend from Hardware to Software

In our view, the concept of a fully programmable SDR, in particular as it relates to wireless terminals, will evolve in a series of incremental steps, starting from what we term *hardware selectability*, progressing to *hardware reconfigurability*, and perhaps at some point in the future leading to complete *software reconfigurability*. During the course of this transition, certain functional aspects of a terminal will experience a greater degree of programmability than others. As such, the important question for us is not a purist view of whether a handset terminal can or cannot be a fully SDR, but rather the *extent* to which it is, and *what* functional aspects of a terminal benefit from this transition. If we examine a broad range of handset terminal designs over the course of the last 10 or 15 years, to a large degree many functions have in fact migrated from rigid, fixed forms to more generic, programmable versions. Certain functionality lends itself naturally to this transition, such as functions associated with channel and speech coding, while others have had a slower transition, such as radio and mixed-signal functions.

3.1.2.3. Increasing Programmability

Although we do not believe that, in the short term, a cost-effective SDR-based terminal can be successfully mass produced, judging by the way that terminals have evolved, it is very likely that in the future the degree of programmability will continue to increase. This is the reason why we strongly believe in the role of programmable hardware, which we have been pursuing over the course of several wireless terminal generations. DSP processors play a critical role in the design of the wireless handset. Before going into details of the Blackfin processor core, which enables the above mentioned migration, we will address the broader topic of general requirements for handset architectures, which in turn greatly impact the baseband processing platform.

3.1.3. Wireless Handset Architecture – The Baseband Platform

3.1.3.1. Elements of the Digital Baseband

The Digital Baseband (DBB) platform has emerged as the key component of the wireless handset having the highest degree of programmability. Depending on different terminal architectures and level of integration, the DBB encompasses a complete set of baseband processing functionality with possible mixed signal functions (digital-to-analog and analog-to-digital converters, as well as numerous auxiliary and voiceband converters) integrated on the chip [2]. DBB architectures for wireless communications typically rely on multicore platforms, including DSP and microcontroller cores, as well as dedicated hardware functionality (accelerators, and coprocessors) for a given wireless standard, and a large amount of internal memory [3]. Dedicated hardware usually covers processing elements that require speed or complexity of operation that is beyond current DSP cores available for handset integration, such as chip-rate processing (e.g. a rake receiver), cell search, path-search algorithms, turbo decoding, and FFT operation, to name a few.

3.1.3.2. Evolution of DSP Architectures

The DSP has always been in the heart of SDR architectures (for historical perspectives see several papers focusing on DSP in [4]). Given the advances in wireless terminal design, particularly for multimode operation, modern DSP architectures have undergone significant changes, including among other things:

- ability of the DSP to execute control-oriented tasks;
- high computational capability;
- greatly increased memory size and I/O bandwidth (both on-chip, and interface for off-chip memory);
- fast interrupt response times to sustain DSP system performance.

Many functions found in channel coding and in speech coding have, over the years, migrated from hard-wired ASIC implementations entirely into software executed on a DSP. This transition has been possible primarily due to advances in DSP technology. For example, only one or two DBB generations ago, trellis-based GSM channel decoding and equalization, which are among the most computationally intensive channel coding functions, required a high degree of hardware acceleration external to the DSP in the form of hard-wired logic. Today, processors such as the Blackfin have sufficient speed and built-in support for these functions to eliminate the need for dedicated hardware blocks. We expect this trend to continue at a rapid pace, utilizing the advances both on the architectural as well as the semiconductor technology side.

The evidence of this evolutionary path has already been seen in wireless standards, such as EDGE. Data computation for channel coding and equalization is far greater than it is in GSM/GPRS, and, in addition, system requirements of link adaptation and incremental redundancy have added additional complexity. The advanced processor such as Blackfin was designed to handle this new challenge entirely in software by providing instruction-set support for most of the channel equalization functions, such as filtering and trellis decoding. Going forward to 3G standards, DBB requirements are, in adddition to more complex modem functionality, driven by a new class of applications such as video. This puts pressure on the DSP to support media operations, and the Blackfin instruction set is an example what can be achieved in this area.

3.1.3.3. Software Downloads

Further functionality for DBB focuses on a broader software framework for multimode terminals. Software downloads are performed today in wireless terminals via the SIM application toolkit, which offers the capability to download simple menus, further extended by the wireless application protocol (WAP) which allows more complex download functionality. A further evolution path is via the mobile station application execution environment (MExE), based on WAP and later on personal Java. In the future, full dynamic ReConFiGuration (RCFG) is expected in 3G systems.

3.2. The Blackfin Core

The Blackfin core is a dual-MAC modified Harvard architecture that has been designed to give good performance on modem, voice and video algorithms. In addition, some of the best features and simplicity of microcontrollers have been built into the Blackfin core. Incorporating microcontroller functionality as a fundamental part of the Blackfin allows for new approaches to system-level partitioning between the MCU and the DSP. Many tasks that previously required complex inter-processor coordination, especially in the physical layer (also referred to as layer 1) are now executed entirely in the DSP domain. Microcontroller functionality in Blackfin also allows for a far greater degree of commonality in this two-processor environment.

3.2.1. Key Features

The DSP features of the Blackfin core include: two 16-bit single-cycle throughput multipliers, two 40-bit split data ALUs and hardware support for on-the-fly saturation and clipping, two 32-bit pointer ALUs with support for circular and bit-reversed addressing, two separate data ports to a unified 4 GB memory space, a parallel port for instructions, and two loop counters that allow nested zero-overhead looping.

The microcontroller features of the Blackfin core include: arbitrary bit manipulation, mixed 16-bit and 32-bit instruction encoding for code density, an extensible nested and prioritized interrupt controller for real-time control, separate user and supervisor operating modes, stack pointers and scratch SRAM for context switching, memory protection and flexible power management.

The multimedia features of the Blackfin core include four auxiliary 8-bit data ALUs and a rich set of alignment independent packed byte operation instructions. These instructions enable the acceleration of fundamental operations associated with video and imaging based applications such as are found in 2.5 and 3G wireless algorithms.

Execution time predictability is achieved with lockable caches that can be optionally configured either as SRAM or cache, simplified static branch prediction, and data-independent instruction fetch and execution.

3.2.2. The Blackfin DSP Architecture

The Blackfin core is a modified Harvard architecture based processor. Instructions and data reside in separate L1 memories, and share common lower-level memories. All addresses are 32-bit, allowing the Blackfin core to address a unified 4 GB address space. Figure 3.1 shows the major sub-blocks in the Blackfin core, and Figure 3.2 shows the detailed data path of the execution unit (also called data arithmetic unit), the data address generator unit (address arithmetic unit), and the control unit.

The data arithmetic unit contains an eight-entry, 32-bit data register file, two 16×16-bit multipliers, two 40-bit split ALUs, one 40-bit shifter, four 8-bit video ALUs and two 40-bit split accumulators. Keeping the accumulators separate from the data registers allows for efficient load/store architecture to coexist with the accumulator-based design of traditional DSPs. The data arithmetic unit is internally pipelined to four stages: data register file read, multiplier array or video units, ALU/accumulate, and register write.

Figure 3.2 also shows a diagram of the address arithmetic unit, which contains two 32-bit address ALUs and an address register file. The address register file consists of six 32-bit general-purpose pointer registers and four 32-bit circular buffer addressing registers. The four circular buffer addressing registers have associated base, length and modifier registers, all of which are 32-bit wide. The address register file also includes a frame pointer to point to the current procedure's activation record and separate user and kernel stack pointer registers.

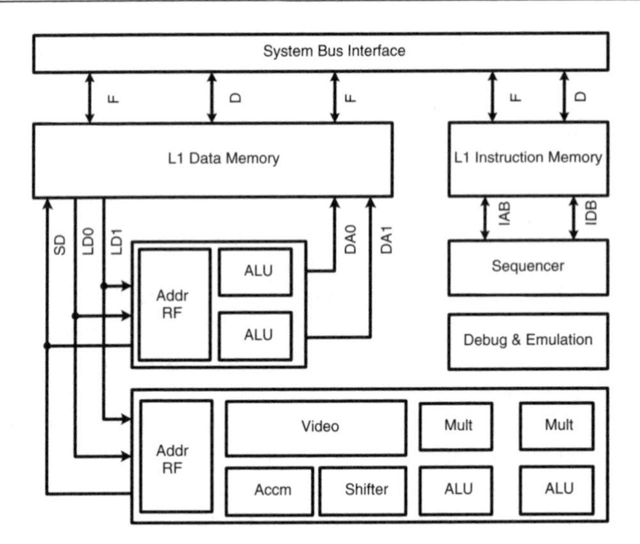

Figure 3.1 Block diagram of the Blackfin core showing major subsystems

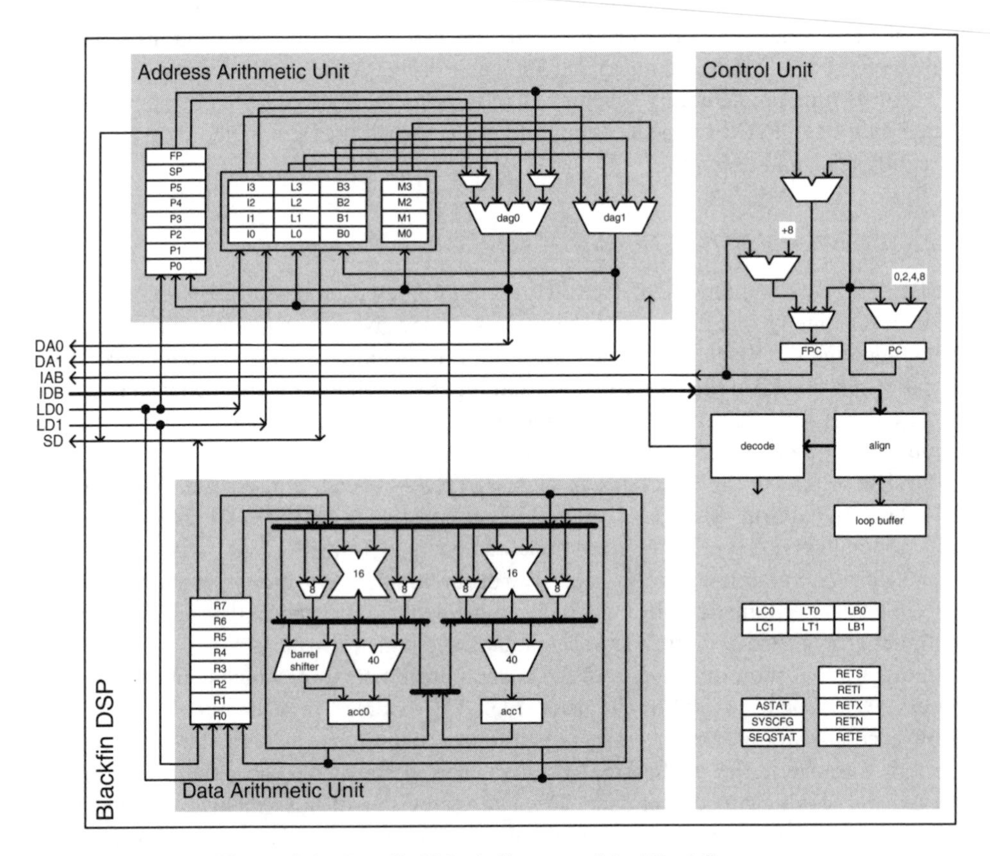

Figure 3.2 Detailed block diagram of the Blackfin core

The sequencer, also shown in Figure 3.2, contains all the control logic, and consists of the event return address registers, arithmetic and system status registers, and two sets of zero-overhead automatic loop counters, LC1 and LC2. The sequencer supports variable length (16/32-bit) instructions with mode-free intermixing and no alignment restrictions. In order to reduce the need for instruction fetches every cycle from the instruction cache, the instruction alignment unit of the sequencer has a four-entry, 64-bit wide instruction buffer. This instruction buffer caches the majority of the zero-overhead loops and significantly reduces instruction fetch bandwidth.

3.2.3. Instruction Set

The instruction set of the Blackfin core has two types of instruction:

- those used primarily for DSP-oriented computation, and
- those used for microcontroller and general tasks.

Specific Blackfin instructions are tuned for their corresponding task, but in general instructions can be intermixed with no restrictions.

Generally, DSP instructions read two 32-bit operands from the data register file, compute results, and either store these results back to the data register file or accumulate in the two internal accumulators. Each MAC unit is capable of computing a 16×16 bit multiplication in one of the following modes: signed, unsigned, mixed signed/unsigned, integer, or fractional. Each ALU unit is capable of a 32-bit operation on two 32-bit inputs, a 16-bit operation on two 16-bit inputs, or two 16-bit operations on a pair of packed 16-bit inputs.

The microcontroller-specific instructions provide the essential instructions needed to perform basic processor control and arithmetic operations. This set of operations includes load/store, arithmetic, logical, bit manipulation, branching, and decision-making functionality. Conditional register move instructions are provided to allow efficient implementation of short if-then-else statements.

The load/store instructions support the following addressing modes: auto-increment, auto-decrement, indirect, circular, bit-reversed, indexed with immediate offset, post-modify with nonunity stride, pre-decrement store on stack pointer.

Each DSP instruction may be issued by itself, or in parallel with two load or one load and one store instruction. The maximum number of operations achievable in one cycle is: two 40-bit adds, two 16×16-bit multiplies, two 32-bit address calculations, and two memory access operations.

3.2.4. Arithmetic Data Path and Programming Model

In the Blackfin core, operands are delivered from the register file to the compute units via two 32-bit operand buses. The compute units either store these results back to the register file or accumulate in the two internal accumulators.

Each MAC unit has four possible combinations of input operands:

- Rm.L times Rp.L,
- Rm.L times Rp.H,

- Rm.H times Rp.L,
- Rm.H times Rp.H.

where Rm.L is the low half, and Rm.H the high half of register Rm ($0 \leq m \leq 7$), and similarly for register Rp. When both MAC units operate in parallel, there are a total of 16 possible combinations of 16-bit input operands. The diagram in Figure 3.3 illustrates the four combinations allowed for MAC0 when the results are accumulated in A0.

An assembly instruction example for a dual MAC operation is

$$A1 += R1.H \times R2.L, A0 += R1.L \times R2.L;$$

This represents two MAC operations: one on Execution Unit 1, where the high half of register R1 is multiplied times the low half of register R2, and added to the contents of accumulator A1; and the other on Execution Unit 0, where the low half of register R1 is multiplied by the low half of register R2, and added to A0. The absence of an arithmetic mode qualifier indicates that the source data types are signed fractional numbers in the 1.15 format (a total of 16 bits, the most significant bit representing the integer part of the number, followed by 15 bits representing the fractional part).

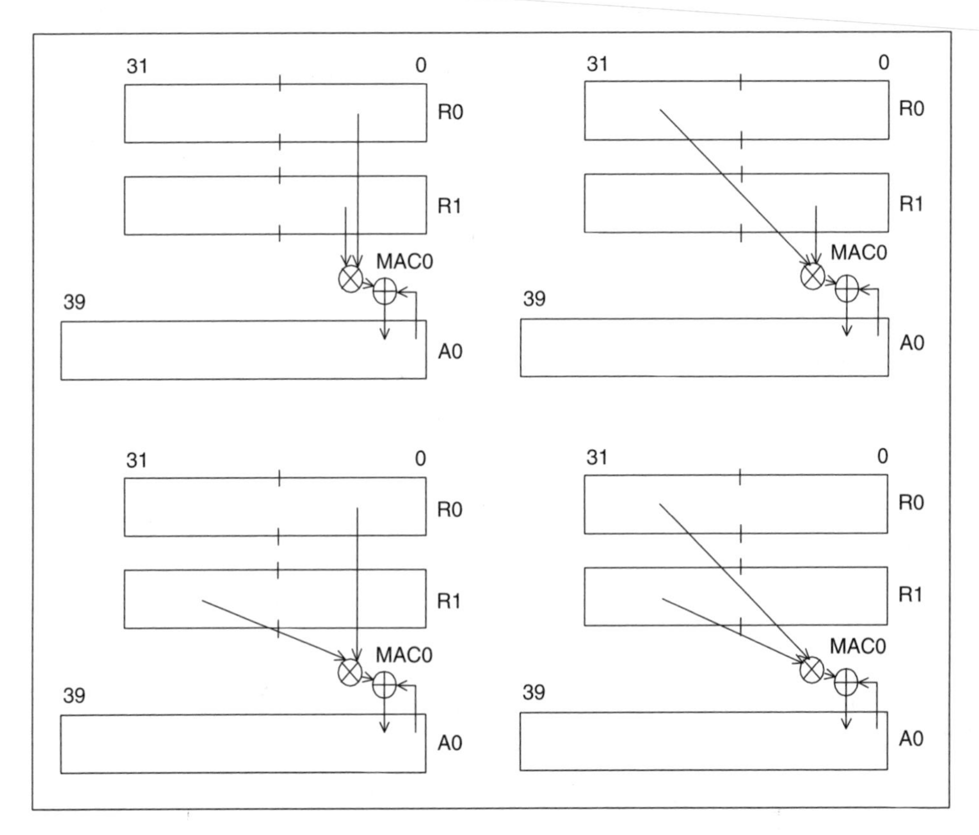

Figure 3.3 Four possible combinations for the 16×16 to 40-bit MAC0 operation. R0 and R1 represent two 32-bit source registers, and A0 one of the accumulators

3.2.5. Memory Architecture

The Blackfin core contains an L1 instruction memory and a separate banked dual-ported L1 data memory as shown in Figure 3.4. This allows two data load operations, or one load operation and one store operation, to occur in parallel with an instruction fetch. Both the L1 instruction and L1 data memories may be configured either as caches or as SRAM. Caches support ease of use and relieve the programmer from explicitly having to manage data movement into and out of the L1 memories. This allows code to be ported to the Blackfin core quickly, with no performance optimization required for the memory organization, without having to understand and program the system DMA controller.

The L1 instruction memory can be configured as either 16 kB SRAM or as 16 kB of lockable four-way set-associative instruction cache. The instruction address bus (IAB) is 32 bits wide and the instruction data bus (IDB) is 64 bits wide. Misses in the L1 instruction memory are sent to the system bus interface using the F port. An external bus master, DMA for example, can use the D port to access the L1 instruction memory when it is configured as SRAM.

The L1 data memory consists of two 16 kB super-banks and a separate 4 kB of SRAM for scratchpad use. Each of the 16 kB super-banks can be configured as either 16 kB SRAM or as 16 kB of two-way set associative data cache. The data address buses (DA0 and DA1) are 32 bits wide. The load data buses (LD0 and LD1) and the store data bus (SD) are 32 bits wide. Misses in the L1 data memory are sent to the system bus interface using the two F ports. An external bus master, DMA for example, can use the D port to access the super-banks in the L1 data memory when they are configured as SRAM.

The memory management unit in the Blackfin core supports both protection and selective caching of memory. In general the memory is divided up into regions in which the memory management rules apply. The Blackfin core supports four different page sizes: 1 kB, 4 kB,

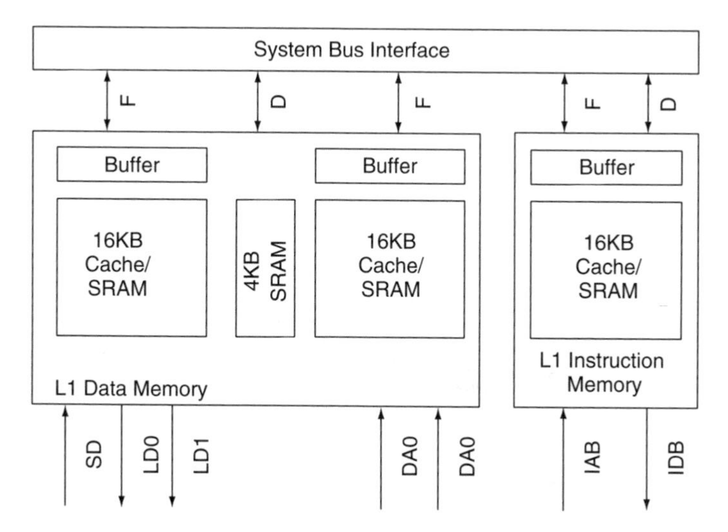

Figure 3.4 Block diagram of L1 memories. The buses shown at the bottom are connected to the Blackfin core

1 MB, and 4 MB. Each page is described by a set of cacheability and protection properties. These properties are stored locally for the most active regions in memory in Cacheability Protection Lookaside Buffers (CPLBs). Each of the three major buses (IAB, DA0 and DA1) has memory management.

3.2.6. *Operating Modes*

As illustrated in Figure 3.5, the Blackfin core has five modes: user, supervisor, emulation, idle and reset. The user, supervisor and emulation modes provide basic protection for system and emulation resources. Application-level code has restricted access to system resources. The system acts on behalf of the user through system calls whenever application-level code requires access to system resources. System resources include protected registers and protected instructions.

The Blackfin core is in supervisor mode when it is handling an interrupt at some level, or a software exception. The Blackfin core enters emulation mode as a result of an emulator event, such as a watchpoint match or an external emulation request.

In the emulator mode, the Blackfin core fetches instructions from a JTAG scannable emulation instruction register. These instructions bypass the memory system and are directly fed to the instruction decoder.

3.2.7. *Interrupt System*

The Blackfin core has an integrated interrupt controller, and in addition, the MSP500 system has a system-level interrupt controller. The interrupt system on the Blackfin core is nested

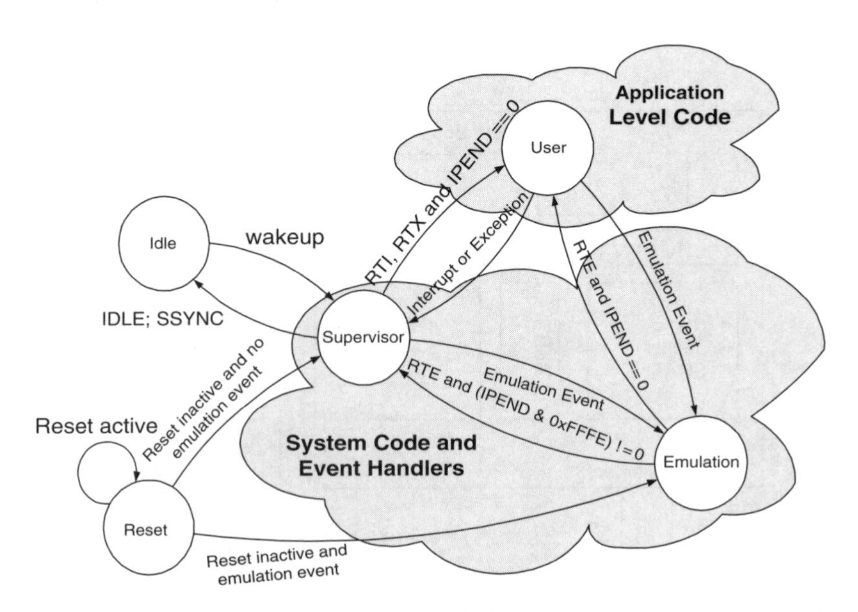

Figure 3.5 Blackfin operating modes. RTI, RTX, RTE are the return instructions from interrupt, exception, and emulator events, and IPEND is the interrupt state register

and prioritized, and, on an interrupt, the processor state is saved on the kernel stack. The interrupt system consists of five basic event types: emulation, reset, nonmaskable interrupt, exceptions and general-purpose interrupts. Each basic event has an associated register to hold the return address, as well as an associated return-from-event instruction. The Blackfin core processes all interrupts and exceptions in the supervisor mode. The emulation event is processed in the emulation mode.

A higher priority event always preempts a lower priority event. General-purpose interrupts may also be configured to be self-nesting at the same priority level. This allows an external interrupt controller to extend the number of available interrupt inputs.

Reset is treated as an event and is lower priority than emulation. Upon exit from reset, the Blackfin core starts execution from the reset service vector in the supervisor mode. A return from interrupt instruction is used to exit the reset service routine and start execution in the user-operating mode. There is also an idle/wakeup mechanism for low-power operation.

3.2.8. Debug Features

The debug features of the Blackfin core include hardware watchpoint registers (sometimes referred to as breakpoints), a trace buffer, performance monitors and cycle counters. The watchpoint unit consists of six instruction address watchpoints and two data address watchpoints. The six instruction address watchpoints may be combined in pairs to create three instruction address range watchpoints. The two data address watchpoints may be combined to create a data address range watchpoint. Each of the watchpoints and watchpoint ranges is also associated with a 16-bit event counter.

The trace buffer consists of a 16-element FIFO that records discontinuities in program flow. Discontinuities due to hardware loops are not recorded in the trace buffer. The trace buffer also has a compression feature that can be used to compress one or two levels of software loop.

The performance monitor unit contains two 32-bit special purpose counters and registers to count the number of cycles or occurrences of an event of interest. There is also a 64-bit free running cycle counter that can be used for code profiling purposes.

3.2.9. Implementation Using the Blackfin Core

A number of systems have been implemented with the Blackfin core, all of which achieve different levels of performance, area, and power consumption, depending on such factors as pipeline† depth, levels of memory/cache integration, and process technology. Some of the high-end systems are capable of speeds in excess of 600 MHz, as in, for example, the ADSP-BF531/32/33.

The Blackfin core implementation built into the MSP500 is based on an eight-stage pipeline as shown in Figure 3.6, and is capable of 300-MHz operation. Due to the extremely stringent constraints on power dissipation of wireless handsets, both in active as well as in standby power, the AD6532 is based on a low-power process.

† *Pipeline*: In order to achieve high clock rates, processors are built with several stages of logic. For example, a load data instruction is divided conceptually into three stages: a stage to load the instruction from memory, a stage to compute the address of the data to be loaded, and a stage actually to load the data.

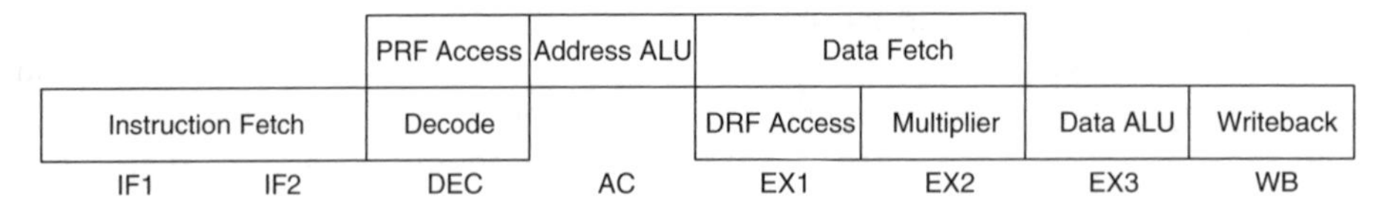

Figure 3.6 Pipeline diagram

Table 3.1 Description of the pipeline stages

	Name	Description
IF1	Instruction Fetch 1	Start L1 instruction memory access
IF2	Instruction Fetch 2	Finish L1 instruction memory access and align instruction
DEC	Decode	Start instruction decode and read Pointer RF
AC	Address calculation	Data addresses and branch target address
EX1	Execution Stage 1	Read Data RF, start access of L1 data memory
EX2	Execution Stage 2	Finish accesses of L1 data memory and start execution of dual cycle instructions
EX3	Execution Stage 3	Execution of single cycle instructions
WB	Writeback	Write architectural states to Data and Pointer register files and process events

The pipeline of the AD6532 is fully interlocked[†]. This allows for the minimum number of stalls to be inserted to maintain program correctness. Table 3.1 describes the operations performed during each stage and Figure 3.6 shows the arrangement of the stages.

This pipeline design allows for the results of load instructions to be forwarded to the execution units without incurring any stalls[‡]. Conventional deep pipelines introduce stalls between operand load and use, and require extensive loop pipelining [5][¶], as well as a relatively large number of registers. The Blackfin core is designed to avoid completely all load-to-use stalls, thus providing high pipeline efficiency and virtual elimination of the need for loop pipelining. The elimination of load-to-use stalls results in code density reduction for the majority of DSP and video algorithms.

[†] *Interlock*: This mechanism ensures that the correct data is delivered when an instruction must wait for data from another instruction. A processor with interlocks will stall the pipeline until the time that data becomes available.

[‡] *Stall*: As instructions flow through a pipeline, often the result from one instruction is required sooner than it is available. These delays in the instruction flow are referred to as pipeline 'stalls.' For example, if a load data instruction (instruction A) is followed by an arithmetic instruction that consumes the incoming data at an early stage (instruction B), then we say that instruction B stalls due to A.

[¶] *Loop Pipelining*: A method used to schedule the instructions in a program to avoid inefficiencies caused by stalls. The side effects of loop pipelining are larger code and higher register pressure.

In addition to the lack of stalls in the load-to-use pipeline, data accumulation operations that use the data ALUs, all MACs, and also the pointer updates that use the address ALUs, are all single cycle operations and incur no stalls. Overall, these pipeline features allow the Blackfin core to achieve a very efficient cycle count on critical inner loops with very small code. Alignment independent byte operation is provided with control signals forwarded from the address ALUs to MUXs in the data register file, and contribute to cycle efficiency and code size in the video algorithms.

3.3. Handset Application Examples and Benchmarks

3.3.1. Finite Impulse Response Filter

Table 3.2 shows a code segment for a finite impulse response (FIR) filter. Software pipelining and loop unrolling techniques are used to adapt the algorithm effectively to the computational pipeline [5]. This loop consists of two nested loops: a two-instruction inner loop (LP1STR, LP1END), and a seven-instruction outer loop (LP0STR, LP0END). Each line of the inner loop has two MAC instructions. The operands for the MAC instructions come from registers R0 and R1, and are accessed as 16-bit register halves: R0.H (high) and R0.L (low). The load instructions use pointers I0 in post-decrement mode (R0.H = W[I0--]) to fetch a single 16-bit delay line input, and I3 in post-increment mode (R1 = [I3++]) to fetch a pair of 16-bit coefficients. The inner loop is executed O(T/2) times and the outer loop is executed O(N/2) times for an effective cycle count of O(NT/2).

The absence of load-to-use latencies between load and compute instructions reduces the need for loop unrolling to only one iteration, improves code size and allows simple assembly programming. Algebraic assembly syntax also contributes to the simple assembly programming.

3.3.2. Viterbi Algorithm

The Viterbi algorithm is used to detect signals in channels with history in the presence of noise, and is commonly used in the channel equalizer of the core GSM physical layer [6, 7]. The example presented here is an eight-state decoder, with binary PSK modulation (each symbol encodes one bit), but this solution extends to any number of states or modulation.

Figure 3.7 shows the Trellis diagram for an eight-state decoder. Each starting state has two possible transitions to two ending states, since a symbol encodes two possible states (a ‘1’ or

Table 3.2 FIR filter implementation on the Blackfin core

```
I0=delay line_pointer;  I1=input_pointer;  I2=output_pointer;
I3=coefficient_pointer;
P0=N/2; P1=T/2-1;  R0=[I1++]  ||  R1=[I3++];  [I0++M0]=R0;
         LSETUP (LP0STR, LP0END) LC0=P0;
LP0STR:  LSETUP (LP1STR, LP1END) LC1=P1;
           A1=R0.H*R1.L,            A0=R0.L*R1.L    ||R0.H=W[I0--];
LP1STR:    A1+=R0.L*R1.H,           A0+=R0.H*R1.H   ||R0.L=W[I0--]||R1=[I3++];
LP1END:    A1+=R0.H*R1.L,           A0+=R0.L*R1.L   ||R0.H=W[I0--];
       R2.H=(A1+=R0.L*R1.H), R2.L=(A0+=R0.H*R1.H)   ||R0   =[I1++]  ||R1=[I3++];
                                                      [I0++M0]=R0;
LP0END:                                               [I2++]   =R2;
```

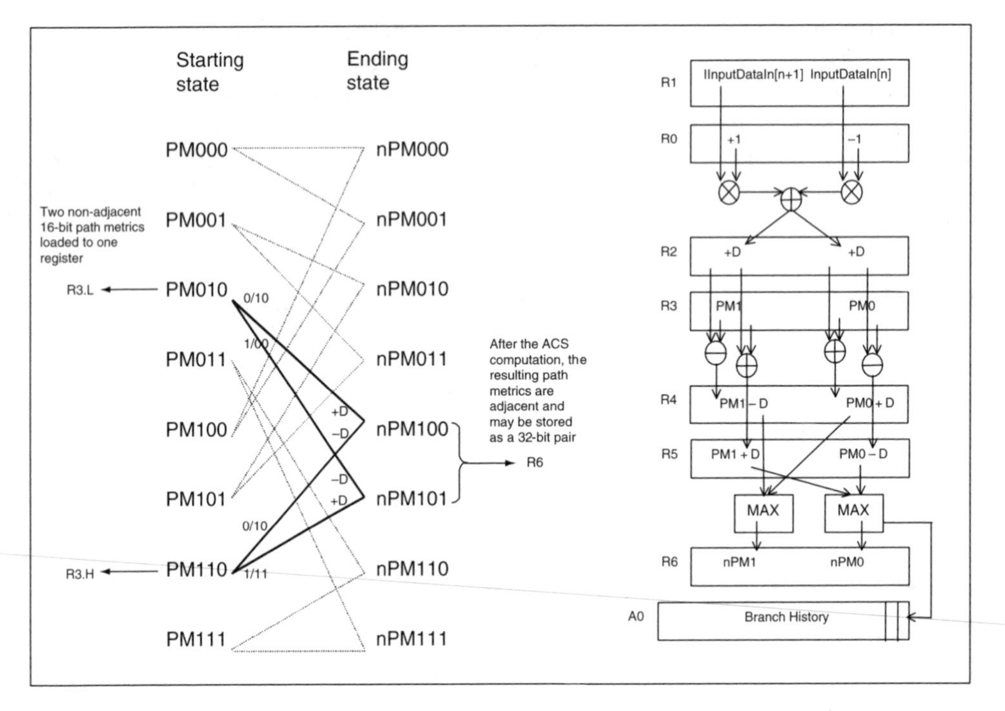

Figure 3.7 Viterbi trellis diagram. The butterfly with bold transitions is computed in one iteration of the inner loop

a '0'). For example, if the encoder is in starting state '010' and receives a '1', the ending state will be '101', assuming that the received symbol is a '00' (indicated as '1/00'). We call the four transitions from two starting states to two adjacent ending states a *butterfly*. In Figure 3.7, one butterfly is represented with bold transitions. For brevity, we omit the description of the encoder, branch metric calculation, and traceback, but additional details can be found elsewhere [6, 7]. We only focus on the computation of the trellis, which is the most time consuming function of the algorithm.

In the Blackfin architecture, a complete Viterbi butterfly is computed with three instructions in three cycles. Figure 3.7 also shows the flow graph for a butterfly computation, and Table 3.3 the assembly code. The first instruction is called a 'add-on-sign,' and calculates the branch metrics (labeled '+D' and '–D') based on the soft decision input data symbols. The structure of the trellis (which corresponds to a particular encoder polynomial) is captured in a precomputed table that stores +1 or –1. For example, the butterfly shown in bold in Figure 3.7 requires branch metrics +1 and –1. Once the branch metric D is computed, it is stored in duplicate in both high and low halfs of a register (in the figure, register R2).

The second instruction is a quad add-subtract, that computes the four possible candidates for maximum selection, namely, old path metrics plus and minus branch metrics PM1 + D, PM0 – D, PM1 – D, and PM0 + D (stored in register R4 and R5).

Finally, the instruction 'VIT_MAX' selects the maximum value of each of these number pairs, as shown in Figure 3.7. This instruction also shifts the contents of accumulator A0, and inserts two bits into bit positions A0[1] and A0[0], which represent the result of both maximum

Table 3.3 Blackfin assembly code for the Viterbi add-compare-select computation

```
        // Load soft decision InputData into R1=(InputData [SymNo*2+1],
        // InputData [SymNo*2]).
        // Load pre-computed branch metrics into R0=(D1, D0).
        // Load starting path metric into R3. L=PM0  (PM0 indicates upper
        // path metric)
        R1=[ P0 ++ ]  || R3.L=W[ I1++ ];
        A0=0  || R0=[ I0 ++ ]   || I2+=M2;
        // Setup inner loop for 6 iterations
        //
        LSETUP (START, END) LC1=P1;
        // Apply sum-on-sign instruction, and load R3.H=PM1 (upper path metric).
        //
START:R2.H=R2.L=SIGN(R0.H)*R1.H+SIGN(R0.L)*R1.L  || R3.H=W [ I2++];
        // Compute add/subtracts of butterfly: R5=(PM1+D, PM0-D) and
        // R4=(PM1-D, PM0+D).
        // Load branch metrics for next butterfly R0=(D1, D0).
        // Store two new path metrics from R6 (nPM1, nPM0) as 32-bit store.
        //
        R5=R3+|- R2 , R4=R3-|+R2  || R0=[ I0 ++ ]  || [ I3 ++ ]=R6;
        // Compute Maximum selection, and store two results in R6=(nPM1 nPM0).
        // This instruction also updates history in A0 and shifts A0 by two bits.
        // Load next upper path metric R3.L=PM0.
        //
END:    R6=VIT_MAX( R5, R4 ) (ASR)  || R3.L=W [ I1 ++ ];
```

comparisons. After 16 iterations (and 32 comparisons), A0 contains the history of 16 butterflies, and is stored for subsequent processing by the traceback function. Hence, the throughput of the Viterbi routine is three cycles per butterfly, including branch metric calculation.

3.3.3. Motion Estimation

Motion estimation (ME) is a technique used for taking advantage of the temporal redundancy inherent in video image sequences. ME has become a fundamental component of virtually all video compression standards, including ISO/IEC MPEG-1, MPEG-2, MPEG-4 and the CCITT H.261/ITU-T H.263 video compression standards [8]. Most ME-based methods are based on variations of block-matching algorithms, which attempt to find the best matching position of a block of a given size in the current frame to a corresponding block in a search region within a reference frame. A reference frame may be temporally before or after the current frame. Typical block sizes are 4×4, 8×8, or 16×16 pixels, although recent advanced video coding techniques use blocks of a size that adapts the image content. The blocks at the current frame are compared at positions that represent candidate motion vectors within the search region. The goal of ME is to find the motion vector which best represents the displacement of the block within the search window of the reference frame. Several fast schemes have been introduced which trade off quality, in terms of PSNR and bit-rate, against computational complexity [8, 9]. The distortion measure between a block in the current frame and a block in the previous frame may be based on several different distance criteria. The sum-of-absolute-differences (SAD) measure is most often used due to its low implementation complexity and high efficiency.

The Blackfin core has four 8-bit video ALUs. These ALUs are combined with split accumulators to accelerate the basic operation of ME, namely the SAD over four consecutive bytes. Most classes of ME algorithms display a high degree of parallelism that can be exposed by removing performance penalties due to data layout in memory. The fast motion search algorithms are a classic example of the need to access multiple operands on byte boundaries since they tend heavily to dominate the execution resources during the encode process. The video pixel operations each use two source register pairs that are coupled with alignment multiplexors. The subtract-absolute-accumulate, (SAA), instruction provides for calculating a block SAD. Figure 3.8 illustrates the alignment capability and the SAA instruction, and Table 3.4 shows a sample assembly code used to compute the motion estimation routine.

The address for the block in the current frame, and the block in the reference frame are allowed to be on byte boundaries. Operand alignment is being provided directly on the source register pairs R1:0 and R3:2 based on the LSBs of the DAG pointers I0[1:0] and I1[1:0]. High throughput is achieved by alternating between the forward and reverse byte ordering of the instruction and ping-ponging the parallel loads between each of the 32-bit registers in each of the source register pairs.

The SAA instruction accepts four pairs of 8-bit operands, subtracts byte-wise, takes the absolute values, and adds each of the four results to separate 16-bit halves of the accumulators, A0.H, A0.L, A1.H, and A1.L. The Blackfin core achieves an effective cycle count of

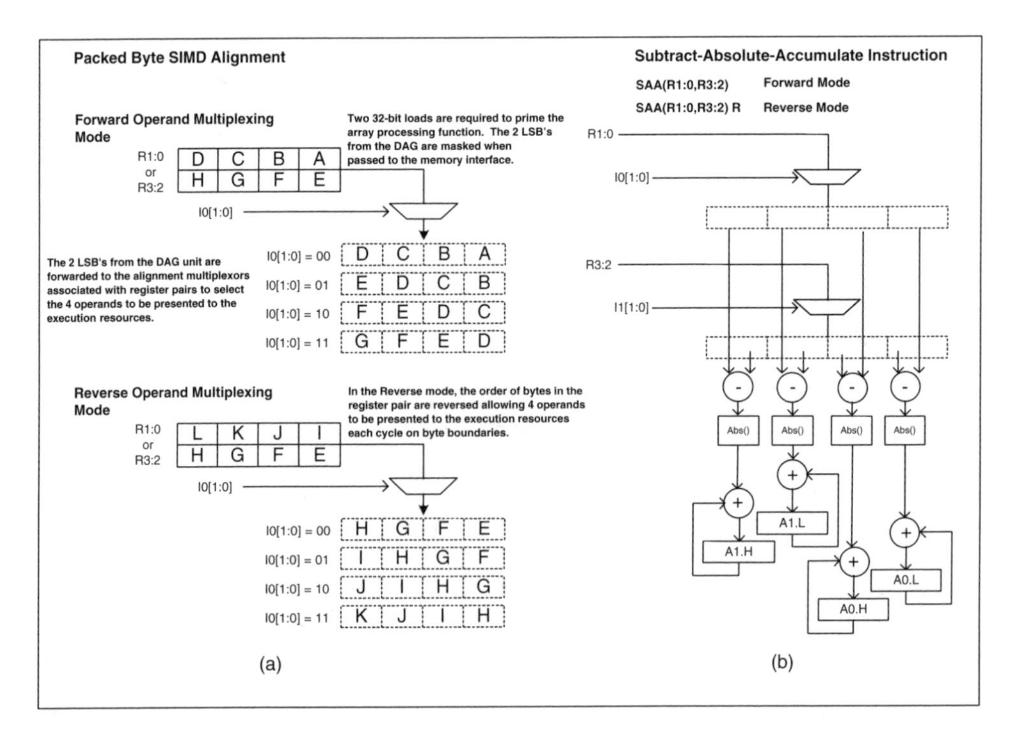

Figure 3.8 (a) Illustration of pixel register pairs and (b) subtract-absolute-accumulate instruction

Table 3.4 Blackfin assembly code for the 16 × 16 block SAD computation

```
       // I0 points to the current frame, I1 points to the previous frame,
       // and M0 holds the
       // image width. The loop is setup for 15 iterations
       //
       LSETUP ( L$0 , L$0e ) LC0=P0;
       // Prime the inner loop by loading 8 pixels from each image array.
       // Mis-aligned access
       // Exceptions are disabled while accessing the image data.
       //
       DISALGNEXCPT               || R0=[I0++]      || R2=[I1++];
       DISALGNEXCPT               || R1=[I0++]      || R3=[I1++];
       // Inner loop is unrolled allowing the stride to be set with the
       // "ImageWidth" parameter.
       // 16 pixel differences (1 row of the macro-block) are accumulated
       // per iteration.
       //
L$0:     SAA(R1:0, R3:2)          || R0=[I0++]      || R2=[I1++];
         SAA(R1:0, R3:2)   (R)    || R1=[I0++]      || R3=[I1++];
         SAA(R1:0, R3:2)          || R0=[I0++M0]    || R2=[I1++M0];
         SAA(R1:0, R3:2)   (R)    || R0=[I0++]      || R2=[I1++];
L$0e:    DISALGNEXCPT             || R1=[I0++]      || R3=[I1++];
       // Perform last row of 16 pixels subtract-absolute accumulations.
       //
       SAA(R1:0, R3:2)            || R0=[I0++]      || R2=[I1++];
       SAA(R1:0, R3:2)   (R)      || R1=[I0++]      || R3=[I1++];
       SAA(R1:0, R3:2)            || R0=[I0++]      || R2=[I1++];
       SAA(R1:0, R3:2)   (R);
       // The partial SADs are collected into R0.
       //
       R2=A1.L+A1.H, R3=A0.L+A0.H;
       R0=R2+R3 (NS);
```

$O(N^2/4+(N-1)+2)$ for $N \times N$ block SAD calculations, independent of alignment of the blocks in either the current or previous frames. The stride through the 2D image array is also supported by incrementing the pointers by the image width at the end of each row.

3.3.4. *Performance Benchmarks*

The balanced execution and data memory bandwidth of the Blackfin core allows for efficient implementation of most DSP inner loop kernels. Table 3.5 summarizes the performance of the Blackfin core on the kernels of typical DSP benchmarks.

In addition to the DSP kernels, the Blackfin core was designed to allow efficient implementations of typical image and video processing applications such as are found in 3G wireless applications. Table 3.6 summarizes the performance of the Blackfin core on the kernels of typical image and video processing benchmarks.

Table 3.5 Performance on DSP algorithms

Algorithm	Cycle count
N sample T tap real FIR filter	NT/2
N sample T tap complex FIR filter	2NT
N sample T tap real LMS adaptive filter	3NT/2
N sample T tap complex LMS adaptive filter	5NT/2
N sample B stage Biquad IIR filter	5NB/2
N sample L section lattice analysis filter	2NL
N sample L section lattice synthesis filter	2NL
N sample T tap convolution	NT/2
N sample maximum or minimum with or without index	N/2
N dimensional vector dot product	N/2
N dimensional vector sum	N
N dimensional weighted vector sum	N
N dimensional sum of squares	N/2
N dimensional weighted sum of squares	N
N dimensional euclidean distance	N
N dimensional weighted euclidean distance	3N/2
Complex FFT Radix-2 butterfly	3
256-point complex FFT (Radix-2 including bit-reversal)	3176
256-point complex FFT (Radix-4 including bit-reversal)	2630
Viterbi for GSM (16 states, 378 soft decision symbols)	6069

3.4. The SoftFone Digital Baseband Architecture

The SoftFone DBB architecture is an excellent example of an evolutionary architecture with the characteristics of an SDR. A number of elements of the architecture are based on the general SDR principle, yet it provides a commercially viable solution in a very competitive marketplace. The SDR personality of the SoftFone architecture enables terminal manufacturers to explore different production options, thus enabling them to differentiate on the market while using a common DBB platform. The wireless handset SoftFone architecture provides a flexible solution at low cost, and at the same time it is expandable and upgradeable. Having a generic hardware architecture allows the SoftPhone to be standard-independent. In addition,

Table 3.6 Performance on image and video algorithms

Algorithm	Cycle count
$N \times N$ Sum of absolute differences (SAD), $N = 8$, 16	$N^2/4 + (N-1) + 2$
$N \times N$ Sum of squared differences (SSD), $N = 8$, 16	$3N^2/4 + (N-1) + 1$
$N \times N$ Motion compensation (1/2X), $N = 8$, 16	$3N^2/8 + (N-1)$
$N \times N$ Motion compensation (1/2Y), $N = 8$, 16	$N^2/4 + (N-1)$
$N \times N$ Motion compensation (1/2XY), $N = 8$, 16	$3N^2/4 + (N-1)$
$N \times N$ Motion compensation (integer), $N = 8$, 16	$N^2/4 + (N-1)$
8×8 Discrete cosine transform (DCT)	284
8×8 Inverse discrete cosine transform (IDCT)	404
$N \times N$ Zig-zag scan (classical), N = 8	68
$N \times N$ Zig-zag scan (vertical), N = 8	72
$N \times N$ Zig-zag scan (horizontal), N = 8	43
Image convolution with 3×3 kernel (per pixel)	5
RGB24-YCrCb 4:4:4 color space conversion (per group)	12

the SoftPhone provides architectural support for modular software, with embedded debugging. The SoftPhone architecture has been incorporated into a number of cellular phone designs from a variety of OEMs (Original Equipment Manufacturers), primarily in Asia and in Europe. We will briefly present the SDR-related features of the SoftFone architecture and illustrate them on the latest family member.

3.4.1. SDR-related Features

3.4.1.1. A RAM-based Architecture

In contrast to earlier implementations of handset chipsets, where much of the software was hard-coded in mask-ROM, SoftFone is a RAM-based architecture. The lower cost of ROM is offset by the difficulty of changing software every time a new feature is added. The RAM-based SoftFone allows a single handset platform to be software configured, and hence to add new features or update the control software in order to keep up with the evolution of wireless standards. Time-to-market and flexibility advantages favor this approach. Moreover, this allows differentiation of the final handset not only by adding the latest features on modem functionality, but also by having the flexibility required to incorporate new and evolving applications on the wireless handset.

3.4.1.2. Modularity

Another important feature of the SoftFone architecture is modularity, in that it allows the substition of the DSP and MCU cores as more powerful cores become available or necessary. The architecture is symmetric, allowing both processors to access either on-chip or off-chip memory with a common, unified, 32-bit memory map. The access to the memory is via bus arbitration, ensuring that the necessary throughput can be achieved, hence enabling seamless core interaction. Sufficient cache memory is included to obtain the best possible performance from each core for time-critical functions.

The modular approach further enhances the SDR capability of the architecture, depending on embedded cores in the system. With the advanced features of Blackfin, traditional partitioning of DSP modem oriented functions (e.g. equalization, channel coding), DSP applications-oriented functions (e.g. audio and video processing), applications for wireless handsets (e.g. games, WAP, etc.) and protocol stack code can be revisited at the system level to ensure efficient realization in the handset.

3.4.1.3. Interfaces

Different applicatons and handset architectures require a variety of interfaces, which is easy to see given the endless combinations of peripheral devices. In order to establish effective support for a multitide of peripheral interfaces, the SoftFone has introduced the generic serial port (GSP), a connectivity device that may be microprogrammed to support a variety of interfaces, including I^2C, I^2S, IrDA, UART, to name a few. In addition, the SoftFone provides support for high-resolution color displays, multimedia card, and secure digital removable media.

3.4.1.4. Coprocessors

The need for specific high-speed operation that is still beyond the capability of DSP or MCU cores can be addressed via coprocessors, either integrated on the chip or as external devices via a coprocessor interface. Examples of this type of functionality include ciphering functions, and chip-rate processing for WCDMA.

3.4.2. The AD6532 DBB Platform

3.4.2.1. Overall Block Diagram

The AD6532 DBB chip from MSP500 family is the latest in the SoftFone family of platforms for wireless handsets. The top level diagram of the AD6532 is shown in Figure 3.9.

At the top right is the Blackfin subsystem, which consists of the DSP core, L1 code and data memories (configurable as cache or SRAM), unified L2 memory, DSP DMA controller, and DSP peripherals (timing and event processor, DSP interrupt controller, high speed logger, Bsport, external coprocessor interface). The Blackfin subsystem is connected with the SBIU crossbar. This subsystem is clocked at the maximum speed in the system, namely 300 MHz. At the lower right is the ARM7TDI subsystem, which consists of the ARM7 core, cache and DMA. The lowest level on-chip memory in the system is called system RAM (this is L3 memory), and is accessible by both the Blackfin and the ARM. The rest of the system consists of general connectivity peripherals for control of most devices present in a wireless terminal, as well as control of the analog baseband (ABB) and radio systems.

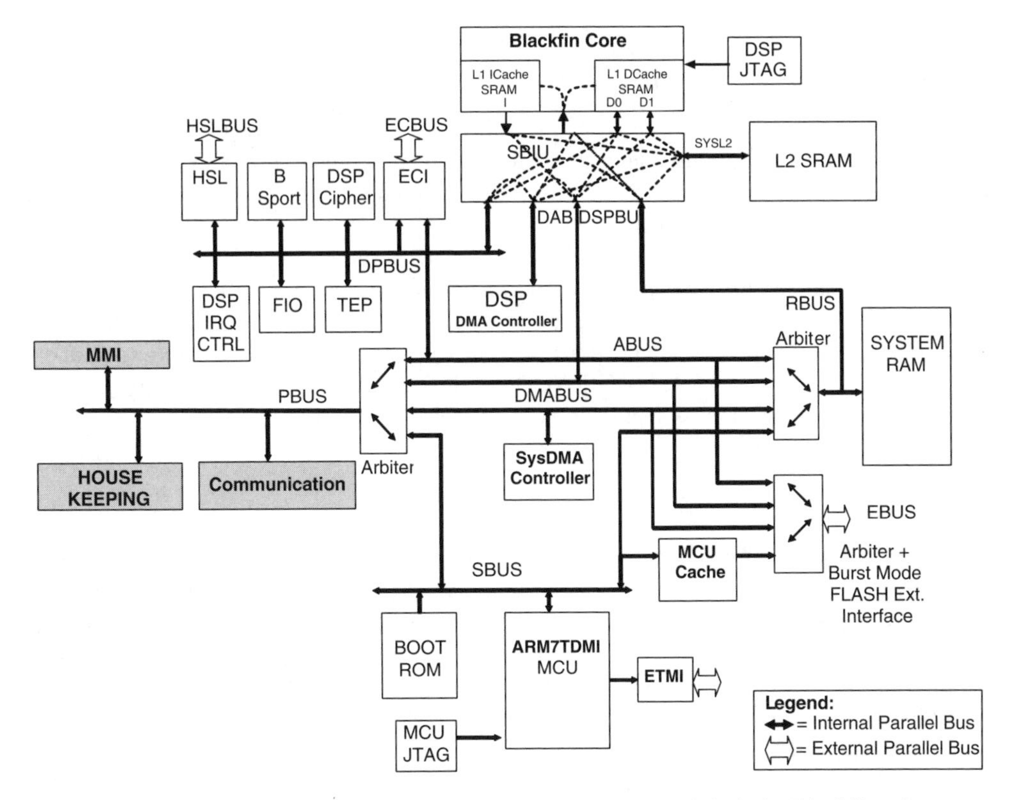

Figure 3.9 Top-level block diagram of the AD6532 system. Top right is the Blackfin subsystem, bottom right is the ARM7TDMI subsystem, and left is the connectivity peripheral subsystem

3.4.2.2. Hierarchical Memory

One of the most important aspects of the AD6532 is that it is based on a hierarchical memory system. From the perspective of the Blackfin, L1 memories provide a limited amount of fast zero-wait state storage. Due to the high speed at which the Blackfin is capable of running, L1 memory is expensive in silicon area, and it is economically feasible to incorporate only a small amount of fast memory. Lower levels of memory provide larger capacity storage, but at slower access times.

3.4.2.3. Caching

In order to support this hierarchical memory system, general purpose processors rely on caching, and in a similar manner, the L1 memory in the AD6532 is configurable either as SRAM or cache both for code and data. Caches are required to maintain a simple programming model, and as software plays a more important role in the development of wireless terminals, we expect that increased use of caching will be prevalent in future generation systems. This cached memory model is very common in general purpose computing, but is relatively new in embedded systems. In addition to internal memories, there are interfaces for external flash and SRAM through a 16-bit interface.

3.4.3. Architectural Extensions for Multimode Operation

3.4.3.1. The Timing and Event Processor

A key peripheral, also shown in Figure 3.9, is the timing and event processor (TEP), which provides support for all timing- and synchronization-related functions required for time-base management. The TEP is a small, multithreaded, microprogrammed processor capable of generating multiple timing signals. The TEP is designed as a multithreaded processor to generate the time bases for the control of several nonsynchronous physical layer protocols in parallel. For example, the TEP is capable of generating the time base of a GSM basestation during a voice connection, as well as the time base to scan WCDMA basestations. The ability to operate across several communications standards is often referred to as multimode operation. In addition to GSM/GPRS/EDGE/WCDMA timing synchronization capability, the TEP can also support other wireless standards, such as IEEE802.11 and Bluetooth.

3.5. Conclusions

In this chapter we presented a new dual-MAC Blackfin core that combines some of the best features of traditional DSPs and microcontrollers. We have also presented the digital baseband SoftFone architecture, showing the extent to which the SDR concept applies to a device in the handset space. While in the past the SDR concept has focused primarily on supporting multiple air interfaces, the dynamics of the handset industry have greatly emphasized the need for a flexible platform that will support the wide variety of applications to be incorporated in a product. The rich set of multimedia instructions in Blackfin make it ideally suited for implementing algorithms found in wireless applications.

In terms of integration and cores, we believe that there are fundamentally two trends. One is the integration of multiple cores into a single DBB component (more than the current two), and the other is an increase in parallelism at the instruction level of future DSP cores. We see these two trends developing at a faster rate than pure increase in clock speeds. In the severely power-constrained environment of wireless handsets, where total device power dissipation cannot increase from one product generation to the next (and hence a degradation of talk and standby times), parallelism at the core and instruction level can provide increases in performance in a more power-efficient manner than clock speed.

Looking into the present state of handset technology, fully SDR based terminals will likely not be economically feasible over the course of the next few years. Instant SDR gratification in newly launched systems will not be feasible, particularly for standards with high processing demands on the air interface such as WCDMA. Over the next decade, we are looking at an uphill battle for optimized SDR handset technical solutions coupled with cost and power constraints. And yes, the history will repeat itself: technology will need time to mature. We may be discussing whether we can accelerate SDR migration, but working hard to establish it while it is still maturing is inevitable.

Acknowledgments

We would like to acknowledge and thank all the individuals who have contributed directly or indirectly to the success of the MSP500 family of processors.

References

[1] J. Reed, *Software Radio: A Modern Approach to Radio Engineering*, Prentice Hall, Upper Saddle River, NJ, 2002.

[2] J. Soerensen, P. Birk and Z. Zvonar, 'New challenges for integrated circuit solution,' *Wireless Personal Communications, An International Journal, Special Issue on the Future Strategy for the New Millenium Wireless World*, 2001, **17**(2/3), 291–302.

[3] Z. Zvonar, J. Fridman and D. Robertson, 'Software-Defined Radio in Wireless Handsets: Evolution versus Revolution', *EE Times*, 2002, August 9.

[4] J. Mitola and Z. Zvonar (Eds), *Software Radio Technologies: Selected Readings*, IEEE Press, 2001.

[5] J.L. Hennessy and D.A. Patterson, *Computer Architecture: A Quantitative Approach*, Second Edition, Morgan Kaufmann Publishers, 1996.

[6] J.D. Gibson (Ed.), *The Mobile Communications Handbook*, CRC Press, IEEE Press, 1996.

[7] H.-L. Lou, 'Implementing the Viterbi algorithm,' *IEEE Signal Processing Magazine*, 1995 (September), 42–52.

[8] P. Kuhn, *Algorithms, Complexity Analysis and VLSI Architectures for MPEG-4 Motion Estimation*, Kluwer Academic Publishers, 1999.

[9] K.R. Rao and J.J. Hwang, *Techniques and Standards for Image, Video, and Audio Coding*, Prentice Hall PTR, 1996.

4

XPP – An Enabling Technology for SDR Handsets

Eberhard Schüler and Lorna Tan
PACT XPP Technologies

4.1. Introduction

4.1.1. The Challenge

Next-generation mobile terminals must provide consumers with a wide host of applications and connectivity to meet the demands of an ever-advancing concept of mobile connectivity. Users increasingly will expect their mobile terminals to connect to any possible network infrastructure anywhere in the world. In addition, the number of transmission standards increases continuously. Therefore, the mobile terminal must be able to adapt dynamically to any wireless infrastructure and to download and run applications offered by the service providers.

Wireless protocols for GPRS, CDMA, UMTS and wireless LANs require processing power that exceeds the capabilities of the conventional microcontrollers and DSPs dominating current generations of wireless terminals. Figure 4.1 gives an indicative comparison of the processing needs of several standards. GSM phones require approximately 10 MIPS. GPRS/HSCSD functionality requires approximately 100 MIPS. For the implementation of EDGE around 1000 MIPS are required. Potentially up to 10 000 MIPS are required for the implementation of third generation UMTS/W-CDMA. Wireless LAN protocols implementing orthogonal frequency division multiplexing (OFDM) require around 5000 MIPS.

4.1.2. The Dilemma

For maximum flexibility, all functions of a mobile terminal should be defined in software. However, because of the computing requirements, implementing them in software overstrains the possibilities of microcontrollers and even sophisticated VLIW DSPs. Furthermore, power consumption, silicon area, and – last but not least – engineering time and development costs

Software Defined Radio: Baseband Technologies for 3G Handsets and Basestations. Edited by W. Tuttlebee
 ISBN: 0-470-86770-1

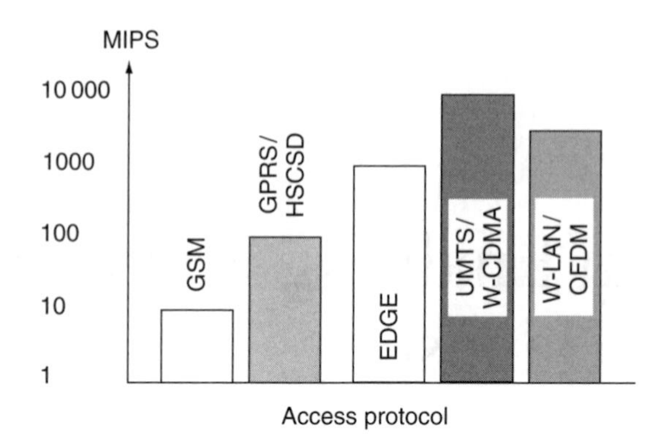

Figure 4.1 Processing requirements for wireless protocols

for new products must be minimized. Several architectural possibilities can be considered to solve these conflicting demands:

- Increase the clock speed of microcontrollers and DSPs and add specialized features and opcodes to their core – or cores in the case of VLIW architectures.
- Improve microcontrollers and DSPs by adding application specific hardware accelerators.
- Utilize reconfigurable hardware or processors as adaptable coprocessors.

Each solution has pros and cons. Increasing the clock speed of microcontrollers and DSP also increases the power consumption. The advantage of sequential processors is that software development is straightforward and the product provides maximum flexibility. Adding specialized hardwired accelerators increases the development time of a product and the silicon area, since specialized hardware is required for each intended algorithm. The product profits from specialized hardware in low power consumption and low production costs. Reconfigurable hardware or processors require adaptation of algorithms for parallel execution on an array of parallel hardware or processing elements. This increases the development effort but promises flexibility, combined with low power consumption.

A rule for a good product design is that each technology should be used within its optimal range for a given application. Tasks for software defined radio are quite heterogeneous, ranging from data flow processing of the baseband signals, error correction on bit level, sophisticated protocol handling, and on top level, the operating system for applications. This implies that several processor technologies should be used for SDR designs and baseband processing in particular.

4.1.3. Partitioning a System

Wireless protocol implementations are complex software systems. Generally, the tasks can be split into two algorithmic categories.

4.1.3.1. Control Flow

Complex decision trees, layered software structures, state machines and communication interfaces between the layers characterize the control-flow part. Sequential processors such as microcontrollers and DSPs are the best choice for processing of sequentially structured control-flow tasks. Software development may follow the beaten path and existing software, typically written in C, C++, System C, or commercially available library functions, can easily be adapted to support new designs.

4.1.3.2. Data Flow

The data-flow part is characterized by extensive arithmetic computations on data streams. The 'inner-loop' of these algorithms typically requires most of the processing needs of the standards in Table 4.1. The algorithms are relatively uniform and provide implicit parallelism. These algorithms are best suited to run on hardware that can exploit the implicit parallelism of the algorithms. This hardware can either be designed to provide fixed functionality for a given job or to utilize reconfigurable hardware that can be adapted dynamically to the required algorithm. Table 4.1 provides examples of typical baseband processing tasks for several standards and the proposed 'best-fit' processor technology.

4.1.3.3. Partitioning Between Processor Types

The partitioning should also minimize the bandwidth needs for data exchange between the processing units, I/O and RAM. Since, in baseband processing, most of the processing units must operate in parallel, the hardware should provide individual communication channels and independent RAM banks for the data exchange.

Table 4.1 Algorithm partitioning for typical baseband processing tasks

Technology	Tasks	Arguments
Microcontroller/DSP	Protocol stacks and management. Portion of channel estimation, control flow type of processing	Complex software structure with large decision trees. Requires little raw arithmetic or not continuous usage Signal processing tasks that do not require continuous processing can be executed
Hardwired functions	FEC (forward error correction) CDMA code generation Analog RF	Viterbi decoder or turbo decoder Hardwired functions require minimum energy and area Currently no programmable solution expected
Reconfigurable processor	Rake receiver, correlations, searcher, header detection, channel correction downsampling, FFT, demodulator, short PN code	These algorithms require most of arithmetic processing power Requires realtime processing of data streams

4.2. The XPP Reconfigurable Processor

4.2.1. The XPP Basic Concepts

Several reconfigurable architectures for baseband processing have been proposed and are available today on the market. In contrast to reconfigurable *hardware* such as FPGAs, which provide fine-grained, hardware-oriented functional units, a reconfigurable *processor* provides a higher level of abstraction. Reconfigurable processors can be configured to adapt their functional units, which are provided in hardware, to the problem or application. In this chapter, we describe a coarse-grained scalable architecture designed not only to provide maximum performance combined with low power consumption, but also to simplify algorithm design and programming tasks.

The XPP idea consists of a well-balanced set of methodologies listed below:

- configurable ALUs are arranged in an array and communicating via a packet oriented communication network;
- mapping of flow graphs to the array of ALUs and RAMs;
- packet-oriented data stream processing with autosynchronization;
- dynamic reconfiguration.

4.2.1.1. Flow Graph Mapping

In contrast to von Neumann-based processors, which are programmed sequentially, the XPP provides structures that allow direct mapping of parallel algorithms to an array of ALUs and RAMs. The configurations are basic parallel calculation modules, which are derived from the data-flow graph of the algorithm. Nodes of the data-flow graph are mapped to the fundamental machine operations such as multiplication, addition, etc. (see Figure 4.2). The graph's edges are the connections between the nodes. As long as data streams through a single configuration,

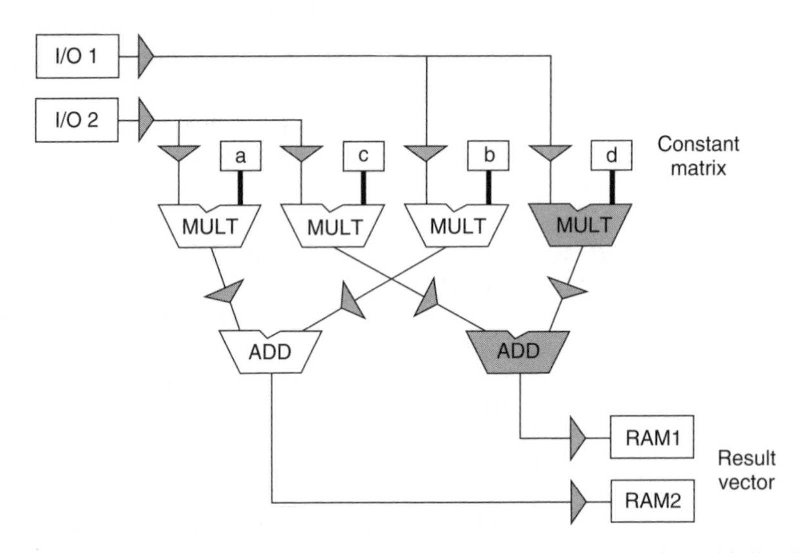

Figure 4.2 A parallel configuration showing a parallel vector matrix multiplication

the graph remains static, i.e. no opcodes and connections are changed. Since data flows are pipelined through the array, all ALUs can be busy thus enabling high parallel usage of the resources.

4.2.1.2. Streams of Data Packets

In XPP, a data stream is a sequence of single data packets traveling through the flow graph that defines the algorithm. A data packet is a single machine word (e.g. 16 or 24 bit). Streams can, for example, originate from natural streaming sources such as A/D converters. When data is located in a RAM, the XPP can generate packets that address the RAM producing a data stream of the addressed RAM-content. Similarly, calculated data can be sent to streaming destinations, such as D/A converters, or to integrated or external RAMs.

In addition to data packets, state information packets are transmitted via an independent event network. Event packets contain one bit of information and are used to control the execution of the processing nodes and may synchronize external microcontrollers. The XPP network enables automatic synchronization of packets. An object (e.g. ALU) operates and produces an output packet only when all input data and event packets are available. The benefit of this autosynchronizing network is that only the number and order of packets traveling through a graph is important – there is no need for the programmer or compiler to care about absolute timing of the pipelines during operation. This hardware feature provides an important abstraction layer, allowing compilers to map programs to the array effectively. To optimize the execution speed, the compiler can automatically perform pipeline balancing, thus eliminating unnecessary wait cycles.

4.2.1.3. Dynamic Reconfiguration

Configuration sequencing virtually extends the application space. Complex algorithms requiring more ALU resources than are available on a given array are split into separate configurations which are executed sequentially. Data to be passed to the next configuration is buffered in internal or external RAMs. Since a full configuration of the XPP requires only some kbits of configuration data, a new configuration can be loaded extremely fast onto the array. Configurations can be cached to make a new configuration available immediately.

Generally, this programming concept replaces the von Neumann instruction stream by a configuration stream where streams of data instead of single machine words are processed (Figure 4.3).

The sequence of configurations is controlled either by XPP's configuration manager or by the host processor when XPP is used as a dynamically adaptable coprocessor.

4.2.1.4. Applications

The XPP is optimized for applications where a large quantity of streaming data must be processed. Consequently, XPP is best suited to a wide variety of applications in addition to wireless baseband processing. It also enables high performance applications such as imaging, video codecs, radar/sonar, real-time visualization and automotive. The XPP is designed to provide a higher abstraction layer as compared with FPGAs. In particular, the data flow synchronization mechanism is completely transparent to the programmer, who does not need

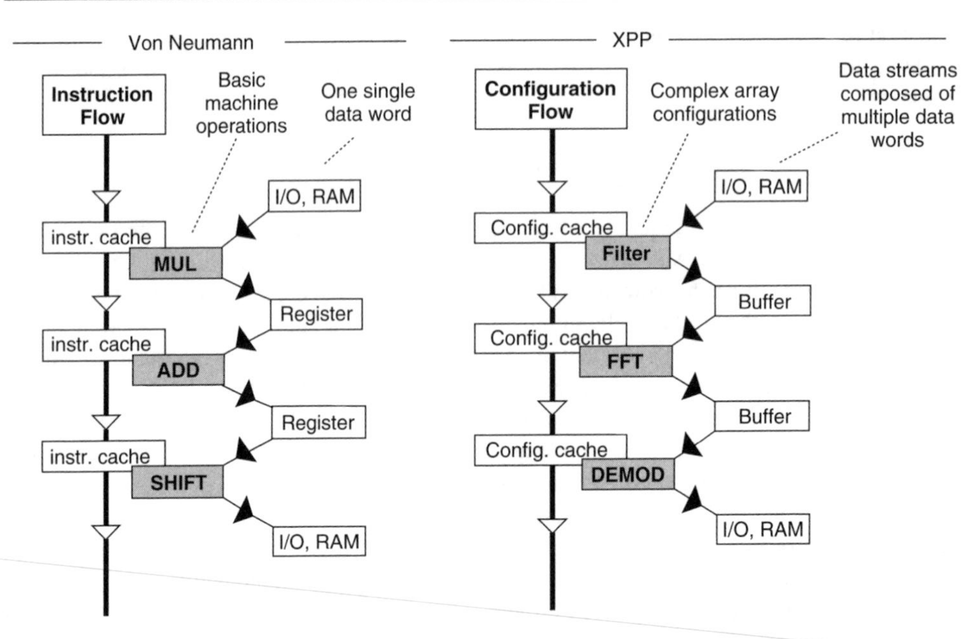

Figure 4.3 From instruction flow to configuration flow

to care about exact pipeline timing and balancing. Pipeline stalls, which often occur if external shared memories or buses are accessed, are handled automatically.

4.2.2. The Processing Array

At the heart of the XPP, an array of ALU processing array elements (ALU-PAEs) and RAM processing array elements (RAM-PAEs) provide the parallel processing resources. The processing array cluster (PAC) is connected to versatile I/O ports. I/O ports allow cascading clusters directly and permit external RAM access through addresses and data. XPP's data and event synchronization mechanism is implemented at the I/O ports by means of handshake signals. In SoC designs, the I/O connects the XPP-internal communication channels with the system buses such as AHB.

An XPP-core is built from a rectangular array of ALU-PAEs, RAM-PAEs at the left and right side of the array and I/O interfaces at the edges. The illustration in Figure 4.4 shows a sample array with 36 ALU-PAEs, six RAM-PAEs and I/O. The data word size of a silicon implementation can be chosen to be 16, 24 or 32 bit (fixed-point). ALU opcodes may be extended by special functions for the intended applications, for example, complex arithmetic for baseband processing.

4.2.2.1. ALU-PAEs

ALU-PAEs (Figure 4.5) are built from three objects and a connection matrix. ALU-PAEs enclose an ALU object featuring a typical DSP-command set including multiplication. ALUs

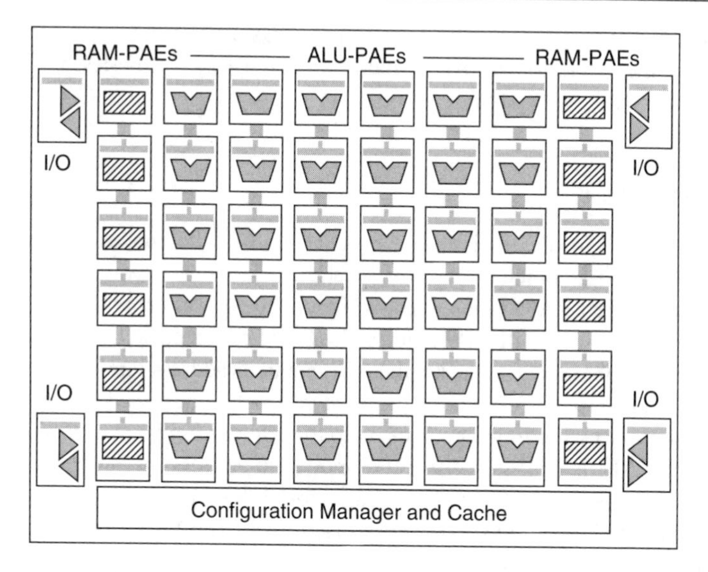

Figure 4.4 An XPP-array with 6×6 ALU-PAEs

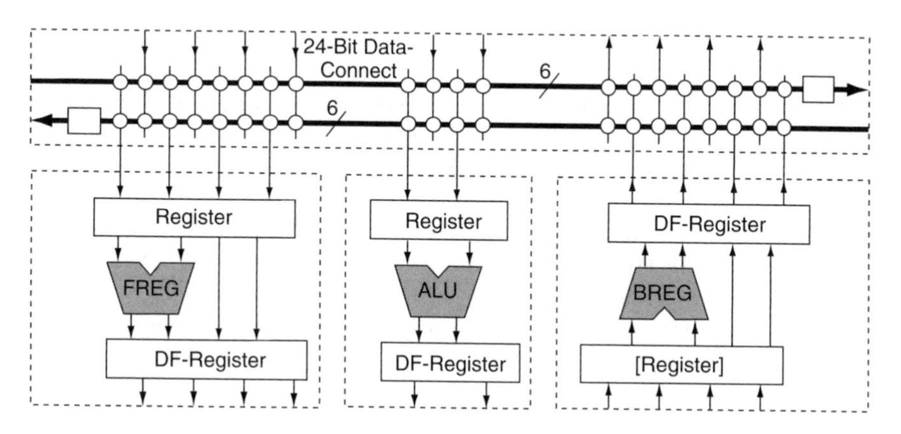

Figure 4.5 ALU-PAE objects

do not require instruction sequencers and caches, since the opcode is statically configured during the lifetime of a configuration. The backregister object (BREG) is used for routing from bottom to top, for arithmetic and normalization. The forwardregister-object (FREG) provides routing channels from top to bottom and a specialized unit with data-flow operators. The objects have input registers that can be preloaded during configuration – in the BREG this register can be bypassed. The output register (DF register, DataFlow register) is able to buffer one packet if the transfer to the next connected object is stalled.

Vertically, each object can be connected to one or more horizontal buses. Configurable switch objects are used for segmenting the communication lines horizontally to the neighboring PAEs. In parallel to the data connections, the similarly designed independent event network enables the transfer of status information from the ALUs. Events can be used to steer the data flow, to control the operation of ALU opcodes. The BREG provides a lookup table for manipulation of several event streams.

4.2.2.2. RAM-PAEs

RAM-PAEs (Figure 4.6) are similar to the ALU-PAEs, with the exception, that a RAM object replaces the ALU object. The size of the RAM ranges from 0.5 kwords to some kwords. The dual-ported RAM object has two independent ports enabling simultaneous read and write operations. As with all XPP objects, the RAM offers packet-oriented data handling. To read from a RAM object, a data packet must be sent to its address input. As a result, the RAM object generates an output packet with the content of the addressed RAM cell. Similarly, writing to RAM requires sending data packets to the address inputs and the data inputs of the write port. If the RAM is configured in FIFO mode, no addressing is required and the FIFO generates output packets as long as packets are stored. RAMs and FIFOs can be preloaded during configuration. This allows them to be used as lookup tables for storing coefficients and initialization parameters. Moreover, events may control read and write operation and inform about the status of the RAM objects if they are concatenated to larger capacity RAMs.

4.2.2.3. I/O Elements

I/O elements (Figure 4.7) are available with different interfacing. The XPP standard I/O interface provides two bidirectional streaming sources for data and events. This interface can be configured to provide direct connection to SDRAMs.

For system on chip (SoC), two AHB-master and two AHB-slave I/O interfaces enable simultaneous data transfers to and from the XPP array. The I/O interfaces perform not only the adaptation to the AHB layers and the AHB clock domain but also the correct alignment of 32-bit RAM accesses from the 16-bit or 24-bit XPP core. Addresses of 24 bit can be generated by preloading the high byte or by 2 × 16-bit calculation within the XPP array.

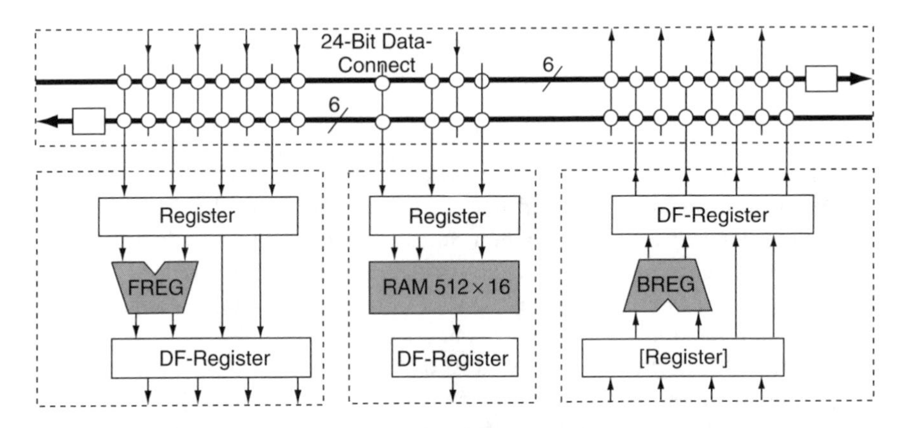

Figure 4.6 RAM-PAE objects

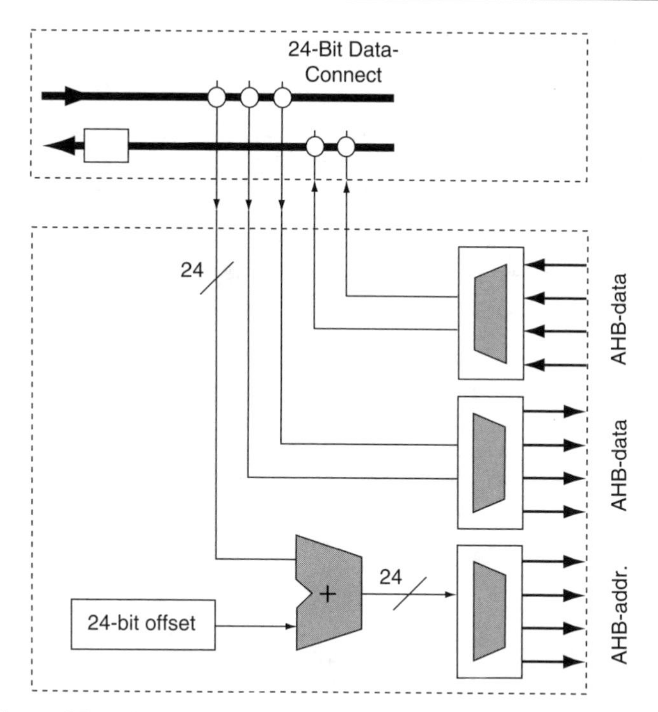

Figure 4.7 The master and slave AHB I/O interface of the XPP core

4.2.3. Packets, Streams and Synchronization

An important part of the XPP technology is its packet oriented communication network. The interconnection network consists of two independent layers. One is used for transmitting data and the other for events. Using these networks, streams of data can be processed, generated, decomposed and merged by utilizing protocols implemented in hardware.

4.2.3.1. Packet Handling

A packet of data is handled as an individual data object. XPP objects operate according to the data-flow paradigm. An operation is performed when all necessary data input packets are available and the previous result has been consumed.

The example in Figure 4.8 illustrates this. Five data packets are processed according to a data flow graph. In step 1, the input packets are consumed by the ALUs labeled '1' and '2.' The operation labeled '3' is not performed in the first step of computation, due to a missing packet at the second input. Four steps are needed to produce a single output packet. Since the communication system is designed to transmit one packet per cycle, all ALUs are busy if the pipeline is full.

4.2.3.2. Pipeline Balancing

The example in Figure 4.9 demonstrates the need for pipeline balancing. Input (b) is connected to ALU 1 and 2. Since ALU 2 must wait for the result of ALU 1 pipeline 2 is stalled until

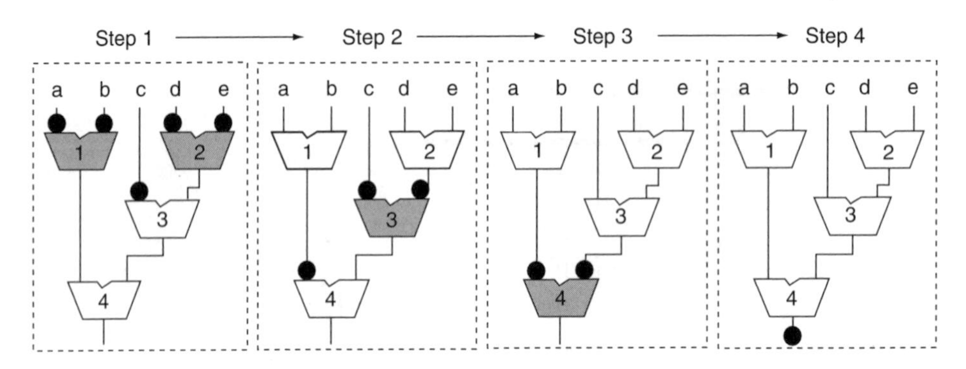

Figure 4.8 Synchronization of data packets

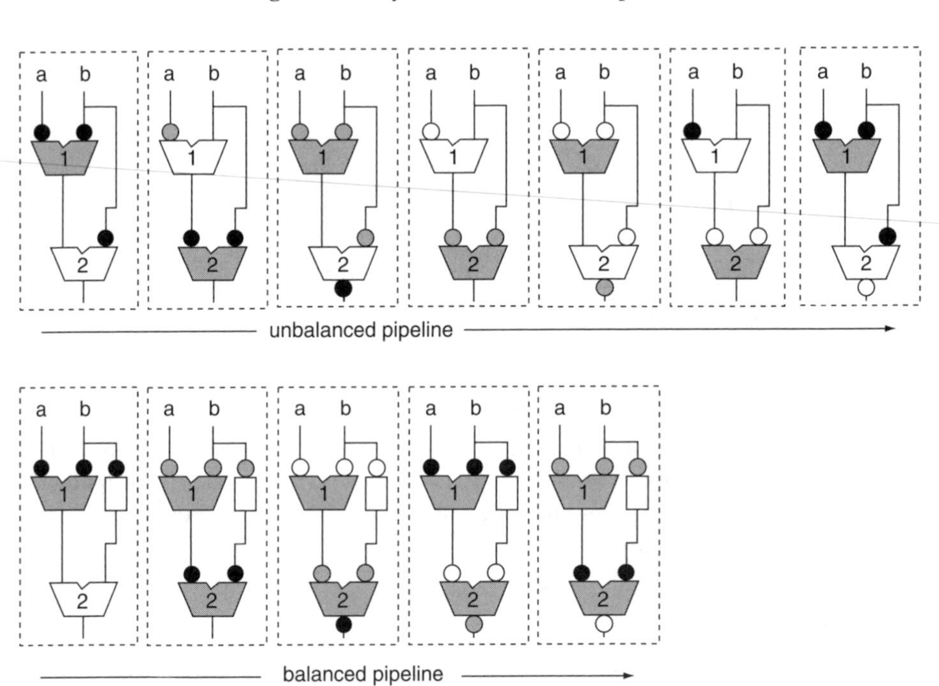

Figure 4.9 Pipeline balancing

ALU 2 has both input packets available. This pipeline can deliver a result only with every second clock cycle. The second pipeline has an extra register in parallel to ALU 1, therefore ALU 2 receives both inputs in parallel and the pipeline (b) is not stalled. This solution delivers a result every clock cycle. For maximum performance, the XPP compilers add those pipeline registers automatically wherever possible.

4.2.3.3. Events

Event packets can be used to control the processing of streams. Events originate from I/O ports or are generated by ALU operations. An event packet can be input to ALUs (e.g. Carry),

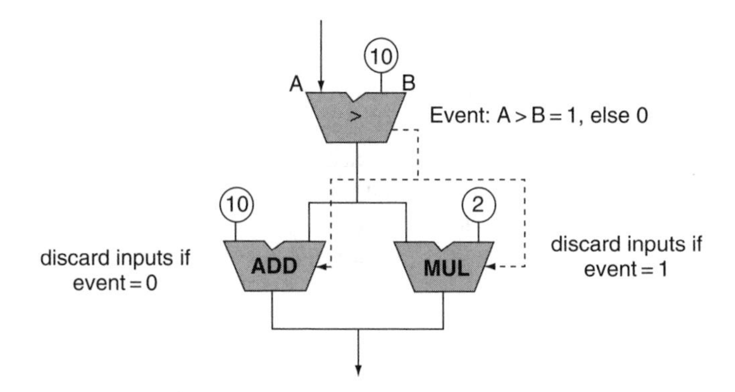

Figure 4.10 Conditional operation controlled by an event signal

can steer the dataflow through special opcodes in the FREG, or can be used to control the consuming or discarding of data at the input of an object.

A simple example demonstrates how conditional operations can be implemented with event packets see (Figure 4.10). The example performs the algorithm:

$$if\, x > 10, y = x + 10, else\; y = x \times 2.$$

First, the input data stream is compared to constant 10. The compare-opcode delivers an event value = 1 if the input is > 10 and event value = 0 if the input is ≤ 10 while the data packet is routed through the ALU without modification. Two further ALUs, which are controlled by the event packets, are configured as adder and multiplier. The adder discards the input packet if the event = 0 and generates an output packet only if the event = 1. In the opposite case, the multiplier multiplies the input by the constant 2 if the event = 0 and discards the input packet if the event = 1. ALUs generate an output packet only if both the condition for the event is met and all input data packets are available. The example shows also that packets are automatically duplicated if more than one object is connected to an output.

Similarly, data streams can be merged, multiplexed and swapped with specialized opcodes in the FREG register. An object can also be controlled by more than one incoming event, which can be AND-ed (all events must have arrived, the event values are OR-ed) or OR-ed (the first arriving event is used). In addition, a lookup table (LUT), which is integrated in the BREG Register, can be used to combine events for more complex decision structures.

4.2.4. Reconfiguration Management

The configuration manager controls the configuration of the array. The state of an object indicates that it is configured, i.e. ready to receive and compute packets, or not configured. An 'unconfigured' object does not send and receive packets and can be reloaded with a new configuration.

Hardware protocols ensure that no data packets are lost during the configuration process. In addition, there exists a special event handling mechanism used to 'unconfigure' a set of connected objects. If an event arrives at that ALU-object input, all connected objects (i.e. the complete network) are automatically set to the 'unconfigured' state. The configuration manager then is allowed to reconfigure these objects. This feature allows the removal of configurations and the freeing up of the occupied resources based on the results of the algorithm.

The configuration manager can preload RAMs or FIFOs and the input registers of all objects. This feature allows constants to be defined and configurations to be initialized. Since loading is also possible with running configurations, the configuration manager (or host processor in a coprocessor environment) can change constants, generate data packets by writing into object ports, or even steer the dataflow by writing event packets into event input ports.

4.2.5. Power Management

One of the main benefits of reconfigurable processors is their inherent low-power consumption. In contrast to sequential architectures, they do not need instruction caches, instruction decoders, or features such as branch prediction. In a sequential processor all those elements are active with every clock cycle and consume energy for tasks that are not directly related to the computation of data. In XPP, data is not stored in intermediate registers but is transferred directly to the destination. This also minimizes the power that is needed for data transfers as compared with the register–register transfers in sequential processors.

Further strategies for power saving are implemented in the XPP. On the first level, objects, which are 'unconfigured', are cut off from the clock tree (clock gating). Therefore, 'unconfigured' objects don't practically consume power. On the second level, objects that are waiting for data packets are set to the power saving idle mode until all data packets have arrived. In idle mode, only the hardware protocols for the interobject communication are active, the rest of the object is isolated from the clock signal. Both methodologies save a substantial amount of electrical energy, since in most algorithms only a fraction of the elements is active.

4.2.6. Development Tools

As outlined in the first section, XPP is typically used in a coprocessor environment and the development tools reflect this. Control-flow sections running on a microcontroller or host processor are coded in C/C++ and compiled with its standard tools. Data flow sections to be executed on the XPP core are separated by the partitioner and compiled by the XPP compilers. A C-API for the host processor provides functions for loading and starting of configurations, for data exchange via DMA or shared RAMs, and for task synchronization. Figure 4.11 shows the typical design flow for coprocessor application development.

4.2.5.1. Compiler for XPP

At the top level, the vectorizing XPP C-compiler is used for the sequencing of configurations, for the mapping of loops and for buffer handling. The compiler supports a subset of C (without structs and pointers) and is able to vectorize automatically the implicit parallelism in the algorithms and to parallelize nested loops. If maximum performance and area saving is mandatory, critical parts should be taken from a library of predefined functions or should be implemented with the native mapping language (NML). NML is an easy-to-learn descriptive language, which directly defines the operations, connections and properties of the XPP-core

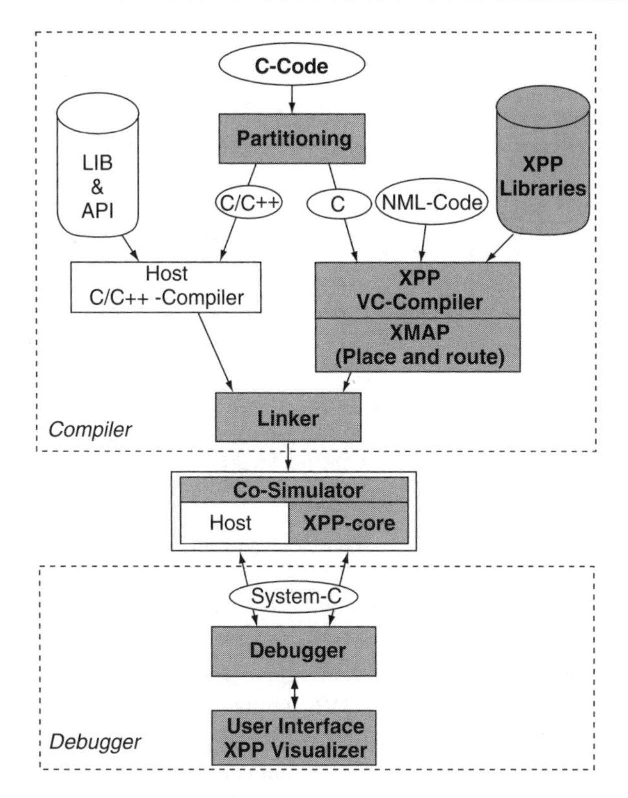

Figure 4.11 Design flow for XPP in a coprocessor environment

and provides all features for area and speed optimized XPP code. The linker combines the host binaries with the XPP configurations to an executable binary. This binary can be executed on the host or accesses transparently the XPP simulation environment.

4.2.5.2. Simulation and Debugging

The clock accurate XPP simulator generates all states, the data flow and the sequence of configurations in the XPP. The functional cosimulation provides all functions of the C-API for data transfer and synchronization enabling functional test and verification of applications partitioned to a microcontroller and an XPP coprocessor. The graphical interactive debugger (Figure 4.12) shows all registers and XPP states and allows manipulation of single values and RAM contents.

The design process and tools do not require completely new skills. Designers who are familiar with multitasking embedded applications feel comfortable with the tools chain. Development of NML applications requires good understanding of the algorithm and experience with hardware oriented programming – however, hardware skills and knowledge of hardware description languages such as VHDL is not required.

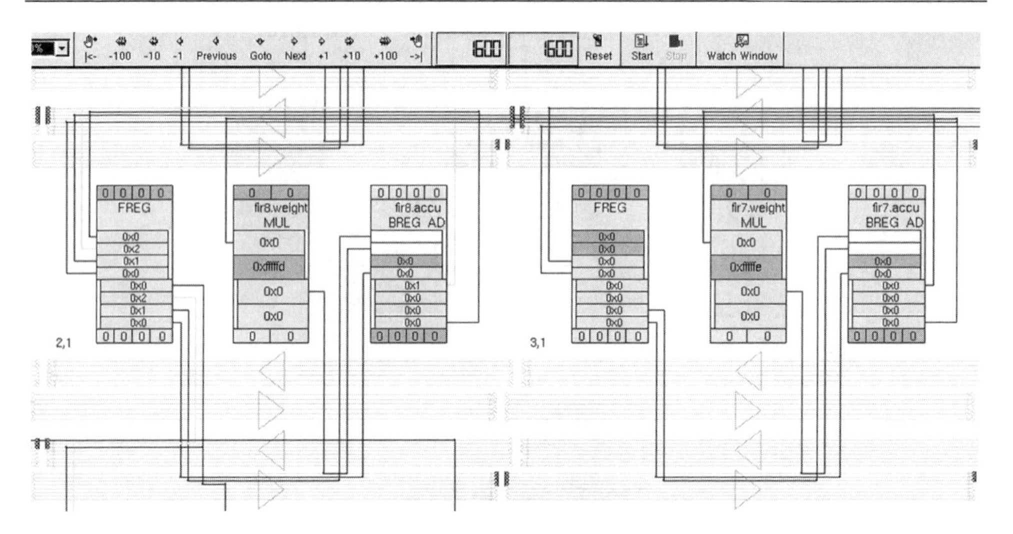

Figure 4.12 Screen shot interactive visualizer and debugger for XPP

4.2.5.3. Hardware Prototyping

Algorithms for the microcontroller and a XPP core can be evaluated, tested and verified on the XPP system development platform (SDP, Figure 4.13). The SDP is a PCI board that incorporates all modules that are required for baseband processing. The microcontroller is a 32-bit MIPS-based processor with RAM, ROM and peripherals. It controls the configuration of the XPP64-A processor and initializes the peripherals and the dataflow on the board. The XPP64-A processor, built in 130 nm CMOS technology [1], incorporates 8×8 24-bit ALU-PAEs, 2×8 RAM-PAEs, a configuration manager and four versatile I/O Interfaces.

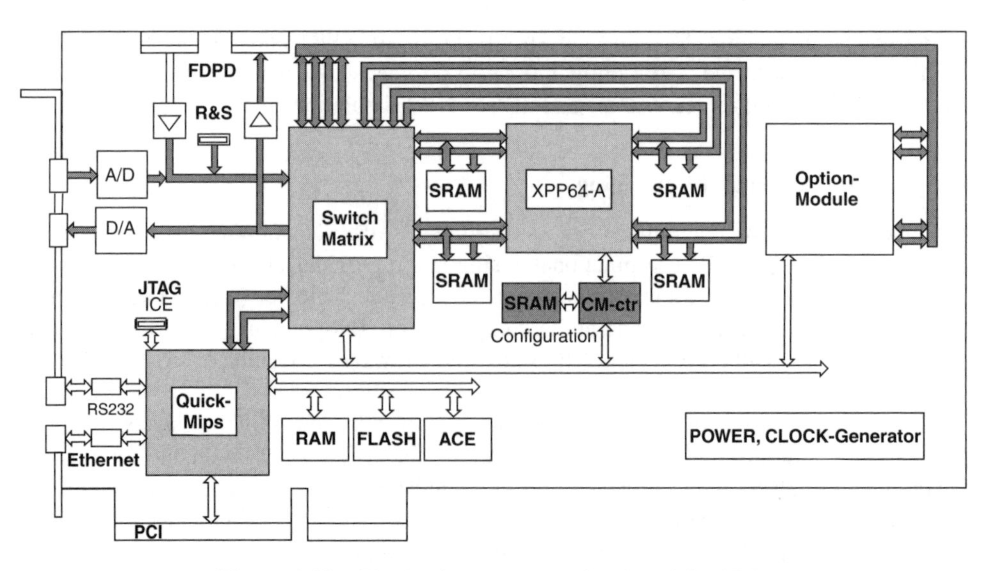

Figure 4.13 The hardware prototyping board for SDR

Each port can be used either for two streaming data channels or can access directly external 512 k × 24 bit SRAMs. A central switch matrix routes all data streams on the board (shaded in Figure 4.13), which provides several interfaces. The link to the development system running on a Linux or Windows2000 PC is done via an Ethernet port. High-speed dual 50 Msamples A/D and 125 Msamples D/A converters allow direct interfacing of the streams to analog baseband circuits. Hardware supporting the FDPD (front panel data port) standard [2] can directly send and receive data streams from and to the board. An optional module with streaming interfaces to XPP and a bus interface for the microcontroller can incorporate additional DSPs or interfaces. Both the XPP development suite with all tools and the SDP hardware are commercially available off the shelf [3].

4.3. Examples for Baseband Processing

Given its reconfigurability, the XPP can adapt from one wireless standard to another. On a GSM system, XPP can be configured to process computation intensive modules like the Viterbi Equalizer for channel equalization. Using the same mobile phone, when the user roams onto a W-CDMA/CDMA system, XPP can be reconfigured to run modules like a rake receiver. In W-LAN hot-spots, an OFDM receiver with an FFT may be loaded. As a result, XPP's dynamic reconfiguration permits uninterrupted connectivity. In this section we illustrate in some detail the application of the XPP to implement complex air interfaces, using the two example cases of W-CDMA and OFDM.

4.3.1. W-CDMA/CDMA Receiver

The UMTS/W-CDMA/CDMA infrastructure uses soft hand-off to transfer a mobile from one basestation to another. This occurs when a mobile user is traveling from one base station to another. In a soft hand-off, the mobile station maintains at the maximum a list of six base stations in its active set. The active set is a list of base stations with which the mobile user can communicate. The mobile uses the pilot channel to detect the base stations and the active set is updated based on the strength of the received pilot signals. This process should be done efficiently to ensure reliable communications as well as to extend the life of the mobile battery. When a mobile is on a cell border, it communicates to at least two basestations for the soft hand-off while monitoring the received strength of the nearby basestations. In a CDMA system, distinct pilot channels are identified by an offset index, which specifies the time offset in multiples of 64 from the zero-offset pilot PN sequence. The length of this PN sequence is 2^{15}. During code acquisition, the searcher tries to locate pilot signals by doing a complex correlation of the received complex signal with the complex pilot PN sequence. The result of the complex correlation is then compared to a predefined threshold. This complex correlation and PN sequence generator for CDMA for base station search can be done inside the XPP. The received pilot signal strength from different multipaths can be combined to give the effective pilot signal strength.

Diversity combining is used in W-CDMA/CDMA to take advantage of the multipaths. A rake receiver is employed to combine the different multipaths. With XPP's reconfigurability, when not in use, the PAEs that have been used by the searcher can be freed and reconfigured to act as a finger in the rake receiver. A simple block diagram of the W-CDMA receiver is shown in Figure 4.14.

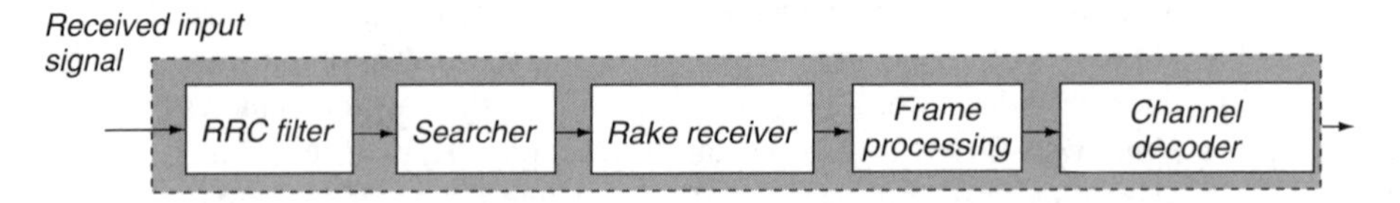

Figure 4.14 Basic blocks of a W-CDMA receiver

4.3.1.1. Searcher

In W-CDMA, the searcher performs several tasks including slot synchronization, frame synchronization, basestation identification and multipath search. For basestation identification, the searcher performs a complex correlation on the received complex pilot signal and the corresponding scrambling code for the basestation. The output of this correlation is processed to detect the multipath. Though it is possible to have more multipaths than the number of fingers in a rake receiver, in this example we assign the paths with the strongest signals to the rake fingers. When not in operation, the searcher can be reconfigured to act as a rake finger. Each finger in a rake receiver performs channel estimation and channel correction for the path assigned to it.

A path searcher performs a correlation of a fixed set of pilot signals over a sliding window to detect the paths with the strongest signal in a multipath environment. The offsets of these paths are stored within a control context and are used to generate the required offsets for the individual rake fingers that descramble and despread the chip rate signals. The path searcher performs a coarse search and later a fine search to track the signals. The tracker is responsible for the tracking and the resynchronization of the paths that are currently being received. The channel estimator calculates the channel coefficients that are used for the channel correction. The channel coefficients are calculated on the basis of a specific sequence of pilot signals.

4.3.1.2. Rake Receiver

The operational scenario for the mobile terminal Rake receiver implementation shown in the following section involves a 'soft' hand-off scenario with up to six basestations, and reception of three multipaths per basestation. The basic partitioning of tasks between a DSP, dedicated hardware and reconfigurable hardware is shown in Figure 4.15. Data-flow oriented tasks that

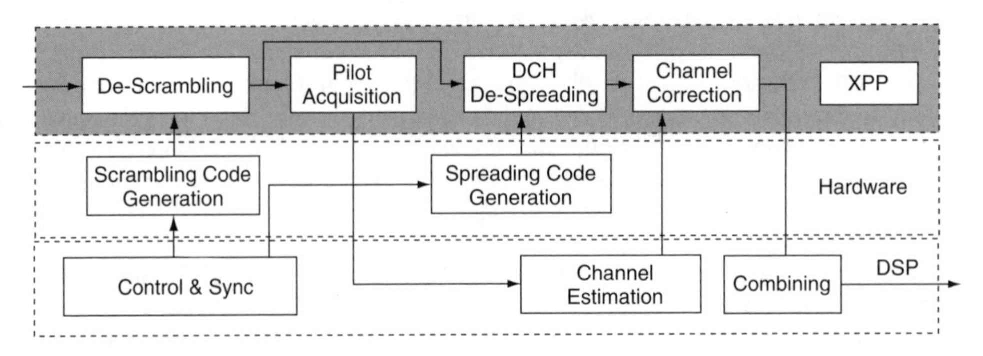

Figure 4.15 An example of partitioning of the searcher and rake receiver

Table 4.2 Rake receiver finger scenarios

Basestations	Channels					
	1	2	3	4	5	
1	18	9	6	4	3	Multipaths
2	9	4	3	2	1	
3	6	3	2	1	1	
4	4	2	1	1	–	
5	3	1	1	–	–	
6	3	1	1	–	–	

operate on a word level, granular, data stream are executed using the reconfigurable hardware. The DSP controls the interaction of the modules and performs the channel estimation based on the results from the pilot acquisition. Bit-level data processing tasks that execute continuously, such as the PN code generation and spreading code generation, are mapped onto dedicated hardware resources.

For this operational implementation, 18 (6×3) rake fingers for the descrambling and despreading operations have been realized. As the UMTS/W-CDMA chip rate is 3.84 MHz, a single physical finger is actually implemented. By repeating the descrambling and despreading operation on a single chip over multiple scrambling and spreading codes and time multiplexing the resulting data stream, the single physical finger thus corresponds to an implementation of 18 rake fingers. The minimum operational frequency of the single finger to accommodate this maximum scenario is 18×3.84 MHz = 69.12 MHz.

Further possible channel, basestation and multipath scenarios are shown in Table 4.2. The scenarios that require the full frequency of 69.12 MHz are shaded. The remaining scenarios need not to run at the full frequency.

The physical finger is implemented in the form of a pipeline on the reconfigurable hardware. The following assumptions are made in the design:

Sampling rate:	3.84 MHz
Data representation:	12-bits for I and Q each
Spreading factors:	4 to 512
Symbol encoding:	Space time transmit diversity (STTD)

The individual components of the rake receiver finger are described in the following. If more hardware resources (i.e. PAEs) are available on a given XPP core implementation, several instances of the rake finger can be loaded, thus increasing the number of available rake fingers.

4.3.1.3. Descrambler

The descrambling operation involves the complex multiplication of the synchronized received data with the scrambling codes. Implementation of the descrambler on the reconfigurable hardware is shown in Figure 4.16.

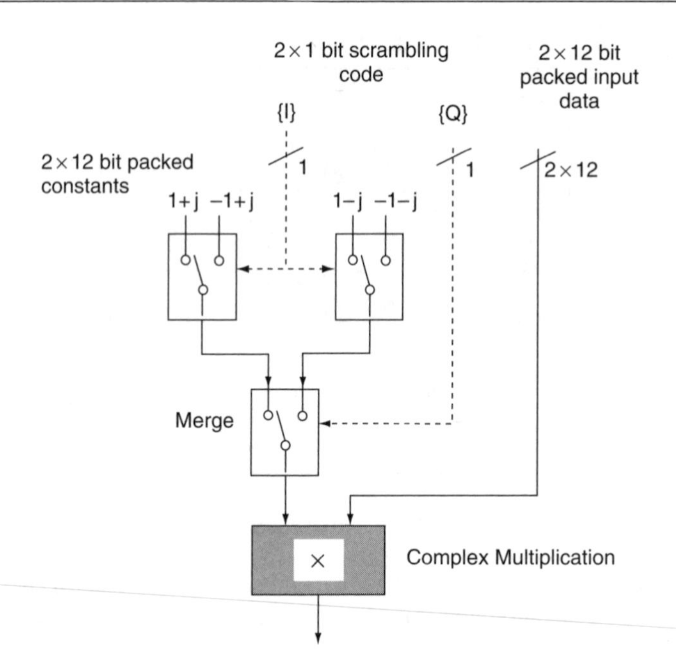

Figure 4.16 XPP implementation of the descrambler

The hardware provides the scrambling code in the form of a 1-bit representation for the {I} and {Q} components. The code is streamed in the form of event signals to the multiplexer stages that provide the complex signals. Since the multiplication has to be done with ±1, the special sign-flip opcode for packed 12-bit complex data can be utilized. This saves significantly energy since the parallel multiplier is not engaged.

4.3.1.4. Despreader

The despreading operation is a multiplication of the corresponding spreading code (OVSF code) with the real and imaginary part of the descrambled data sequence followed by a summation over a length equal to the spreading factor. The spreading factor in the downlink can range from 4 to 512 chips.

Figure 4.17 shows the block diagram implementation of the despreader on the XPP. The input data streams are time multiplexed with basestations, channels, multipaths and channels as annotated. The FIFO buffer stores the data packets in order to resolve the time-multiplexing scheme.

4.3.1.5. Channel Correction Unit

Figure 4.18 shows the implementation of the channel correction unit. In addition to the actual channel correction, the unit also performs the STTD decoding of the symbols. In STTD encoding, the symbol stream is divided into two streams, each with half the transmit frequency. Each stream is transmitted over a locally separate antenna. The first symbol stream

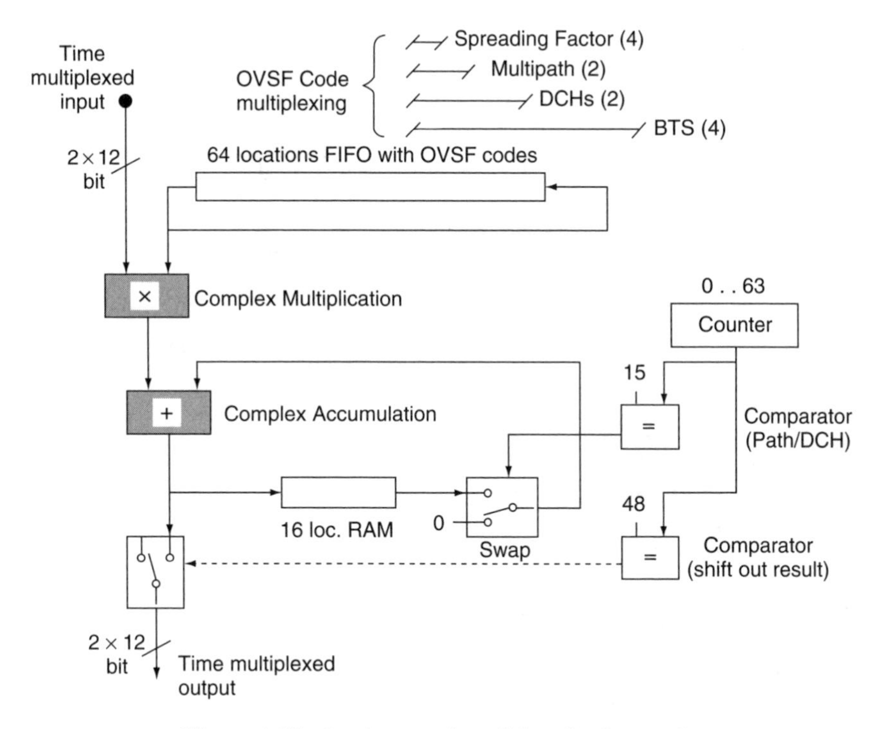

Figure 4.17 Implementation of the rake despreader

remains unchanged. The second symbol stream is reordered and the conjugate complex of the symbol is transmitted. The antennas are located sufficiently far apart that each stream has its own channel coefficient, but close enough for both symbols to arrive at the receiver at the same time.

Using the pilot sequences, the DSP calculates the channel coefficients (weight 1 and weight 2), which are then transferred to the configuration. The upper FIFOs shown in Figure 4.18 store the channel coefficients for the finger. The channel corrector takes two symbols from the despreader at half the symbol rate then to execute the STTD decoding and channel correction. Computation of the channel weights consists of a complex multiplication and the addition of two subsequent symbols for each finger. The symbols arrive from the despreader in a time-multiplexed manner; therefore the accumulation requires a 16 location FIFO for buffering of intermediate results.

4.3.2. OFDM Decoder

The high-bandwidth wireless LAN standards IEEE 802.11a and Hiperlan/2 are orthogonal frequency division multiplexing (OFDM) systems. Hereby, symbols are modulated and spread over 48 low-bandwidth carriers, with an additional four carriers containing pilot signals. The standards define various modulation schemes and code rates, which specify data rates from 6 up to 54 Mbit/s. Figure 4.19 shows the required modules and their mapping onto the DSP, the dedicated hardware and the reconfigurable hardware.

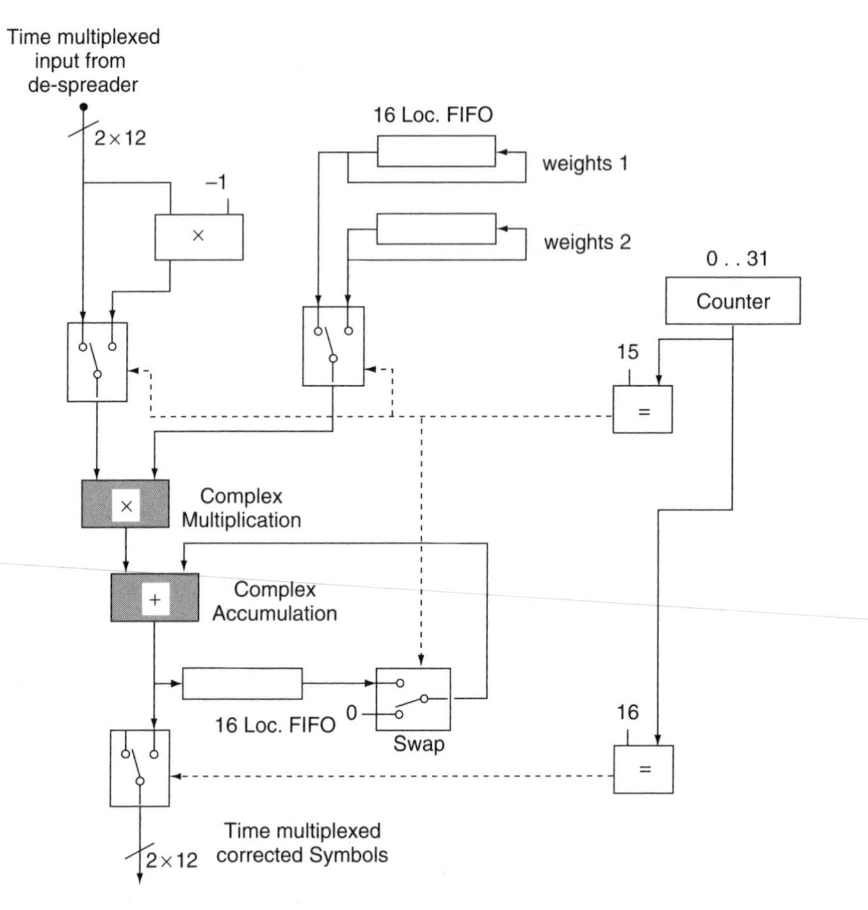

Figure 4.18 Block diagram of the channel correction unit

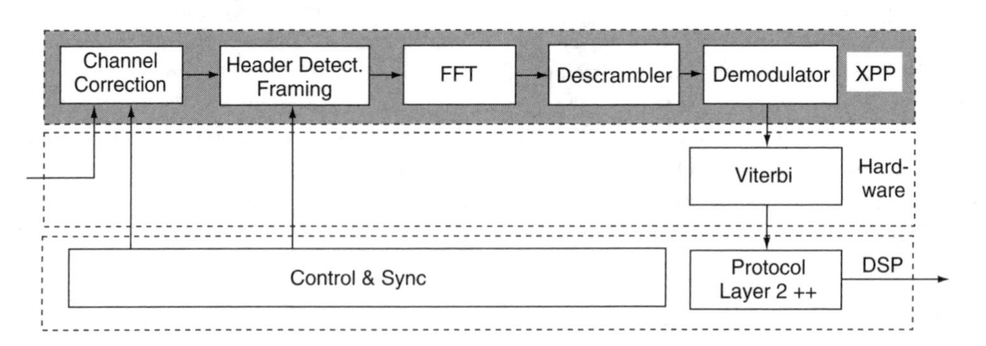

Figure 4.19 Partitioning of the OFDM decoder tasks

The complex input samples are down-sampled, interpolated and then propagated to the preamble detection for framing and synchronization. The FFT64 is used to transform the resulting OFDM symbol from the time domain into the frequency domain. Individual symbols are acquired by demodulating the resulting carrier signals. A Viterbi decoder is used for the forward error correction before propagating the data points for higher layer protocol processing.

4.3.2.1. Implementation of the FFT64

The FFT64 uses the radix-4 approach. The radix-4 computation is performed in a pipeline delivering a result value every clock cycle. The block diagram in Figure 4.20 shows the implementation of the FFT64 with the radix-4 kernel.

Read-and-write addresses are stored in circular lookup tables, which are implemented as preloaded FIFOs. Twiddle factors for all three stages of the FFT64 are also stored in a lookup table. Initially, 64 samples stream into the data RAM. The output of the RAM is multiplied with the twiddles and then streams to the radix-4 module. The output is read back to the dual-ported data RAM for the next iteration. After the third iteration, the transformed data is available at the output multiplexer. A simple counter and comparator control the multiplexer stages.

Figure 4.21 shows details of the radix4 kernel that is mapped to the XPP. The kernel receives an input stream and delivers one result per clock cycle sequentially. The accuracy of the complex input signal is 10 bit. With every stage a scaling (2-bit right shift) is required to prevent overflow. For three stages of the FFT64, we get a 4-bit precision in the result.

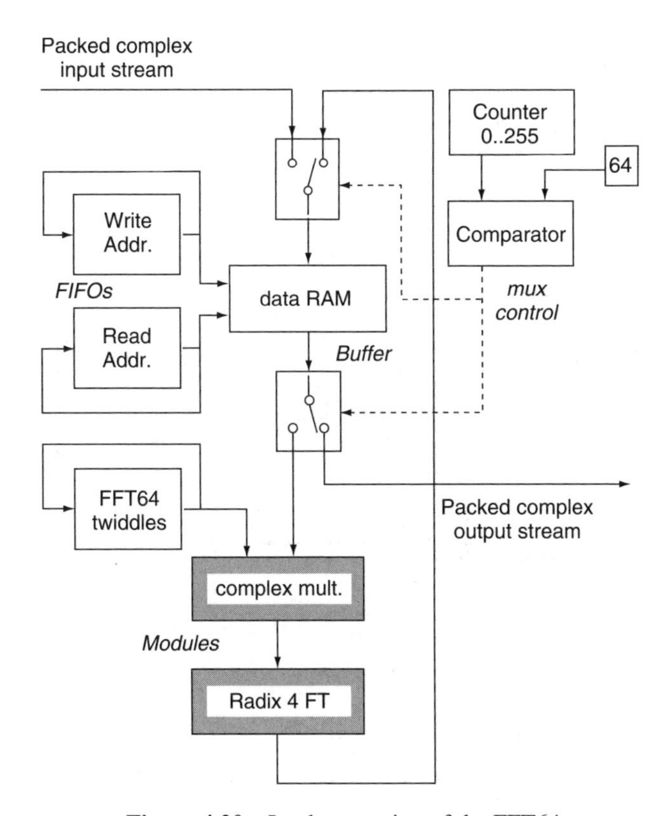

Figure 4.20 Implementation of the FFT64

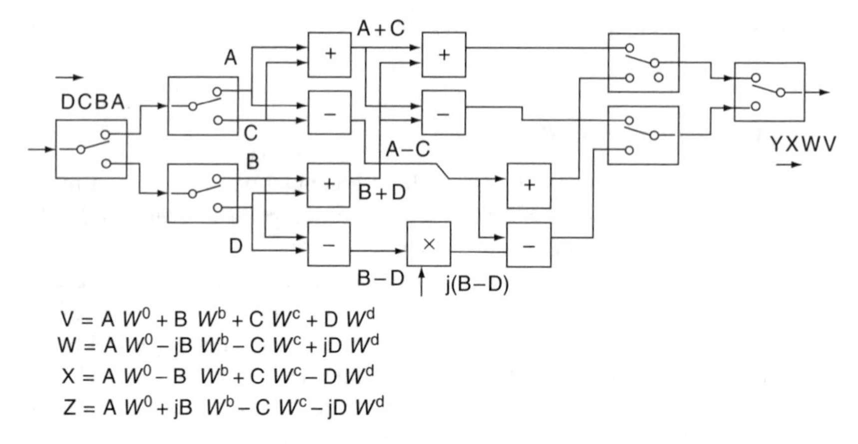

Figure 4.21 The FFT radix-4 kernel mapped onto complex-arithmetic ALUs

4.4. Software Defined Radio Processor SDRXPP

4.4.1. Structure

As outlined, software defined radio requires flexible processors and dedicated hardware support for an optimal design. These modules are implemented on a system on chip (SoC) that can be the heart of a mobile terminal. Figure 4.22 shows a proposed design for the flexible baseband processor SDRXPP with an integrated XPP core.

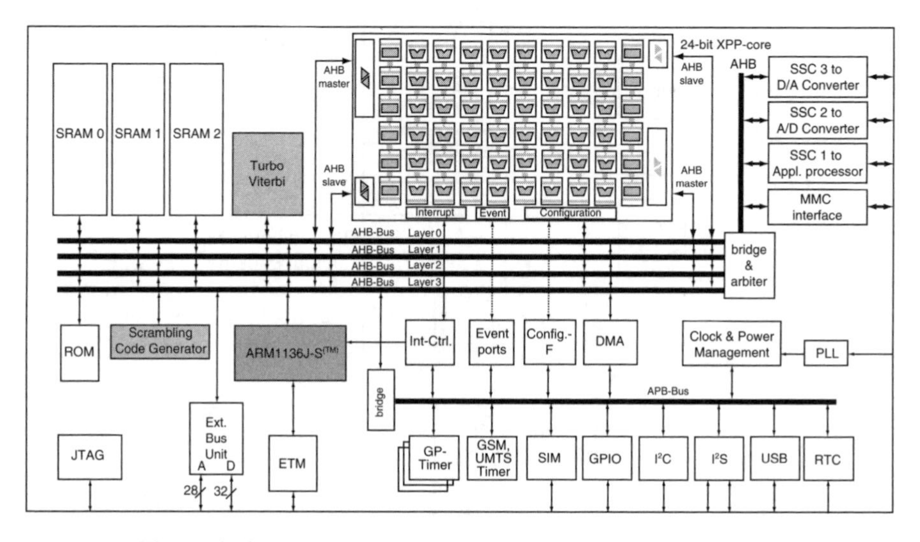

Figure 4.22 The SDRXPP system on chip for mobile terminals

The SDRXPP integrates an ARM1136E microcontroller, peripherals, a reconfigurable XPP core and high-speed interfaces. The SDRXPP operates as a stand-alone wireless modem processor that performs the digital baseband processing and protocol stack handling. The ARM11 manages the system and real-time demands, processes the protocol stacks and initializes the XPP coprocessor, which handles the high bandwidth data flow and signal processing tasks.

The components are arranged around a multilayer AHB-Bus that allows simultaneous transfers between the high bandwidth components and independently accessible RAM banks. The 24-bit XPP array is built from 6×8 ALU-PAEs, 2×6 RAM-PAEs, two AHB master interfaces and two AHB slave interfaces. This design enables simultaneous data transfers between the A/D, D/A ports and the XPP core. For example, this allows streaming data from the A/D converters directly to the array and simultaneous storage of results in a RAM bank. Modules for the forward error correction (Viterbi decoder, turbo decoder) and scrambling code generators are implemented, area and speed optimized, in hardware. The ARM11-core provides ARM's DSP extensions, which allow low bandwidth tasks, such as channel estimation, to be processed without support from XPP.

Algorithms for the SDRXPP have been designed and tested with the XPP system development platform (XPP-SDP) and software. Though this evaluation hardware does not exactly reflect the proposed design, the important interfaces, communication structures and task synchronization mechanisms can be tested and verified before the SoC silicon implementation is started. An SoC, which is similar to the SDRXPP design, is expected to be integrated on a multistandard mobile platform in 2004. The calculated power consumption of a configured and active PAE will be about 25 μW/MHz.

4.4.2. Processor Extensions for Baseband Processing

Tasks such as channel correction or demodulation require complex arithmetic with limited accuracy. Therefore, additional opcodes have been implemented in the XPP ALUs.

One such opcode is complex multiplication of two 12-bit half-words. With this feature, complex multiplication can be done with two ALU-PAEs, halving the area count and power consumption as compared with full 24-bit calculation. The other set of opcodes provides complex correlation of bit streams without utilizing the multipliers. Spreading code correlation can be implemented efficiently with these opcodes.

4.5. Conclusions

This chapter has sought to outline a new stream-based approach to signal processing that offers significant efficiency enhancements over conventional approaches. Detailed implementation of this approach within the XPP (co-)processor has been described and example implementations of the physical layer processing for W-CDMA and OFDM have been presented. XPP is well placed to be used as part of an SoC solution to provide an effective SDR processing engine, the SDRXPP, to provide an SDR capability for mobile terminals, requiring flexibility and efficiency. The inherent ability of the XPP to support fast reconfiguration means that such an SDR engine could support emerging requirements for handset air-interface reconfiguration whilst roaming.

Acknowledgments

Lorna wishes to acknowledge Wolfgang Gerner for his valuable support in her work and for being instrumental in the writing this chapter. She also wishes to acknowledge Guenter Zeisel for his support, insight and advice. She also would like to acknowledge Prashant Rao.

References

[1] *The XPP64-A Reconfigurable Processor Datasheet*, PACT XPP Technologies AG, 2003.
[2] http://www.fpdp.com/what_is_fpdp.html
[3] PACT XPP Technologies AG. *www.pactxpp.com*

Bibliography

[1] A.F. Molisch, *Wideband Wireless Digital Communications*, Prentice Hall, 2001.
[2] Roke Manor Research, *Roke Manor Research Business & Technology Review*, 2001.
[3] G.H. Bruck and P. Jung, *Software Defined Radio in drahtlosen Endgeräten*, Universität Duisburg, 2001.

5

Adaptive Computing as the Enabling Technology for SDR

David Chou, Jun Han, Jasmin Oz, Sharad Sambhwani and Cameron Stevens

QuickSilver, Technology, Inc.

5.1. Introduction

Definitions

Adaptive Computing: Algorithmic elements are mapped to dynamic hardware resources creating the exact hardware needed, for that moment in time, as often as needed, i.e. software becomes hardware in real time.

SilverC Software Programming Language: A high-level, architecture-independent system language that augments the ANSI C language with temporal and spatial extensions, and that is optimized for highly parallel, multiprocessing computing platforms.

Immediate Benefits of Adaptive Computing

The QuickSilver Adaptive Computing Machine (ACM) as discussed elsewhere [1] meets system requirements for next-generation mobile/wireless devices with high performance, low power consumption, architecture flexibility, and low cost. In doing so, the ACM overcomes the limitations of today's conventional integrated circuit (IC) solutions, such as ASICs, DSPs, RISC/microprocessors, and FPGAs.

The ACM's heterogeneous array of adaptive processing nodes brings together the combination of DSP programmability and ASIC-class power/performance. These combined attributes in a single IC give the ACM the needed characteristics for next-generation mobile and wireless devices to offer multifunctionality, including software-defined radio (SDR).

The ACM combines a high-speed network, a fractal topology, and a heterogeneous set of highly specialized adaptive, or programmable, processing nodes. The ACM architecture comprises a tree structure with clusters of processing nodes at the leaf level. The cluster can have any mixture of processing nodes, with four nodes to a cluster; 16-node clusters combine

Software Defined Radio: Baseband Technologies for 3G Handsets and Basestations. Edited by W. Tuttlebee
 ISBN: 0-470-86770-1

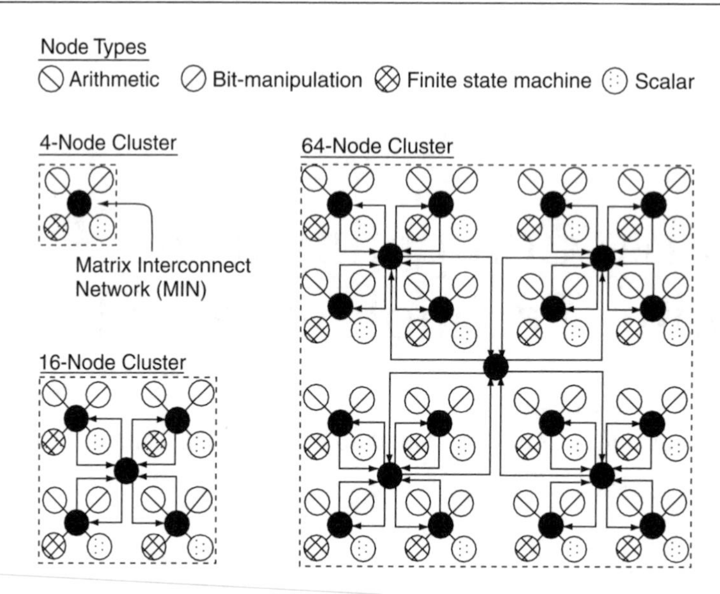

Figure 5.1 A fractal architecture

four four-node clusters. The fractal architecture is therefore scalable and allows for expansion on-chip and off-chip, creating an ACM system with a variable number of nodes. The ability to combine different types of nodes is shown in Figure 5.1 above. The node types, discussed later in this chapter, are also adaptive in nature and are each designed to perform well in a relatively narrow and specific application area.

The ACM adapts, or changes, its hardware in real time to perform what the algorithmic element requires at any point in time. The result is an efficient mapping of silicon resources to the algorithm, and therefore, higher performance and lower power consumption as compared with conventional IC implementations.

5.2. Algorithmic Evaluations in Communications

5.2.1. Algorithmic Subsets

Current solutions for meeting the requirements of the modern communications world are to mix DSPs, ASICs, and general-purpose processors to form a system. This combination requires a highly difficult and costly design process, but offers little to no flexibility for keeping pace with the ever-changing protocols in today's communications world. An improved approach to solving these problems lies in re-examining the constituent parts of the communications protocols – the underlying algorithms and operations.

Algorithms can be viewed as a grouping of mathematical operations to form a desired result. Since both the base mathematical operations and operands are of a varying nature, it is impractical for a singular processor to perform the required computations across all algorithms in an efficient manner. Because computations are of a *heterogeneous* nature, the key then is to develop an architectural solution that best exploits the heterogeneous nature of these algorithms – a mixture of heterogeneous compute engines.

A network of heterogeneous compute engines is grouped together through a normalized structure of input information and data and output information and data. Thus the interconnection between compute engines can be viewed as being that of a *homogeneous* nature.

The optimal implementation for any algorithmic element in performance, power dissipation, and area, is to build a customized ASIC to match exactly these required attributes. Therefore, the value of any other solution should be compared directly to the ASIC 'gold standard'.

These concepts form the foundation for the following discussions.

5.2.2. Algorithm Space

During the initial development of adaptive computing in the late 1990s, it became clear through mainstream research, that FPGA-based reconfigurable computing systems had considerable limitations. Conventional reconfigurable computing technology approaches the problem with a large homogeneous array of micro compute engines (e.g. lookup table, flip-flop, exclusive OR), that is, FPGA-based reconfigurable computing systems have far too fine an architecture and do not work at the level of an entire algorithm or algorithmic element. In reality, it is critical to consider the problem at this higher level.

Consider just how many different core algorithmic elements there are and for what purposes they are used. As an example, a natural separation in the wireless space is to divide computations into those with word orientation and bit orientation. Computing bit-oriented problems with a word-oriented architecture is extremely inefficient. The cost of using word-based machines to compute bit-based problems wastes resources, increases power consumption, and unnecessarily complicates the solution.

For example, consider the number of elements used in word-oriented algorithms, such as time division multiple access (TDMA) algorithms employed in digital wireless transmission (see Figure 5.2). Variants, such as Sirius, XM Radio, EDGE, and so forth, form a set for this algorithmic class. Therefore, a flexible architecture that can handle high-end TDMA algorithms will also be able to handle its less sophisticated cousins.

Once word-oriented algorithms have been evaluated, consider their bit-orientated counterparts, such as those in wideband code division multiple access (W-CDMA) – used for wideband digital radio communications of Internet, multimedia, video, and other capacity-demanding applications – and sub-variants such as cdma2000, IS-95A, and so forth. Other algorithms to consider comprise various mixes of word-oriented and bit-oriented components, such as MPEG image processing, and voice and music compression. The ACM architecture is able to cover this very large problem space and all the points in between.

5.2.3. The True Nature of Algorithms

Increasing wireless service demands require constant improvement in capacity and data-rate capability. Examples are W-CDMA, cdma2000 and IEEE 802.11. These standards require complex air interfaces, as well as video, audio, and multimedia capabilities, and support for the ever-evolving service features. At the same time, the mobility of terminals introduces physical requirements, such as longer battery life, reduced size, and lighter weight. Another consideration is that third-party developers should find it easy to understand and develop applications.

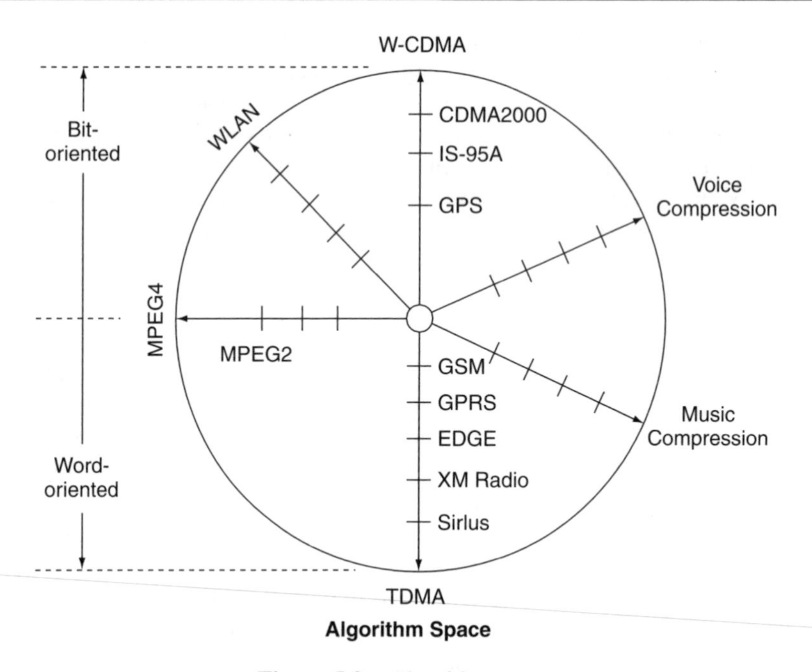

Figure 5.2 Algorithm space

As wireless technologies move from 2G to 3G and beyond, they are becoming extremely complex, requiring much higher processing power than before. For example, the wideband CDMA technology (W-CDMA), at a 5-MHz bandwidth, requires chip rates of about 3-times its narrow band counterpart TIA/EIA-95. The higher chip rate also means that more multipaths will be resolved and, therefore, more rake receiver fingers may be necessary to coherently combine the received multipath energy. Further complicating the matter is the lack of a single worldwide standard and the nonharmonized band allocations around the world. It is foreseeable that an operator would require a 3G phone capable of operating in multiple standards, such as W-CDMA, cdma2000, GSM and TIA/EIA-95. In addition to the increased complexity of the wireless functionality, designers are rapidly incorporating new features into the mobile terminal that utilize the increased bandwidth provided by the latest standards.

Another major factor to consider is that the evolution of wireless standards causes changes in design implementation that are usually painful and costly to manufacturers. New algorithms and techniques require time to mature, and the first cut at a new standard normally falls short of the mark. Unfortunately, this process results in the implementation of a cellular phone being outdated upon introduction to the market, or shortly thereafter.

In addition to physical layer processing constraints, more high-rate communications protocols are requiring the MAC layer to have increased and stringent performance requirements such as the 802.11x family. This places an additional burden on the computational fabric and pushes what typically would be handled by a low throughput microprocessor into the main computational engine.

5.2.4. Classification of Algorithms

Most communications protocols have algorithms that can be grouped into themes of common processing: source coding, channel coding, interleaving, scrambling, modulation, domain transformations, various filtering, and on the receive side, much more complicated DSP algorithms to reverse the effects of the air interface channel impairments. These algorithms in turn have common algorithmic elements that lend themselves to a further grouping. Algorithmic elements are chosen as this point of demarcation because they offer the best combination of performance and flexibility. Coarse-grained reconfigurable computing architectures that work directly with algorithms or applications are not able to cover the myriad communications protocols in existence today. Fine-grained architectures operating at the arithmetic level will face the arduous task of creating an efficient group of operations and instructions that cover all algorithms across all protocols. A more optimal solution is to break algorithms into common algorithmic elements. Hardware can be efficiently designed to operate on groups of algorithmic elements.

Algorithmic elements contain different operations and manipulate different data types.

- **High-rate filtering**: W-CDMA pulse-shaping filter can require up to 1.5 GMultiplies/s, depending on implementation. Other applications such as a telematics receiver can also approach these numbers.
- **Parallel arithmetic processing**: Many communications algorithms fall into this traditional DSP category, such as FFTs, lower-rate filtering operations, convolutions, correlations, linear, and nonlinear operations. Most vectored operations can be categorized as parallel arithmetic processing.
- **Byte processing**: Source and channel decoding, interleaving.
- **Bit processing**: Source and channel encoding, scrambling, encryption.
- **Small control structures**: Efficient control structures and processing implemented as finite state machines.
- **Large control structures**: Bookkeeping and general control most suited to a general-purpose, traditional *instruction set architecture* processor.

Difficulties not only lie in the heavy computation requirements, but also in the complicated control structures intertwined in the processing. Each of the groups of algorithmic elements must support elaborate control mechanisms. The underlying fabric that connects these groups of algorithmic elements must also have the support and capability to meet these requirements.

Algorithms have a locality of reference nature, meaning they always move from one mathematical step to the next immediate mathematical step, and data is only passed to that next step. This can involve several steps in parallel, but they are still the next logical data movement. This means that the wiring needed between the next logical steps should be very dense, but the converse is not true. Unlike an FPGA that has very dense wiring along its whole XY plane, a better architecture will reduce its wiring structure by understanding that algorithms do not communicate, or spread out, to the far ends of the silicon area. The algorithm always moves data to the next logical step, creating a fractal organization. The further away any two blocks of processing are, the less wiring there is between them. The closer two blocks physically reside, the more wiring there is between them.

By partitioning the problem into algorithmic element groups, it is now conceivable to imagine a solution consisting of a heterogeneous architecture that fully addresses the heterogeneous nature of the algorithms, and is designed to run each group in a most efficient manner in terms of area, performance, and power while attaining flexibility to modify existing and future algorithms. The balance of communication and computation between processing elements and the underlying interconnection can be achieved on a single device given the perspective above.

5.2.5. The IC Gold Standard

Currently, the challenges of implementing computationally intensive tasks in modern communications protocols are met through a tedious system design that comprises DSPs and ASIC accelerator blocks. At a very early stage, the designer must consider the trade-offs in choosing what goes into the DSP and what goes into the ASIC. Typically, the designer will allocate as many tasks to the DSP as possible, thus minimizing the cost of the IC and increasing flexibility. However, there are functions that the DSP lacks the power to perform, e.g. Viterbi decoder and despreader. In contrast, ASICs have higher efficiency in power consumption than DSPs, and therefore, allocating tasks to an ASIC has the potential of prolonging battery life in a wireless/mobile terminal.

ASICs have long been considered the 'gold standard' of ICs for performance, power consumption, and silicon area. For a single algorithm or computation, an efficient ASIC design will yield the highest performance and lowest power consumption of any implementation. Such a design represents the ideal implementation of an algorithm in hardware. An ASIC design may also represent the most efficient implementation in terms of silicon area, provided the algorithmic computations are reasonably well balanced and all circuitry is utilized at, or near, its peak capability.

This standard of performance cannot be matched by microprocessors, DSPs, or FPGAs due to the overhead of their control structures and/or inefficiencies in mapping an algorithm to the rigid elements within their design. However, the costs associated with achieving this gold standard are high, including lengthy design and verification times, low silicon efficiencies when an ASIC element is not used full time, and the impossibility of changing the design without a costly and time-consuming chip re-spin. Nevertheless, ASIC implementations represent an accepted standard by which other IC technologies can be judged.

One of the goals of adaptive computing is to make SDR possible by delivering ASIC-class performance in a flexible architecture, such as the adaptive computing machine (ACM). The power consumption and silicon overhead of DSPs and FPGAs limit their use in mobile/wireless handsets. In contrast, adaptive computing seeks to overcome these obstacles by greatly reducing the control overhead associated with programmable computing, while providing architecture flexibility for efficiently mapping computational resources to algorithmic problems in real time.

5.3. Solving P^3 – Performance, Power Consumption and Price

5.3.1. Today's Design Issues

The baseband processing in a wireless device today typically comprises three computational elements physically fabricated on one or more conventional semiconductor components, as follows.

5.3.1.1. General Purpose Processor

The main control processor, typically a RISC, is responsible for supporting the user interface, running user applications such as a web browser, managing external connectivity, and handling the protocol stacks of the air interface.

5.3.1.2. Digital Signal Processor (DSP)

DSPs are incorporated for handling the number crunching associated with both user applications and air interfaces. A DSP provides flexibility in its ability to be reprogrammed. However, DSPs are power hungry and typically do not offer sufficient processing power for a wireless system design to run complex algorithms in software, such as Viterbi decoders, channel decoders, and fast Fourier transform.

5.3.1.3. Application Specific Integrated Circuit (ASIC)

ASICs offer high performance and consume very little power. However, ASICs present several design limitations:

- ASICs can only be fully tested as a finished IC;
- at a very early stage, the designer must consider the trade-offs in choosing which portions of the design are to be assigned to the DSP and which portions are to be assigned to ASIC material;
- ASICs are most often obsolete at design time because of their fixed-function architecture which does not allow for changes or bug fixes during the development cycle;
- high cost is associated with long-lead design, build, and test cycles;
- each function implemented in ASIC material adds to the die size, which means more cost is incurred for the silicon.

5.3.2. Field Programmable Gate Arrays (FPGAs)

FPGAs are not considered to be a practical option for mobile/wireless applications since they consume enormous amounts of power and require large amounts of programming data, placing a burden on system memory. While an FPGA can be changed, the rapid reconfiguration necessary to satisfy the variety of required circuits of a complex air interface does not make its use practical on a dynamic, real-time basis.

Typically, a designer will allocate as many tasks to the DSP as possible, taking advantage of its cost effectiveness and programming flexibility. However, as stated earlier, the processing power of a DSP is no match for that of an ASIC. To bridge the performance gap and to meet the efficiency requirements of next-generation mobile/wireless devices, hybrid system-on-chip (SoC) designs are emerging that use a mixture of RISC, DSP, and ASIC material.

In order to keep up with the evolving complexity of wireless standards, a growing amount of ASIC material on the chip is needed. Due to the inherent conflict between the use of ASIC material and the desire to implement new algorithms as quickly as possible, multiple design teams are required, working in parallel at different stages of the design. Even so, at least 18 months is still required from concept to market for a new feature.

In addition, ASIC design in the deep-submicron region is becoming more problematic and costly due to:

- the need for multiple iterations between logical and physical synthesis to accommodate process effects;
- increased complexity as more transistors are placed onto a fixed die size;
- integration of heterogeneous functionality on the same die, and
- pressure for faster time to market.

The average cost of a 0.13 µm or 0.09 µm process design is $4 to 8 million and $10 to 15 million, respectively. Furthermore, the cost to test a transistor nearly approaches the cost of manufacturing it. All this implies a need for increasingly large volumes in order to amortize the design and fabrication costs and reduce the cost per unit.

Based on the above discussions, there is a clear need for an innovative IC approach that can meet the requirements of increasingly complex technologies and design considerations for SDR communications of today and the future.

5.3.3. Heterogeneous Processors for Heterogeneous Algorithms

Since real-world problems consist of heterogeneous complex algorithmic elements, low-level hardware building blocks must be heterogeneous. The architecture of the ACM integrates a number of compatible, high-performance, scalable, heterogeneous computing elements, memories, input/output ports and associated infrastructure to provide efficient implementations that target broad and diverse sets of applications. The heterogeneous nodes are adaptive elements that exchange data via a homogeneous network.

The principal components of the ACM architecture are:

- Heterogeneous nodes;
- Homogeneous matrix interconnection network (MIN):
  - IMC (internal memory controller)
  - XMC (external memory controller)
  - IOC (input/output connection controller)

The nodes are designed to address diverse problems:

- both high and low vector content;
- both streaming and nonstreaming data;
- data flow;
- complex control;
- arithmetic on various data types;
- bit manipulations;
- both parallel and sequential dominated algorithmic elements.

The desired goal is to make the node fit the algorithms within a pre-defined problem space. To accomplish this, there are special hardware features on the ACM to support the needs of the specific problem space.

The ACM is a data-flow machine. Figure 5.3 shows a typical configuration of an ACM with 32 nodes. More than one ACM can be connected together to form an ACM network.

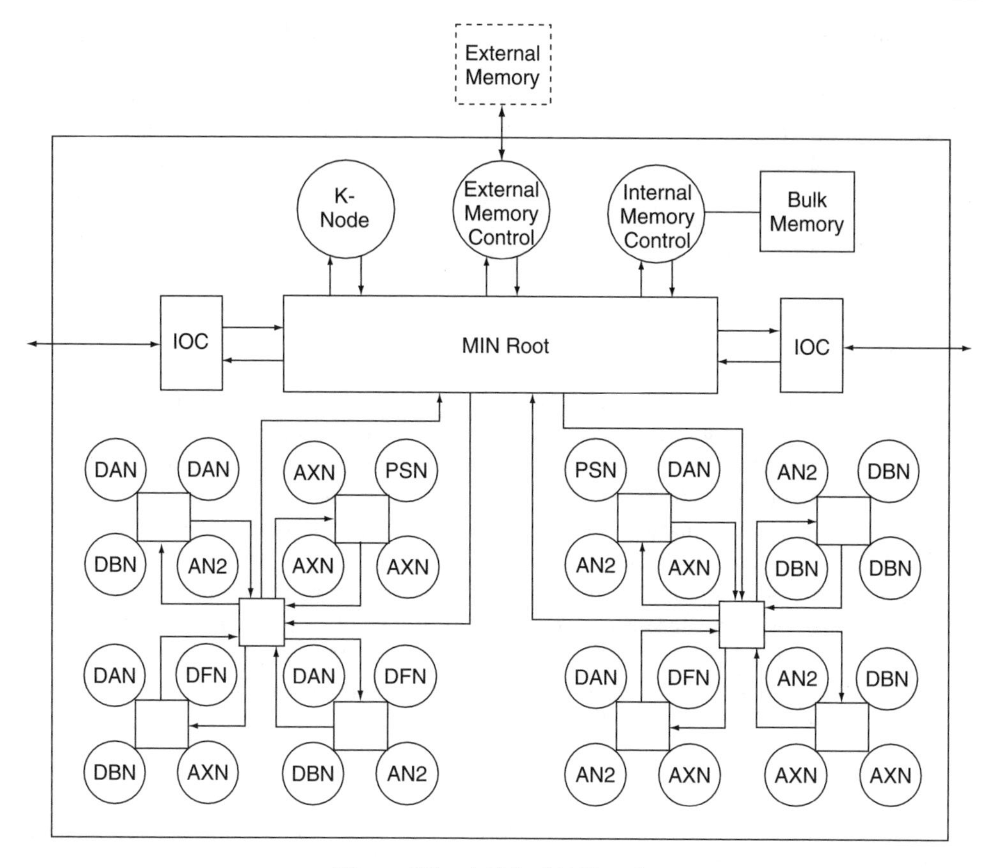

Figure 5.3 ACM with 32 nodes

5.3.4. Node Types and Features

The ACM comprises several node types, all with different algorithm capability and power consumption. Examples of node types are the following:

- AN2: adaptable arithmetic node with one MAC unit;
- AXN: fully adaptable node with four MAC units;
- DAN: domain arithmetic node;
- DBN: domain bit-manipulation node;
- DFN: domain filter node;
- PSN: programmable scalar node.

Currently, each node is assigned with a certain amount (16 kbyte) of local nodal memory. When tasks require more local memory, there are two ways to meet this requirement. First, aggregate adjacent node(s) to double or quadruple the nodal memory. Alternatively, establish connections between the IMC/XMC to the node, which allows the nodes to access data directly from internal/external memory.

The ACM's combination of a homogeneous network with heterogeneous nodes provides the framework for developing system solutions for targeted applications spaces. These nodes can be combined in numerous ways and communicate from one node to another via the matrix interconnect network (MIN).

At a macro level, the ACM advantage is the ability to incorporate multiple standards to implement true SDR, such as GSM/GPRS, W-CDMA, and cdma2000, in a single chip. Further, within the same chip, the ACM also supports multifunctionality, such as an MP3 player, digital camera, or PDA, without any impact to performance, power consumption, or silicon size. The ACM's reuse of gates during a single task for different activities enables tremendous reductions in silicon area (cost) and power consumption.

5.3.5. Homogeneous Infrastructure

The heterogeneous nodes – AN2, AXN, DAN, DBN, DFN, PSN – are interconnected by the MIN, using a by-four fractal architecture to support scaling to any number of nodes. Each node has a highly efficient hardware operation system that enables multiple tasks to be executed on a single node. A node comprises an execution unit (EU) for the computational work and infrastructure composed of memory, node wrapper, and so forth. The node wrapper is the interface between a node and the network.

Most applications are dominated by memory. In order for the architecture to be scalable, the memory and node wrapper must be uniform for all node types. The EU is node specific, the memory and node wrapper are not. Therefore, the ACM has a basically homogenous 'tree' infrastructure and the heterogeneity is expressed only at the leaf level.

5.3.6. Scalability

The ACM's architecture is fractal in nature, comprised of nodes that are interconnected by a highly scalable network that exploits the locality of reference of the problem space. This scalability gives the designer the ability to expand processing capability as needed. The basic element in the fractal hierarchy is the 4-node cluster formed by four nodes and connected via a MIN. A 16-node cluster is formed from four 4-node clusters linked by their own MIN, while a 64-node cluster is formed from four 16-node clusters linked by their own MIN, and so forth. Of significant note is that conventional IC technologies do not use a fractal architecture, and therefore are not scalable.

Integrated into each node is a complex mixture of memory, logic elements and interconnect that enables the ACM to perform a wide variety of operations on a wide variety of data widths. The operations and data widths are fully programmable. In addition, the interconnect between the nodes allows them to be closely coupled for work on larger problems, assembling even larger structures than what may be possible within a single node.

5.3.7. Real-time Adaptability

A key advantage of the ACM's architecture is that any node can be adapted on-the-fly to perform a new function. Rather than passing data from function to function, the data remains resident in a node while the function of the node changes. Temporal sequencing allows the ACM to step through the various algorithms used for solving a particular problem, and to adapt the ACM for solving an entirely different problem(s) at another instant in time.

The ACM maps algorithms onto dynamic hardware (silicon) resources. The term 'dynamic hardware resources' refers to the logical functions inside an ACM that are adapted at high speed, at run time, while the mobile/wireless device remains in operation. Unlike a fixed-function ASIC implementation, the ACM adapts, or changes, in nanoseconds so that only those portions of an algorithm that are actually being executed need to be resident in the chip at any one time. This reuse of gates enables efficient use of the silicon, and thereby reduces the silicon area and power consumption.

The dynamic hardware resources themselves are adaptive, in that they can be configured to solve a broad range of problems over a broad range of data sizes. The ACM is capable of executing a number of different problems simultaneously; in contrast, an ASIC has dedicated resources for each problem executing in parallel.

5.4. Design Advantages

5.4.1. Solving Hardware/Software Codesign

ACM designs are represented in the SilverC™ language, which augments the ANSI C language with temporal and spatial extensions. Applications are developed in SilverC, and then compiled and expressed as downloadable applications in SilverWare™ binary components.

ACM technology eliminates the very difficult problems associated with hardware/software codesign because the entire system is initially represented as software. Having said this, it is important to understand that SilverWare is not executed by the ACM in the same way that machine code is processed by a DSP, i.e. executing a long stream of instructions. Instead, SilverWare is used to dynamically adapt the ACM on-the-fly to create the exact hardware needed to perform whichever algorithmic tasks are required at any particular time. Complex algorithmic elements can be thought of as the smallest operators, and many of these complex algorithmic elements are temporally or spatially combined to form an application.

5.4.2. The SilverC Language

SilverC is a software programming language designed for expressing dataflow-oriented algorithms and complete applications. It is a high level, architecture-independent system language that is optimized for highly parallel, multiprocessing computing platforms.

SilverC augments ANSI C with extensions that are necessary for expressing task-level parallelism. These extensions include the *module* and *process* constructs, as well as the concept of *streams* as expressed using *pipes*. SilverC also adds support for DSP programming through the inclusion of fixed-point data types, fixed-width integers, and specialized pointers for efficient array access.

The fundamental unit of computation in SilverC is the *module*. Each module is a code container that performs a specific logical function. For example, one module may implement a FIR filter and another a Vocoder computation. Modules may also be parameterized, thus providing a generalization mechanism similar to C++ templates[†].

[†] It should be noted, however, that SilverC is not rightly considered an object-oriented language. While a SilverC module does encapsulate both data and operations, SilverC provides no mechanisms to support such concepts as polymorphism.

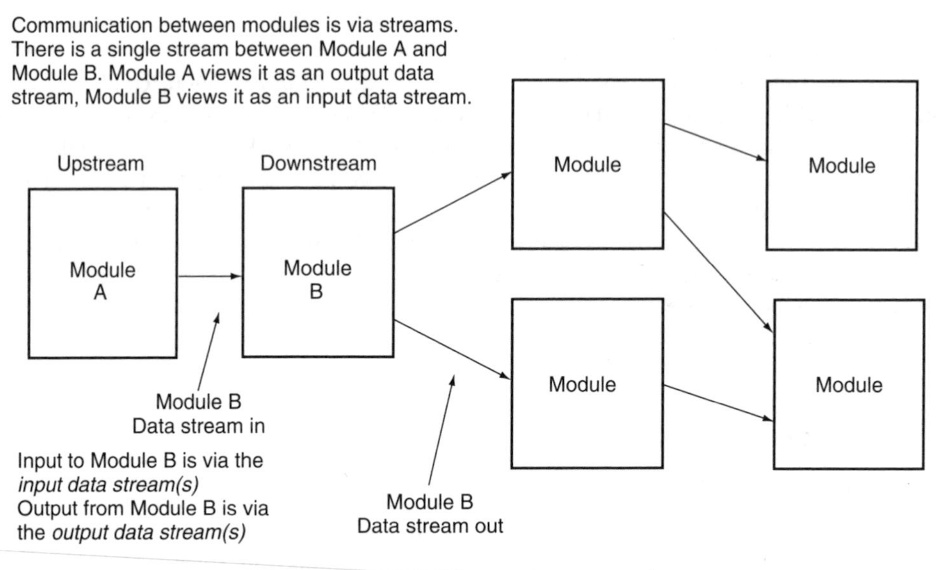

Figure 5.4 A dataflow-style computation

Modules are connected via data streams that move data values from one module to the next. Figure 5.4 illustrates several sample modules connected by a number of data streams. The module producing the data stream values is considered to be upstream of the module that is consuming the values. The module pointed to by the arrowhead, is referred to as the downstream module.

In addition to exposing task-level parallelism, the module construct provides a high degree of encapsulation. Modules can be treated as 'black boxes' that are completely specified in terms of their parameters and stream I/O characteristics.

Modules may be archived and redistributed as *libraries*, exposing only module declarations while hiding the details of the underlying implementation. Together with parameterization, this allows for a high degree of *code reuse*.

Each SilverC module aggregates one or more processes that implement the actual computation, typically by reading data values from one or more input streams, performing some computation on those values, and then sending the results to one or more output streams (see Figure 5.5). The bodies of these processes closely resemble ANSI C. Unlike C functions, processes are not invoked explicitly. They are declared with specific firing conditions and are executed whenever these firing conditions are true. Processes take no arguments and have no return type. SilverC processes can also be implemented using assembly language.

The remainder of this section will further elaborate on the SilverC language by means of a simple example – the implementation of a min-k filter. This filter takes k data values as its input and generates the smallest of the values as its output. For example, a min-k filter can be written in ANSI C using the function shown in Figure 5.6.

A SilverC module implementing a min-k filter may be represented as a dataflow diagram as shown in Figure 5.7. The SilverC code implementing the module is shown in Figure 5.8.

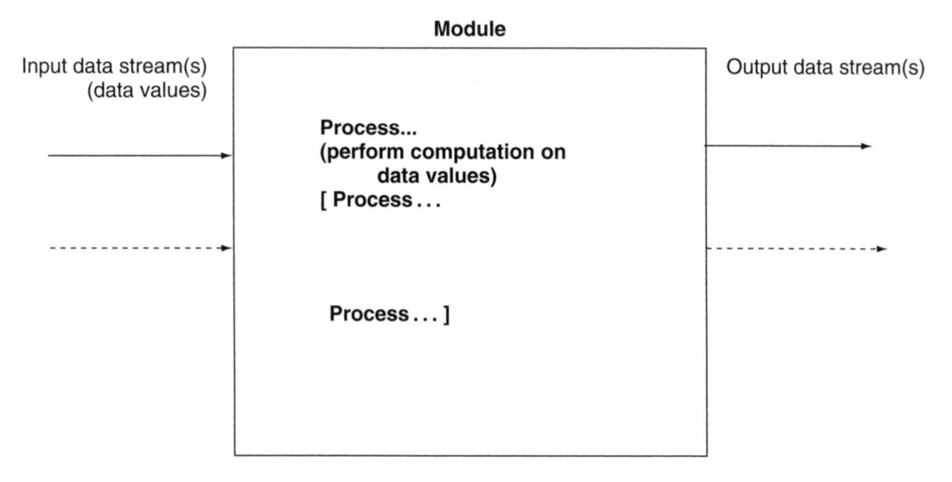

Figure 5.5 A module with processes

```
1:  double mink (const int k, const double data []) {
2:        int i;
3:        double minValue, nextValue;
4:
5:        minValue = data [0];
6:        for (i=1; i<k; i++) {
7:              nextValue = data [i];
8:              if (nextValue < minValue) {
9:                    minValue = nextValue;
10:             }
11:       }
12:       return minValue;
13: }
```

Figure 5.6 An ANSI C min-*k* module

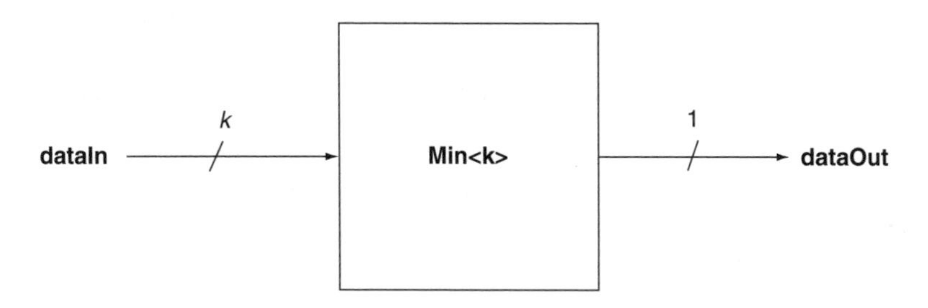

Figure 5.7 A sample abstract min-*k* module

All modules in SilverC are declared using a **module** statement (line 1). The **module** statement is used to name the module, specify its instantiation-time parameters, and open a scope that contains the module definition. A module statement may optionally identify its hardware target.

```
1:  module Min<const int16 k> {
2:      pipes:
3:      inpipe<fract16, 2*k> dataIn;
4:      outpipe<fract16> dataOut;
5:
6:      processes:
7:      process runMinK when
8:          (ready (dataIn, k) && ready (dataOut, 1)) {
9:
10:         static pointer<fract16, dataIn, 1> dataInPtr;
11:         fract16 minValue;
12:         fract16 newValue;
13:         int16 i;
14:
15:         minValue=*(dataInPtr++);
16:         for (i=1; i<k; i++) {
17:             newValue=*(dataInPtr++);
18:             if (newValue<minValue) {
19:                 minValue=newValue;
20:             }
21:         }
22:
23:         dataOut=minValue;
24:
25:         release (dataIn, k);
26:         notify (dataOut, 1);
27:     }
28: }
```

Figure 5.8 A SilverC MIN-*k* module

Each module definition consists of declarations that are organized in distinct sections. There are two sections that every module must contain:

- the first defines the stream interface of the module – the input and output data streams that the module expects;
- the second defines the processes of the module, which perform the actual computation.

These sections are indicated by the **pipes** and **processes** keywords (lines 2 and 6). Section headers appear at the top level of the module scope and are followed by a colon.

Pipes are the conduits for transmitting data streams between modules. Each pipe of a module may be one of two varieties:

- input pipes that bring a data stream into a module;
- output pipes that convey data out of a module.

These pipes are specified using the **inpipe** and **outpipe** keywords, respectively. Pipes are parameterized by the type of element they transfer. The SilverC programming model assumes that data buffering will be performed on the input side of the pipe, therefore, the number of elements that should be allocated for the pipe buffer is also specified for an input pipe.

The *Min* module has a single input pipe and a single output pipe, which are declared on lines 3 and 4. The module pipes are declared to transmit elements that are 16-bit fractional fixed-point values, designated by the **fract16** keyword.

In this example, the input pipe is declared to have $2 \cdot k$ elements of storage to provide a *double-buffering* effect – while this module is processing *k* values, its upstream neighbor can fill the other *k* locations in the buffer with new values.

The body of a SilverC process contains local variables and imperative C-style code, like a traditional C function. The *Min* module contains a single process that runs the min-*k* filter (lines 7–27).

The code declares a process named *runMinK* (line 7). It is declared to fire based on the readiness of the input and output pipes of the module. Firing conditions are specified using the SilverC **ready()** function. The firing condition requires that *k* values be present in the buffer corresponding to the input data stream, *dataIn*, and that the downstream module which *dataOut* is connected to must have sufficient buffer space to store a single output value (line 8). These conditions are designed to ensure that the process never stalls or accesses inappropriate values due to inadequate input data or lack of output buffer space.

A SilverC *pointer* is the primary method of accessing data from within a SilverC input pipe buffer. This variable type is similar to a C pointer and can be used to walk through memory in regular patterns. The constraints on SilverC pointers allow them to be mapped to efficient hardware address generation units. SilverC pointers are declared using the **pointer** keyword, and take a number of parameters that specify their behavior. SilverC pointers may also be nested.

The SilverC pointer used to access the data buffer implementing the *Min* module input pipe is declared at line 10. The declaration creates a static SilverC pointer named *dataInPtr*. The first parameter in the pointer specification indicates that it points to values of type **fract16**. The second parameter indicates the legal range of values for the pointer. In this case, the use of an input pipe identifier indicates that the range of the pointer is constrained to refer to the buffer space used to implement that pipe. The last parameter indicates the size of the step that is taken when the pointer is incremented. In this case, the value '1' indicates that the pointer steps from one **fract16** element to the next in memory.

Because it is defined in terms of an input pipe, *dataIn*, the pointer is initialized to refer to the first element of the *dataIn* buffer. This variable is marked using the **static** keyword, indicating that its value is preserved from one invocation of the process to the next.

5.4.3. Specifying the Computation

The computational code is defined in lines 15 through 21. It strongly resembles the original C code on which it is based. The primary difference is the change from an indexed array access notation to the use of a SilverC pointer that walks through the input data stream.

The pointer is dereferenced and incremented just as a traditional C pointer would be (line 15). Because *runMinK* is specified only to execute when *k* data values are ready in the *dataIn* buffer (line 8), the *k* pointer dereference/increment operations performed by the module are guaranteed to refer to legal values within the buffer. Because *dataInPtr* is declared to be static, it maintains its state across process invocations and continues where it left off, reading the following *k* items of data when they become available. Since *dataIn* was used to specify the limit expression of the pointer, the pointer automatically wraps back around to the beginning of the buffer when it is incremented past the end.

5.4.4. *Writing to Output Pipes*

Having computed the minimum value of *k* inputs, the result is written to the output pipe of the module by using an assignment to the output pipe (line 23). This assignment causes *minValue* to be placed on the network and delivered to the downstream module, where it is stored in the corresponding input pipe buffer of the module. This operation is guaranteed to function correctly because the *runMinK* process is defined with a firing condition that states that there is sufficient space ready for one element in the output buffer before the process begins (line 8).

5.4.5. *Pipe Synchronization*

A SilverC process finishes by synchronizing with the upstream and downstream modules to which it is linked. All data written to an output pipe must be acknowledged by *notifying* the downstream process that new data has arrived. Once the process is done with values stored in its input pipe buffer, it must *release* those values so that their buffer space may be reclaimed for use by the upstream process.

These operations are performed using the built-in SilverC **notify()** and **release()** functions. These function calls appear at lines 25 and 26 of the *Min* module. This specifies that this process has finished with the *k* oldest values in its input buffer, and has generated one new value for its downstream neighbor.

These calls are the mechanism for causing the readiness conditions of a process to become true: neither the actual use of input pipe data nor assignment of data to an output pipe has any effect on the readiness of a pipe.

The relative order or placement of these calls within a process can be changed depending on the priority placed on synchronizing with its upstream and downstream neighbors.

5.4.6. *Instantiating Modules*

In SilverC, modules are instantiated by using C-style declarations within a top-level C function named *main*. Figure 5.9 illustrates a *main* function that utilizes the *Min* module in a simple application.

```
1:   #define K 16
2:
3:   void main () {
4:          const fract16 testValue=-1;
5:
6:           MinProducer<K, testValue> myProducer;
7:           Min<K>                    myMinFilter;
8:           MinConsumer<testValue> myConsumer;
9:
10:          link (myProducer.dataOut, myMinFilter.dataIn);
11:          link (myMinFilter.dataOut, myConsumer.dataIn);
12:
13:          constrain (SAME_NODE, myProducer, myConsumer);
14:  }
```

Figure 5.9 A simple SilverC application

Traditional C notation is used to *#define* a constant *K* and to declare a constant variable *testValue* (lines 1 and 4). These constants are used as the actual parameters for the module instantiations. Both values could have been declared using either mechanism; this example uses both constant-specification mechanisms for the purposes of demonstration.

Next, the *Min* module is instantiated along with two fictitious test bench modules, providing each with their expected parameters. At this point, all of the modules have been instantiated and are ready for use once their input pipes and output pipes are linked together.

5.4.7. Linking Modules

Modules are linked together using a built-in **link()** function. This call connects the output pipe of one module to the input pipe of another. Lines 10 and 11 of the sample application link the three modules together. The result is a dataflow system that is linked as shown in Figure 5.10.

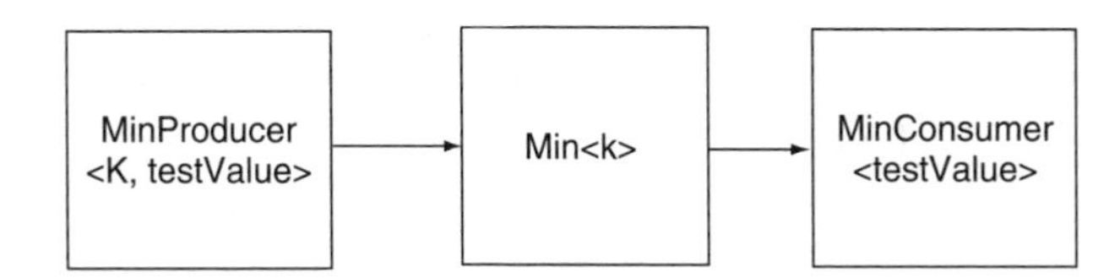

Figure 5.10 The test bench architecture for the MIN-*k* module

5.4.8. Constraining Modules

The SilverC **constrain()** function allows users to specify additional constraints that should be applied to the SilverC application. A single constraint type is currently supported: **SAME_NODE**. This constraint type may be used to force multiple module instances to be allocated to the processor. A **SAME_NODE** constraint is used at line 13 of the sample allocation to force the *myProducer* and *myConsumer* modules to be allocated to the same processor.

5.4.9. SilverC Support for Spatial and Temporal Extensions

The spatial and temporal segmentation (SATS) process of the ACM enables SDR and multi-functionality to occur. Unlike conventional IC technologies, the ACM architecture adapts to the problem at hand, enabling time sharing, or spatial and temporal segmentation. SATS is the process of mapping algorithms for a given task to dynamic hardware resources, then rapidly performing various portions of an algorithm in different segments of time (temporal) and in different locations in the adaptive fabric (spatial) of the ACM.

With SATS, the ACM's gates are rapidly reused, bringing into existence for the exact amount of time needed – clock cycle by clock cycle – the hardware an algorithm requires, and then efficiently running any number of different algorithms on the hardware engine.

The SilverC language allows the programmer to expose the inherent spatial and temporal segmentation of their application. The *module* construct supports spatial segmentation by providing an appropriate unit of code to map onto the adaptive fabric. The programmer may further control the spatial segmentation using the *constrain* function, or allow the compiler to perform this task automatically.

The SilverC *pipe*, *process* and *ready* constructs allow the programmer to efficiently express task-level parallelism and thereby expose the temporal segmentation of their application.

5.5. ACM Benchmarks

The application module library (AML) is an application function library for the ACM. The AML contains a wide range of math, DSP and application specific routines that are hand-assembly optimized on the nodes available within the ACM.

5.5.1. General Benchmarks

In this benchmark example, algorithms ranging from multiplication intensive FIR filters to bit-manipulation intensive interleavers and linear feedback shift registers are presented in the form of ACM modules that can execute on the ACM simulation platform (ASP). The target ACM nodes are the AXN and DBN nodes. Eight modules that operate on 16-bit data have been selected from the AML for benchmarking purposes:

- general real FIR filter;
- general complex FIR filter;
- 64-point complex FFT/IFFT;
- 16-point complex Hadamard transform;
- general bit interleaving;
- general bit puncturing;
- general bit repetition;
- W-CDMA downlink scrambling code generator.

In Table 5.1 the performance metrics of the eight selected modules, including clock cycles, memory requirements, and latency are shown. All the metrics are measured from the ASP simulations, including system and module set-up overheads. Note that in the memory requirement column, double input buffering is used to allow pipelining of modules at the system level. Also, inside each module, pipelining and prolog/epilog optimization are applied to reduce overhead. New features of the ACM architecture are being added to further reduce the overhead and increase the efficiency of the code. With new tool releases, the three DBN bit repetition modules will be merged into one.

In the above AML modules, 16-bit data are packed into 32-bit words. Each memory access transfers one 32-bit word between the nodal memory and execution unit (EU) data path, which translates to one complex data or two real data samples. The state variables, such as the filter coefficients in the real/complex FIR filter modules and twiddle factors in the FFT/IFFT module, are stored in the state section of the nodal memory. The multitasking capability of the nodes makes it possible to implement different tasks in the same node, reducing the number of active nodes and the data movement across nodes. All the modules that are implemented on that node share the state section.

Seven of the listed modules are generic, which are self explained. Of particular note is the W-CDMA downlink scrambling code generator, which generates a complex binary sequence

Table 5.1 Performance metrics of selected AML modules

Module	Target application	Target node	Benchmarks	Clock cycles	Memory requirement (bytes)	Latency (cycles)	Comments
Real FIR filter	General	AXN	Compute N output samples using filter of length M	$9 + N/4(M+5)$	$8N+2M$	$M+57$	$N \geqslant 4$, must be multiple of 4 $M \geqslant 12$, must be multiple of 4
Complex FIR filter	General	AXN	Compute N output samples using filter of length M	$15 + N(M+1.5)$	$16N+4M$	$2M+9$	$N \geqslant 2$, must be multiple of 2
64-point complex FFT/IFFT	General	AXN	Compute 64 complex output samples	FFT: 526 IFFT: 592	1816	FFT: 448 IFFT: 499	Cycle efficient implementation
Complex Hadamard transform	3GPP W-CDMA	AXN	Compute magnitude of 16-point complex Hadamard transform	234	488	226	
Bit interleaving	General	DBN	Perform bit interleaving of 481-bit input buffer within 30 column permutation table	563	1476	39	16-bit indexes are computed by different task (presumably on PSN node) and packed into 32-bit word
Bit puncturing	General	DBN	Perform puncturing of 576-bit input buffer with rate 2/3 and 110110011011puncturin g pattern	435	1540	39	16-bit indexes are computed by different task (presumably on PSN node) and packed into 32-bit word

Table 5.1 Continued

Module	Target application	Target node	Benchmarks	Clock cycles	Memory requirement (bytes)	Latency (cycles)	Comments
Bit repetition	General	DBN	Bit repetition rate of 2 for an input buffer size of 512 32-bit words	10242	2368	16	
			Bit repetition rate of 4 for an input buffer size of 256 32-bit words	9733	1344	15	
			Bit repetition rate of 8 for an input buffer size of 128 32-bit words	8695	876	15	
Scrambling code generator	3PGG W-CDMA	DBN/ PSN	Compute 256 complex chips	526	768	68	Maximum cycle budget (for a 200 MHz clock) is 13333 cycles. For this clock rate 10 such generators can be implemented on a single DBN

that is used to scramble a particular downlink W-CDMA channel. This module is mapped to three ACM tasks distributed across 1 PSN and 1 DBN.

5.5.2. W-CDMA Baseband Transceiver Benchmarks

The following example of a W-CDMA baseband transceiver system mapping onto an ACM, and corresponding benchmark metrics, depicts the flexibility of the ACM and how it achieves performance efficiencies. The main subsystems of the transceiver are shown in Figure 5.11. Figure 5.12 is a block diagram that shows the W-CDMA downlink searcher.

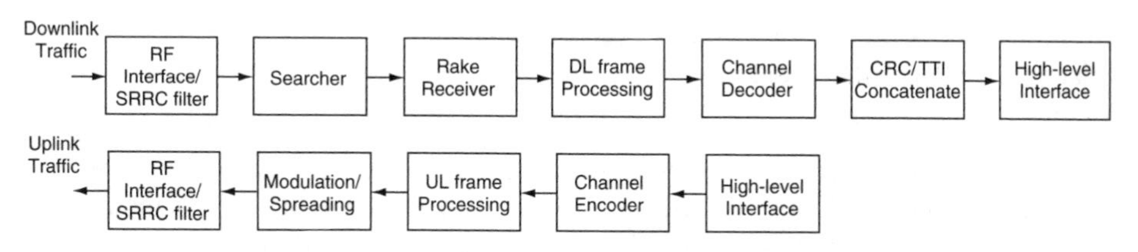

Figure 5.11 W-CDMA transceiver key subsystems

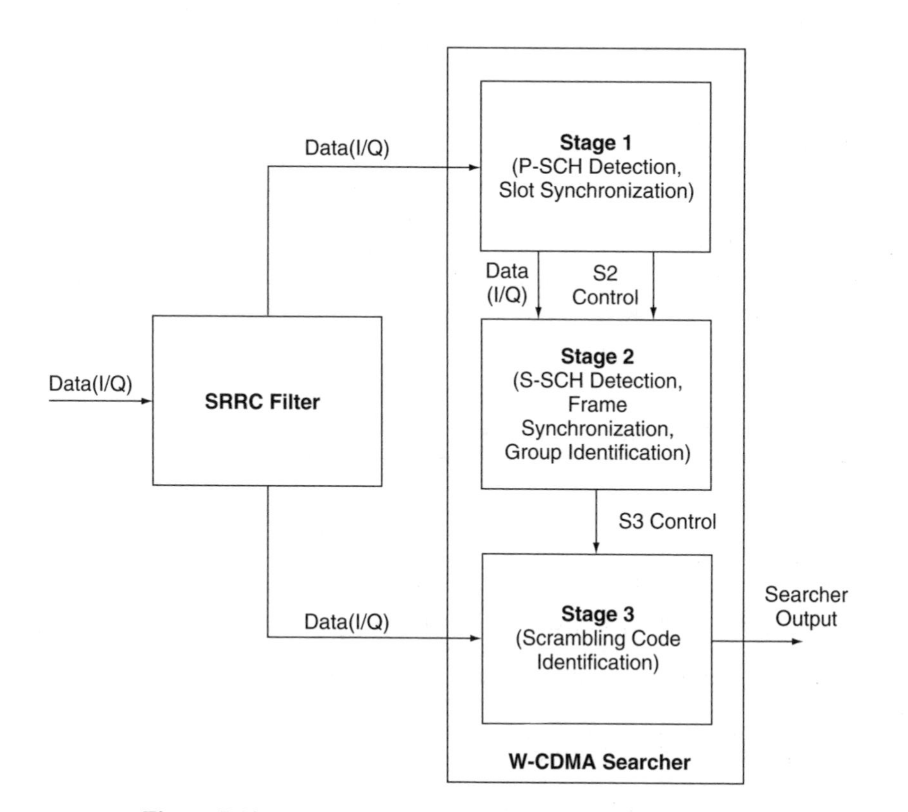

Figure 5.12 Block diagram of W-CDMA downlink searcher

The downlink receiver contains the following baseband subsystems:

- Digital front end: square root raised-cosine (SRRC) filter, AGC, A/D, AFC;
- Searcher (3 step search algorithm): used to perform initial cell search and target cell search;
- Rake: de-spreader, channel estimation-Tx diversity, AFC, DLL, finger assignment used to demodulate dedicated and common channels;
- Downlink power control and SIR estimation;
- DL frame processor: inter/intra de-interleaver, rate de-matching;
- Channel decoder: Viterbi and turbo decoder including turbo-interleaver.

The uplink contains the following baseband key subsystems:

- Digital front end: SRRC filter;
- UL frame processor: inter/intra interleaver, rate matching;
- Channel encoder: Viterbi and turbo encoder;
- Modulation/spreading.

The system parameters are as follows: chip rate 3.84 Mchips/s, data rate 384 kbps, SRRC over-sampling factor 8, input/output data bit width is I and Q 8 bit, search inputs are the SRRC outputs decimated by 4. Table 5.2 and Table 5.3 show the downlink and uplink subsystem input/output data for the key subsystems. For example, for the receiver SRRC filter output (8-bit precision), with an 8×over-sampling factor, the data rate is 3.84 Mchips/s× 8 samples/chip×2 byte/sample×word/4 byte=15.36 MW/s, where the word has a 32-bit width.

A simple example of ACM mapping is the SRRC filter. The SRRC filter can be implemented on a single AXN node, as a single-stage polyphase implementation of a 104-tap FIR filter, in which each sub-filter has 13 taps corresponding to an interpolation factor of 8. The performance characteristics of this filter are conformant to 3GPP standards, with the following parameters: roll-off factor 0.22, input sampling rate 3.84 MHz, output sampling rate 30.72 MHz, 5 MHz stop-band attenuation of 33 dB, etc. Note that this AXN module can be used in both uplink transmitter and downlink receiver. The input data are 256 samples of complex data, with I and Q parts 8-bit each. The ASP simulation shows that the SRRC filter utilizes 91% of one AXN node's processing power, and the nodal memory is sufficient to hold all the required data and program.

Table 5.2 Downlink subsystem data rate at 384 kbps

Subsystem	Input	Output	Comments
RRC filter	N/A	15.36 MW/s	Over sampling factor = 8
Searcher	3.84 MW/s	30 kW/s	Operate on 2 × oversampled data
Rake receiver	15.36 MW/s	0.96 MW/s	Operate on 8 × oversampled data
DL frame processing	0.96 MW/s	0.96 MW/s	Second de-interleaver, first de-interleaver, rate matching
Channel decoder	0.96 MW/s	20 kW/s	Decoder output = hard bit
CRC and TTI	20 kW/s	20 kW/s	Frame = 10 ms

Table 5.3 Uplink subsystem data rate at 64 kbps

Subsystem	Input	Output	Comments
SRRC filter	1.92 MW/s	15.36 MW/s	3.84 Mbyte complex
Mod/spreading	7.5 kW/s	1.92 MW/s	I channel SF = 16 Q channel SF = 256
UL frame processing	6 kW/s	7.5 kW/s	First interleaver, radio frame segmentation, rate matching, second interleaver
Channel encoder	2 kW/s	6 kW/s	Source generates 1280 bits every 20 ms

The goal of the W-CDMA downlink searcher algorithm is to detect a multipath from a cell station (there are 512 different cells in a W-CDMA network), or to detect the start of a cell station's radio frame in the received stream of data for a mobile receiver. This means that the mobile device needs to determine jointly the identity of the cell station, as well as the start of its radio-frame timing within the received data. Figure 5.12 is a diagram of the searcher where the joint detection may be obtained in three major steps [2]:[†]

Stage 1: Detect the start of a slot of an unknown cell station in the received data stream.
Stage 2: Detect the unknown cell station's group, as well as the start of its radio frame in the received data stream.
Stage 3: Detect the cell station's identity.

In any algorithm mapping onto an ACM, it is essential to first define all the required tasks to implement the overall algorithm. An ACM task mapped to a particular node is defined by the inputs it processes and the outputs it provides to the rest of the ACM. There are at least several candidate mappings of the searcher to the ACM. Each mapping differs from the others in size, power consumption, and scalability points of view. Tradeoffs are considered between the different mappings. There are a total of 11 tasks needed to implement the searcher algorithm [3]; they are listed as follows:

(i) Stage 1 Golay correlator;
(ii) Stage 1 Accumulate and peak detector;
(iii) Stage 2 Decimator and binary modulator;
(iv) Stage 2 Hadamard transform;
(v) Stage 2 S-SCH detector;
(vi) Stage 2 Pattern search;
(vii) Stage 3 Multi-dwell control;
(viii) Stage 3 Downlink scrambling code generator control;
(ix) Stage 3 Downlink scrambling code generator run;

[†] In practice, an additional Stage 4 may be included to allow for additional multipath search around the path detected by Stages 1, 2 and 3. However, for purposes of illustrating how synchronization of tasks is attained in the ACM, three stages are sufficient.

(x) Stage 3 Scrambling code de-spreader;
(xi) Stage 3 RSSI computation.

Figure 5.13 shows the node connections of one candidate mapping for a W-CDMA searcher, and Table 5.4 and Table 5.5 list the memory/clock and MIN bandwidth required by this searcher mapping. Compared with other mapping candidates, this mapping balances among three nodes, and hence any other application that requires DBN and AXN can overlay with the searcher. Also, the DBN can accommodate multiple de-spreaders. The AXN can also handle

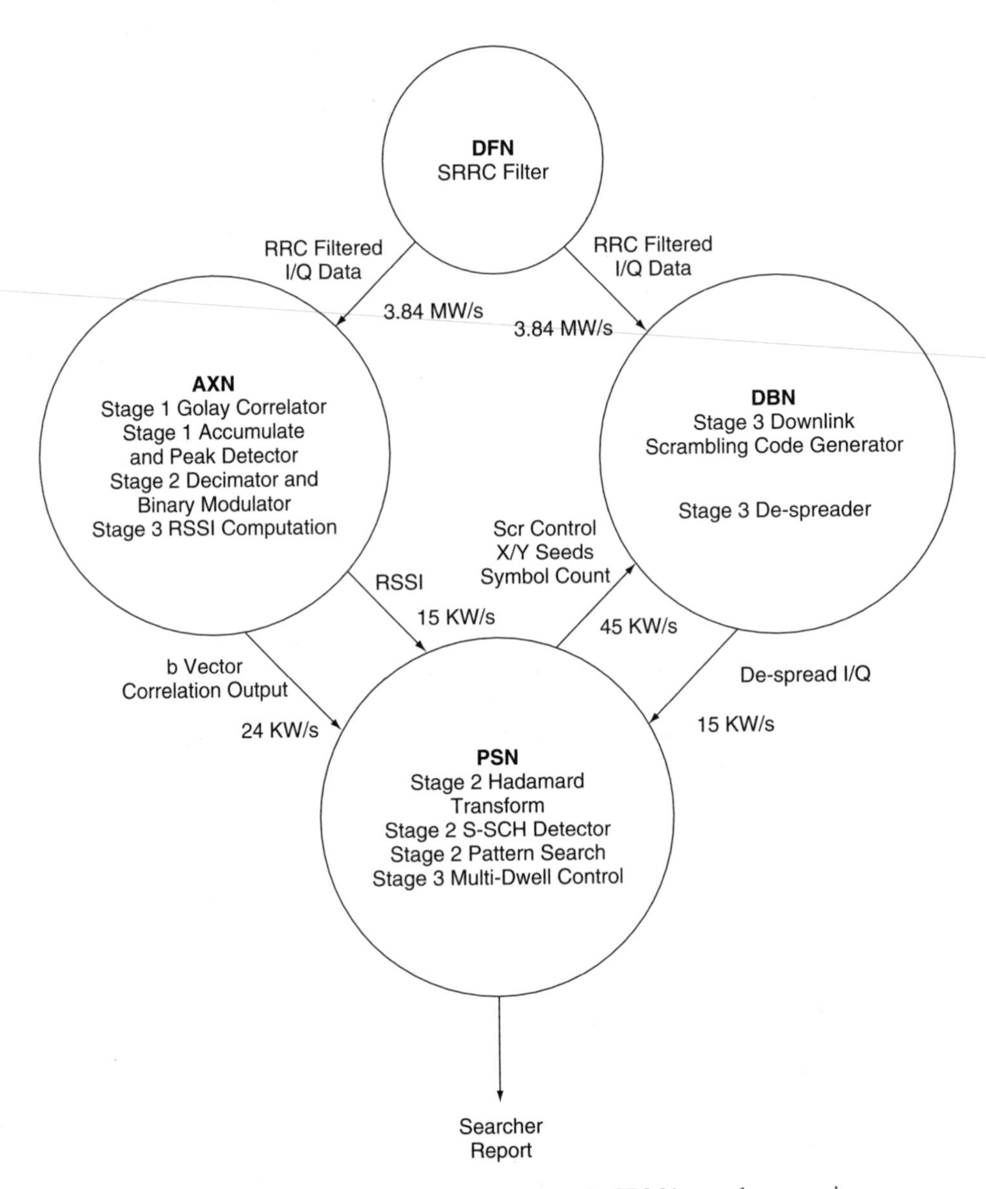

Figure 5.13 Node connection diagram for W-CDMA searcher mapping

Table 5.4 Memory/clocks required by searcher candidate mapping

Node/tasks	Node memory utilization (32-bit words)	Processor clocks consumed per second (MOPS)	Percentage of node's processing power
	AXN		
Stage 1 Golay correlator	793	57.93	28.97
Stage 1 Accumulate and peak detect	1426	6.345	3.17
Stage 2 Decimator and binary modulator	558	0.435	0.22
Stage 3 RSSI computation	16	4.8	2.4
TOTAL AXN:	2793	69.51	34.77
	PSN		
Stage 2 Hadamard transform	80	0.9	0.45
Stage 2 S-SCH detector	60	0.48	0.24
Stage 2 Pattern search	300	9.705	4.85
Stage 3 Multi-dwell control	128	0.005	0.0025
TOTAL PSN:	568	11.09	5.545
	DBN		
Stage 3 Dl Scr code gen control	128	5000	0.0025
Stage 3 Dl Scr code gen run	16	7.2	3.6
Stage 3 De-spreader	844	4.8	2.4
TOTAL DBN:	988	12.005	6.0025

multiple de-spreaders; however, since the scrambling code generators lie in the DBN, there will be additional data transfer of 0.24 MWps per de-spreader. If the number of de-spreaders is greater than seven, then the data transfer of 3.84 MW/s to DBN is still justified, since the net data transfer is minimized. The downside of this mapping is that the real-time data from the RRC filter is fed to both DBN and AXN and hence, interconnect bandwidth is higher. However, if the number of de-spreaders is greater than eight, then the benefit of utilizing the DBN for de-spreading outweighs this point.

In the following we show a mapping of the W-CDMA space-time transmit diversity (STTD) mode rake receiver algorithm to the ACM platform. Figure 5.14 is the diagram of the STTD mode rake receiver with three data channels and six fingers, or paths. The node connection, in Figure 5.15, is a preferred one because it appears to be the most efficient mapping from size, power consumption, and scalability perspectives.

Table 5.5 MIN bandwidth requirement by candidate searcher mapping

Source node to destination node	Total bandwidth
Host → AXN	3.84 MW/s
AXN → PSN	0.039 MW/s
PSN → DBN	0.045 MW/s
DBN → PSN	0.015 kW/s
PSN → Host	Negligible (500 W/s)
Host → AXN	3.84 MW/s
TOTAL	7.69 MW/s

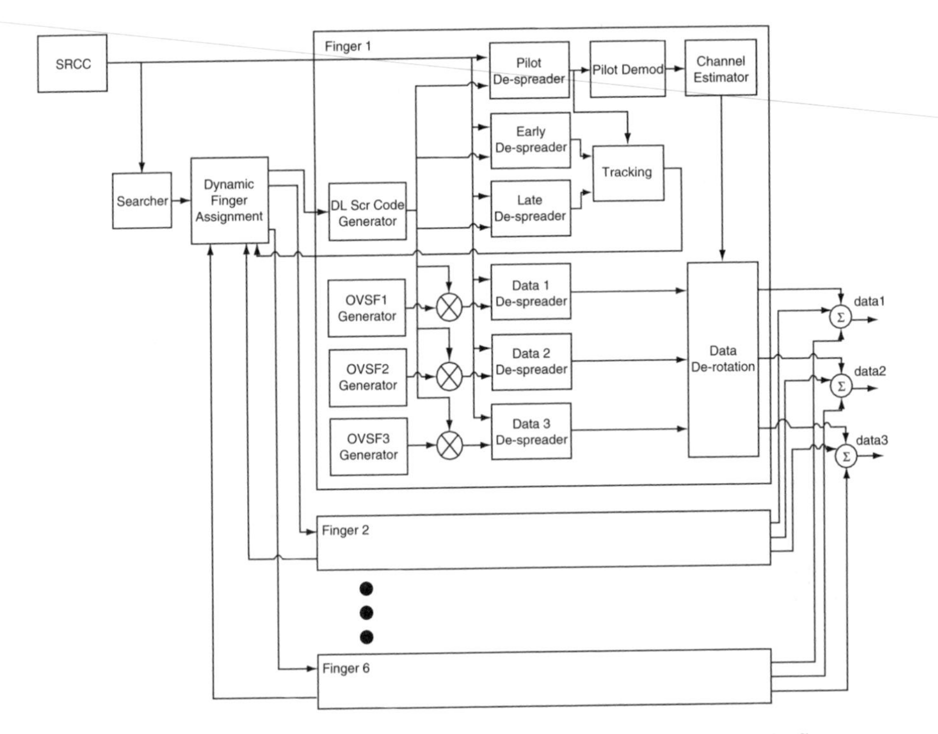

Figure 5.14 STTD-mode rake receiver with three data channels and six fingers

The analysis presented here is based on the following assumptions:

- The receiver is capable of handling up to three data channels received from a maximum of six different fingers paths (i.e. the receiver can process a maximum of 18 fingers).
- The receiver can operate in both STTD and non-STTD mode.

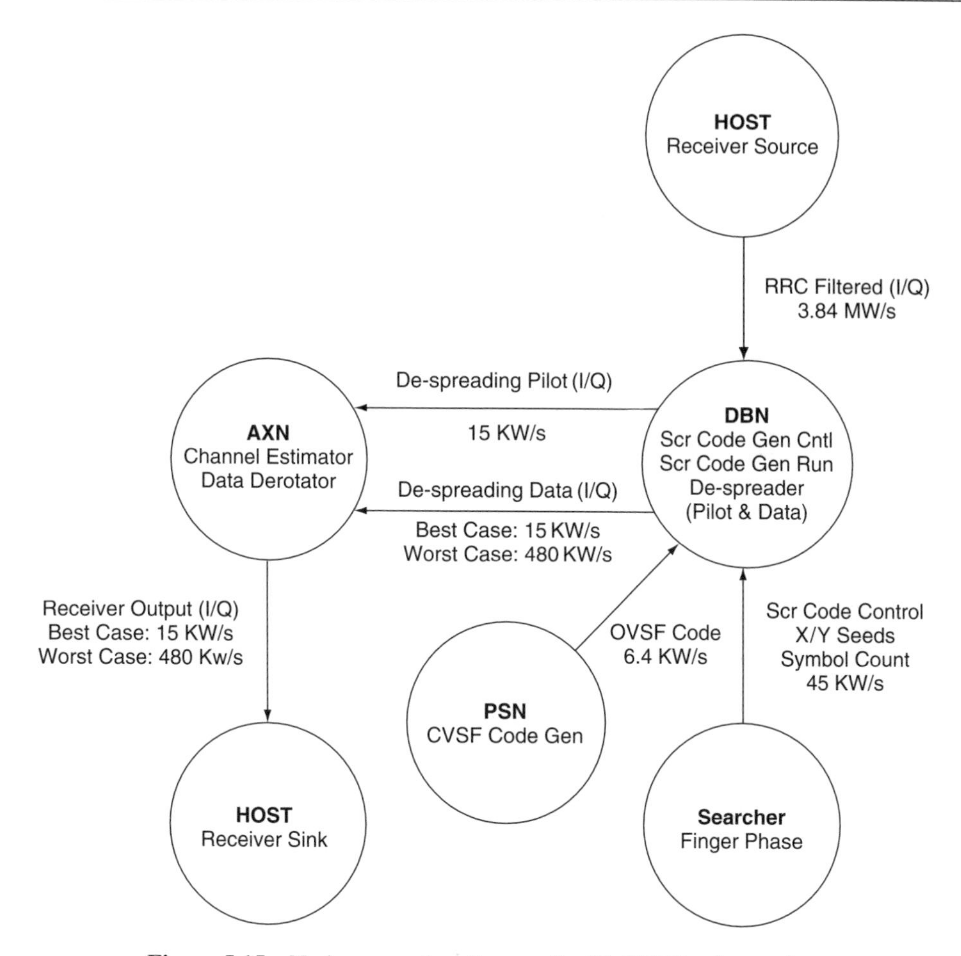

Figure 5.15 Node connection diagram for W-CDMA rake receiver

The ACM tasks for the rake receiver are given below [4]:

- downlink scrambling code generator control;
- downlink scrambling code generator run;
- channelization (OVSF) code generator;
- de-spreader;
- pilot generator;
- channel estimator;
- data de-rotator.

Table 5.6 and Table 5.7 list the memory/clocks and MIN bandwidth requirements for the case when the spreading factor is 8. A 16-tap FIR filter for channel estimation is assumed.

The searcher and the rake receiver are among the most complicated mappings in the W-CDMA transceiver. Most of the other subsystems require lesser nodes and fewer node types. Table 5.8 shows the number of nodes and percentage of node usage required for each subsystem of the transceiver [4], which conforms to the 3GPP standards [5–12]. In the table, node usage is

Table 5.6 Memory/clocks required by proposed RAKE mapping (18 fingers)

Node/tasks	Node memory usage data only (32-bit words)	Processor clocks consumed per second (MOPS)	Node's processing power (%)
	PSN		
OVSF code generator	270	0.459	0.2295
TOTAL PSN:	270	0.459	0.2295
	AXN		
Pilot demodulation	24	0.09	0.045
Channel estimator	120	0.81	0.405
Data de-rotator	530	4.32	2.160
TOTAL AXN:	674	5.22	2.61
	DBN		
DL Scr code gen-ctrl	128	0.03	0.015
DL Scr code gen-run	16	34.8	17.4
Pilot de-spreader	1088	100.8	50.4
Data de-spreader	1088	100.8	50.4
TOTAL DBN:	2320	236.43	118.215

Table 5.7 MIN bandwidth requirement by candidate rake mapping (18 fingers)

Source node to destination node	Total bandwidth (MW/s)
Host → DBN	3*3.840
DBN → AXN	18*0.495
PSN → DBN	18*0.0064
AXN → Host	3*0.495
TOTAL	22.0302

defined as the maximum nodal memory usage over the total memory of that node, and the cycle requirement over the total cycles available on a 200-MHz ACM. Note that since the ACM nodes have multitasking capability, some tasks can share the nodes with other tasks. For example, all the uplink transmitter tasks can fit in 1 AXN, 2 DBN and 2 PSN nodes. Also, the total

Table 5.8 Node assignments/usage in W-CDMA transceivers

Subsystem	AXN node numbers/ (node usage)	DBN node numbers/ (node usage)	PSN node numbers/ (node usage)
Uplink transmitter total	1 (91%)	2 (31%, 0.5%)	2 (7%, 43%)
Downlink receiver			
SRRC Filter (104-tap)	1 (91%)	0	0
RAKE receiver	1 (17%)	2 (51%, 68%)	1 (7%)
Searcher	1 (89%)	1 (5%)	1 (6%)
Frame processing	0	2 (100%, 100%)	1 (80%)
Channel turbo decoder[a]	0	2 (65%, 72%)[b]	0
CRC/TTI concatenate	0	1 (10%)	0

[a] Turbo decoder is a max-Log-MAP decoder, iterations = 8.
[b] Four nodes are aggregated to form two pairs, each with double nodal memory.

number of nodes needed in the implementation of a downlink receiver is much smaller than the total number of nodes listed. Furthermore, only three node types are listed in the table; the actual number of node types is larger, which gives more flexibility in the mapping designs.

5.6. Marketplace Benefits

With adaptive computing, SDR is now possible, and can be designed and implemented in less time, with less design risk, and in less silicon area as compared with its ASIC and other conventional IC counterparts. The software-to-hardware capability of the ACM technology enables designers and original equipment manufacturers (OEMs) of mobile/wireless devices to achieve faster time-to-market, lower cost of development and higher margins, as well as the ability to extend the life of products, increase revenues, react immediately to market trends, create product differentiation, and build brand loyalty by offering a wide range of add-on features and functionality after the initial product sale. Service providers can go beyond their limited price-per-minute business model, and add new revenue streams for added features and services.

Designers are the first to realize the benefits of adaptive computing. For example, a developer can rapidly move from design concept to silicon implementation for a product. Because engineers are working in software with the SilverC language, the development process is easier and faster then the hardware design of ASICs – hours or days versus several months or years, based on the number of functions. Working in software also enables developers to make changes at any time during the design cycle, as well as after product shipment. For example, if updates or bug fixes are needed, turn around can quickly occur in software rather than going through the long lead-time and costly re-spin cycle of an ASIC.

Ultimately, the consumer marketplace wins. The ACM is the enabling technology for a single mobile/wireless product to perform a variety of functions, rapidly changing from an

SDR terminal, to a digital camera, to streaming video, to data retrieval, e-mail, Internet and intranet access, a global positioning system, or an MP3 player. The applications are limited only by the imagination. With SDR capability, users can access any number of protocols and will experience seamless roaming throughout the world, staying 'connected' via the same single mobile device. Today's handsets will essentially become mobile communicators with media rich (data/voice/image/video) applications and the needed features to call, page, e-mail, and stay connected – at any time, anywhere in the world.

In the tradition of other disruptive technologies, such as vacuum tubes, transistors, and integrated circuits that spawned tremendous changes and advantages for the engineering world, adaptive computing brings a fresh approach to how ICs are designed, implemented, and employed. In turn, this stimulates engineering innovation and the development of new applications in the commercial arena, ultimately changing and improving the way in which we do business, communicate, and enjoy entertainment.

5.7. Technology Status

Like any substantially new technology, adaptive computing will take time to find its way into the mainstream consumer product development. At the time of publication, commercial releases of development tools and sample silicon will be available. Early adopters of the ACM technology have been working with early versions of software development tools, simulators and emulators for over a year, and the concepts are well on their way to being proven in a commercial setting. It can be expected that consumers will see ACM-based products in 2005.

Just as important, whether under the banner of 'adaptive computing' or by another name, the concepts of dynamic hardware reconfiguration and the elimination of fixed ASIC-based designs are becoming mainstream. There are extensive research efforts in universities around the world on all aspects of this technology, including architecture, development tools, operating systems and adaptive algorithms. Adaptive computing has evolved from a handful of start-ups evangelizing the technology in the late 1990s to a crowded field of over 50 start-ups and established high-tech companies in Europe, North America and Asia.

References

[1] W.H. Tuttlebee, *Software Defined Radio: Origins, Drivers and International Perspectives*, John Wiley & Sons, Inc., New York, 2002.
[2] Y.-P.E. Wang and T. Ottosson, 'Cell Search in W-CDMA', *IEEE Journal on Selected Areas in Communications*, 2000, **18**, 1470–82.
[3] *W-CDMA Searcher Mapping to ACM*, QuickSilver Technology, Inc. white paper.
[4] *W-CDMA Receiver Mapping to ACM*, QuickSilver Technology, Inc. white paper.
[5] *W-CDMA Uplink Tx Benchmark Test Items*, QuickSilver Technology, Inc. white paper.
[6] 3GPP TS 25.306, *UE Radio Access Capabilities (Release 99)*, V.3.4.0 (2001–12).
[7] 3GPP TS 25.101, *UE Radio Transmission and Reception (FDD) (Release 99)*, V.3.8.0 (2001–12).
[8] 3GPP TS 25.211, *Physical channels mapping of transport channels onto physical channels (FDD) (Release 99)*, V.3.9.0 (2001–12).
[9] 3GPP TS 25.212, *Multiplexing and channel coding (FDD) (Release 1999)*, V.3.8.0 (2001–12).
[10] 3GPP TS 25.213 *Spreading and modulation (FDD) (Release 1999)*, V.3.7.0 (2001–12).
[11] 3GPP TS 25.214, *Physical layer procedures (FDD) (Release 1999)*, V.3.9.0 (2001–12).
[12] 3GPP TS 25.215, *Physical layer – Measurements (FDD) (Release 1999)*, V.3.9.0 (2001–12).

6

The Sandbridge Sandblaster Communications Processor

John Glossner, Erdem Hokenek and Mayan Moudgill

Sandbridge Technologies, Inc., Whiteplains, NY

As applications converge to multimedia systems, user terminals will increasingly be required to support voice, data, and video applications. Further, convergent devices will increasingly roam seamlessly across multiple communications systems. Traditionally, wireless communications systems have been single mode and implemented in hardware. To avoid excessive hardware costs, a software defined radio (SDR) approach offers a programmable and dynamically reconfigurable method of reusing hardware to implement the physical layer processing. From a processor architecture perspective, support for signal processing, control code, and Java execution on a common SDR platform – a convergence architecture – offers a route to both power and cost reduction for such devices.

In this chapter we describe such an architecture applicable to reconfigurable baseband processing. The architecture supports all datatypes appropriate to convergence devices, efficiently executing Java, digital signal processing (DSP), and control code. It is programmed in C or Java, is executed on a multithreaded processor in real time and is capable of executing physical layer processing of multiple waveforms completely in software. Architectural features that reduce power dissipation and enable real-time processing are described, along with the software development tools, both compiler and simulator capabilities, that support the architecture. Finally, results are presented for implementation of 2 Mbits/s WCDMA, IEEE802.11b, and GSM/GPRS physical layer processing.

6.1. Rationale for SDR Processors

Performance requirements for mobile wireless communication devices have expanded dramatically since their inception as mobile telephones. Consumers are demanding convergence devices with full data and voice integration, as well as a variety of computationally intense features and applications such as web browsing, MP3 audio, and MPEG4 video. Moreover,

Software Defined Radio: Baseband Technologies for 3G Handsets and Basestations. Edited by W. Tuttlebee
 ISBN: 0-470-86770-1

consumers want these wireless subscriber services to be accessible at all times anywhere in the world. Such complex functionality and features require high computing capability at low power consumption; adding new features requires adding computing capability.

6.1.1. The Challenge

The technologies necessary to realize true broadband wireless handsets and systems present unique design challenges if extremely power efficient, yet high-performance, broadband wireless terminals are to be realized. The design trade-offs and implementation options inherent in meeting such demands highlight the extremely onerous requirements for next generation baseband processors. Tremendous hardware and software challenges exist to realize convergence devices.

Power dissipation constraints are requiring new techniques at every stage of design – architecture, micro-architecture, software, algorithm design, logic design, circuit design, and process design. With performance requirements exploding as bandwidth demand increases, power conscious design becomes more difficult. System-on-chip (SoC) integration and low-voltage process technologies will contribute to lower power SoC integrated circuits (ICs) but are insufficient as the only solution for streaming multimedia.

Convergence applications are fundamentally DSP applications. A large number of standards exist or have been proposed for the wireless and wired communication markets. Such a diversity of standards necessitates a programmable platform for their timely implementation. In second generation mobile communications, GSM and IS-95 data rates were limited to less than 15 kbit/s. Future third generation (3G) systems may provide data rates more than 100-times the previous rates. Escalating communication rates are accelerating DSP processing requirements.

Traditional communications systems have typically been implemented using custom hardware solutions. Chip rate, symbol rate, and bit rate coprocessors are often coordinated by programmable DSPs, but the DSP processor does not typically participate in computationally intensive tasks. Even with a single communication system the hardware development cycle is onerous, often requiring multiple chip redesigns late into the certification process. When multiple communications systems requirements are considered, both silicon area and design validation are major inhibitors to commercial success. A software-based platform capable of dynamically reconfiguring communications systems enables elegant reuse of silicon area and dramatically reduces time to market through software modifications instead of time consuming hardware redesigns.

Digital signal processors (DSPs) are now becoming capable of executing billions of operations per second at power efficiency levels appropriate for handset deployment. This has brought software defined radio (SDR) to prominence as technology begins to allow the designer to address a new and hitherto difficult portion of the software radio implementation space.

6.1.2. Tiers of Software Radio

The SDR Forum [1] defines five tiers of software radio:

Tier 0: Traditional radio implementation in hardware.

Tier 1: Software controlled radio (SCR), implements the control features for multiple hardware elements in software.

Tier 2: Software defined radio (SDR), implements modulation and baseband processing in software but allows for multiple frequency, fixed function RF hardware.

Tier 3: Ideal software radio (ISR), extends programmability through the RF with analogue conversion at the antenna.

Tier 4: Ultimate software radio (USR), provides for fast (millisecond) transitions between communications protocols in addition to digital processing capability.

6.1.3. The Benefits

The advantages of a reconfigurable SDR solution versus hardware solutions are significant:

- First, reconfigurable solutions are more flexible, thus allowing multiple communication protocols to execute dynamically on the same transistors, thereby reducing hardware costs. Specific functions such as filters, modulation schemes, encoders/decoders, etc., can be reconfigured adaptively at run time.
- Second, several communication protocols can be efficiently stored in memory and coexist or execute concurrently. This significantly reduces the cost of the system for both the end-user and the service provider.
- Third, remotely reconfigurable protocols provide simple and inexpensive software version control and feature upgrades. This allows service providers to differentiate products after the product is deployed.
- Fourth, the development time of new and existing communications protocols is significantly reduced, providing an accelerated time-to-market. Development cycles are not limited by long and laborious hardware design cycles. With SDR, new protocols are quickly added as soon as the software is available for deployment.
- Fifth, SDR provides an attractive method of dealing with new standards releases while assuring backward compatibility with existing standards.

In the subsequent sections we describe the Sandbridge approach to delivering these benefits. The derivation of this solution outlines the major challenges to the technology and of the market, particularly the need to future proof designs against continually evolving standards and air interfaces.

6.2. Processor Architecture

6.2.1. Definitions and Architecture Evolution

The *architecture* of a computer system is the minimal set of properties that determine what programs will run and what results they will produce [2]. It is the contract between the programmer and the hardware. Every computer is an interpreter of its *machine language* – that representation of programs that resides in memory and is interpreted (executed) directly by the (host) hardware. The logical organization of a computer's data flow and controls is called the *implementation* or *microarchitecture*. The physical structure embodying the implementation is called the *realization*. The architecture describes what happens while the implementation describes how it is made to happen. Programs of the same architecture should run unchanged on different implementations. An architectural function is *transparent* if its implementation does not produce any architecturally visible side effects. An example of a nontransparent function is the load delay slot made visible due to pipeline effects. Generally, it is desirable to have transparent implementations. Most DSP and VLIW implementations are not transparent and therefore the implementation defines the architecture.

The requirement for execution predictability in DSP systems often precludes the use of many general-purpose design techniques (e.g. speculation, branch prediction, data caches, etc.). Instead, classical DSP architectures have developed a unique set of performance enhancing techniques that are optimized for their intended market (e.g. 0-overhead loop buffers, visible memory, and exposed pipelines). These techniques are characterized by hardware that supports efficient filtering, such as the ability to sustain three memory accesses per cycle (one instruction, one coefficient, and one data access). Sophisticated addressing modes such as bit-reversed and modulo addressing may also be provided. Multiple address units operate in parallel with the data path to sustain the execution of the inner kernel.

In classical DSP architectures, the execution pipelines were visible to the programmer (i.e. not transparent) and necessarily shallow, to allow assembly language optimization. This programming restriction encumbered implementations with tight timing constraints for both arithmetic execution and memory access. The key characteristic that separates modern DSP architectures from more classical DSP architectures is the focus on compilability. Once the decision was made to focus the DSP design on programmer productivity, other constraining decisions could be relaxed. As a result, significantly longer pipelines with multiple cycles to access memory and multiple cycles to compute arithmetic operations could be utilized. This trend has yielded higher clock frequencies and higher performance DSPs.

In an attempt to exploit instruction level parallelism inherent in DSP applications[†], modern DSPs tend to use VLIW-like execution packets [3–9]. This is partly driven by real-time requirements, which require the worst-case execution time to be minimized. This is in contrast to general purpose CPUs, which tend to minimize average execution times. With long pipelines and multiple instruction issue, the difficulties of attempting assembly language programming become apparent. Controlling instruction dependencies between upwards of 100 in-flight instructions is a nontrivial task for a programmer. This is exactly the area where a compiler excels.

A challenge of using VLIW DSP processors includes large program executables (code bloat) that results from independently specifying every operation with a single instruction. As an example, a VLIW processor with a 32-bit basic instruction width requires four instructions, 128 bits, to specify four operations. A vector encoding may compute many more operations in as little as 21 bits (for example – multiply a four vector, saturate, accumulate, saturate).

Another challenge of VLIW implementations is that they may require excessive write ports on register files. Because each instruction may specify a unique destination address and all the instructions are independent, a separate port must be provided for targets of each instruction. This can result in high power dissipation, which is unacceptable for handset applications.

A challenge of visible pipeline machines (e.g. most DSPs and VLIW processors) is interrupt response latency. Visible memory pipeline effects in highly parallel inner loops (e.g. a load instruction followed by another load instruction) are not interruptible because the

[†] The inherent parallelism of DSP algorithms is discussed in detail in the chapter by Mohebbi *et al.*

processor state cannot be restored. This requires programmers to break apart loops so that worst case timings and maximum system latencies may be acceptable.

DSP processing requires support for filtering and highly compute intensive functions to enable software execution of baseband physical layers. DSPs have traditionally implemented one or more multiply accumulate (MAC) units to perform these functions. DSP operations typically operate on fixed point (fractional) saturating datatypes. Because saturating arithmetic is nonassociative, parallel execution of multiple data elements may result in different results from serial execution. This creates a challenge for high-level language implementations that specify integer modulo arithmetic. Therefore, most DSPs have been programmed using assembly language.

Embedded control processing can be implemented using a standard reduced instruction set computer (RISC); this is also amenable to high-level language compilation.

In addition to embedded and DSP processing, Java processing may be required. Future wireless systems will make significant use of Java; indeed, a number of carriers are already providing Java-based services and all 3G terminals could potentially be required to support Java [10].

6.2.2. *Microarchitecture*

Figure 6.1 shows the architecture of a multithreaded processor capable of executing DSP, embedded control, and Java code, in a single compound instruction set, optimized for handset radio applications developed by Sandbridge Technologies [11, 12]. This design overcomes the deficiencies of previous approaches by providing substantial parallelism and throughput for high-performance DSP applications while maintaining fast interrupt response, high-level language programmability and very low power dissipation.

The design includes a unique combination of modern techniques such as a SIMD vector/ DSP unit, a parallel reduction unit, a RISC-based integer unit, and instruction set support for Java execution. Instruction space is conserved through the use of compounded operations that are combined in a single instruction for execution. The resulting combination provides for efficient control, DSP and Java processing execution.

The processor is partitioned into a program function unit (PFU), a load/integer unit (LIU), a vector SIMD parallel unit (VPU), an instruction cache unit (ICU), and a data memory unit (DMU), see Figure 6.2. Figures 6.1 and 6.2 may be roughly overlaid on each other and depict the classes of instructions run in each portion of the microarchitecture.

6.2.2.1. Program Function Unit (PFU)

The program function unit (PFU) is shown on the left side of both Figure 6.1 and Figure 6.2. The PFU manages program flow control and execution and is the 'brain' of the processor. It provides instruction decoding, instruction fetch and jump address calculation, and interrupt request processing. One compound instruction per cycle is fetched from the instruction cache. The compound instruction is then decoded and dispatched to the integer (INT) and the vector (VPU) units. Some compound instructions may be the target of instructions fetched as a result of unconditional jumps, taken conditional jumps, or other call operations.

For each cycle the PFU calculates either the sequential or taken branch path addresses. Note that the instruction fetch mechanism has to go through address translation in the

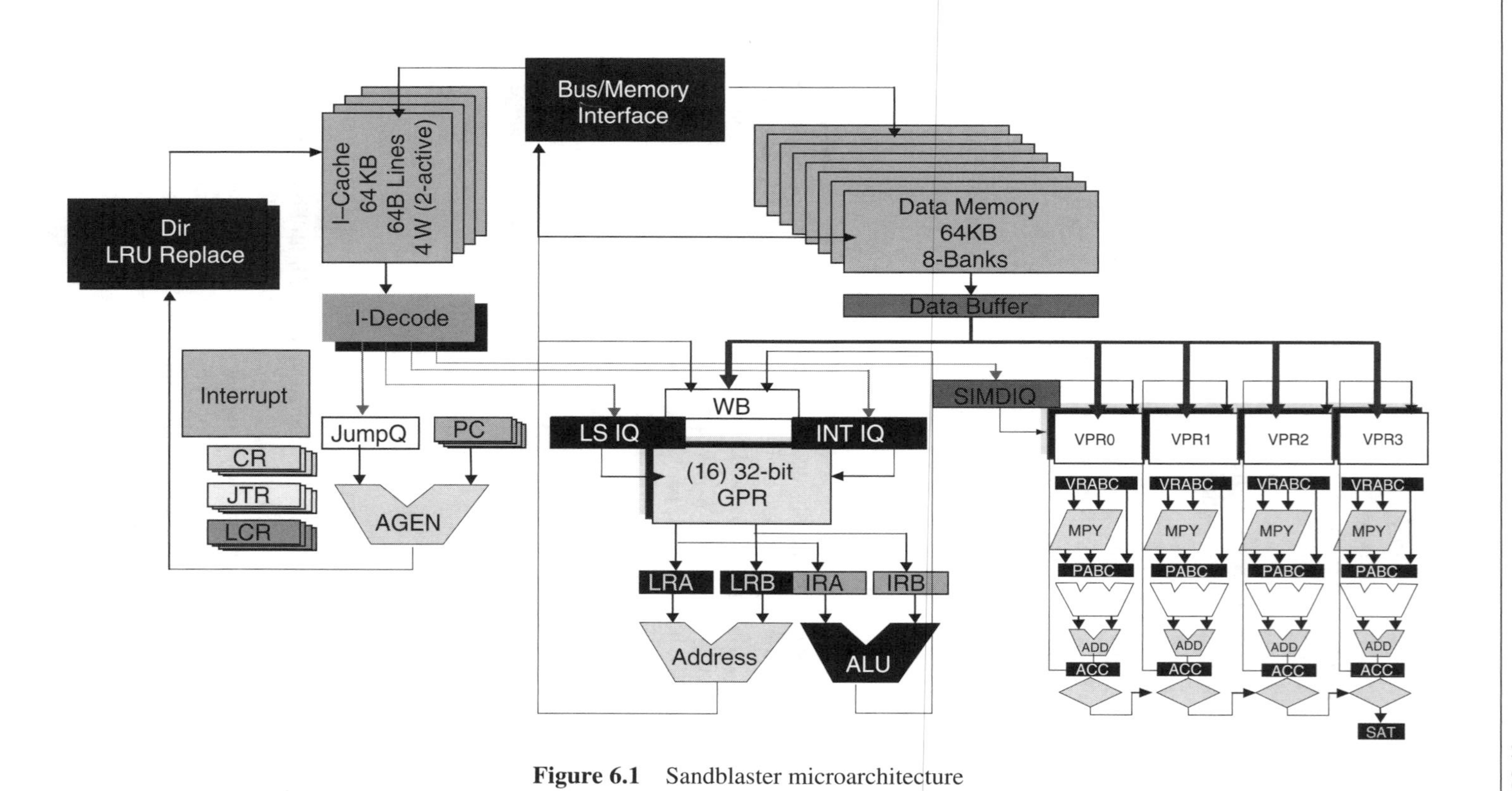

Figure 6.1 Sandblaster microarchitecture

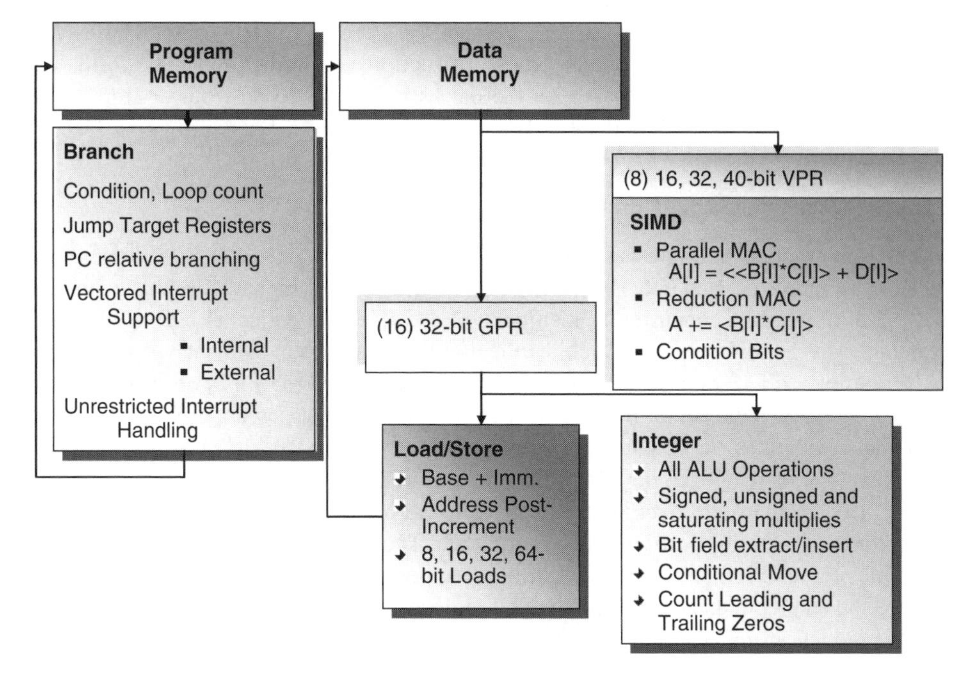

Figure 6.2 Functional partitioning of the processor architecture

instruction cache directories. The PFU also controls interrupt processing by checking the available threads for context switching. The interrupt vector addresses are generated in the PFU. The PFU manages other internal processor resources such as loop counts, jump target addresses, condition flags, timer decrements, and processor status registers.

6.2.2.2. Load/Integer Unit (LIU)

The load/integer unit (LIU), or load-store/integer unit, is shown in the middle of Figure 6.1 and provides integer data manipulation, special register operations, and load/store processing. All integer instructions, arithmetic, logic, shift, bit-extract and insert operations are executed by this unit. All addresses for memory access instructions, such as load/store are generated by the integer arithmetic logic unit (ALU). Since the integer unit can access the general purpose registers (GPRs), all copies to/from special register instructions will be co-executed in concert with other units using GPRs either as the target or as the source for copies.

The integer unit (INT) receives its instructions from the instruction decode/dispatch unit and executes all integer and memory access (load/store) instructions that include execution of the register-to-register operations (RR ops) for arithmetic (also data address generation), compare, logic, shift, extract, and insert instructions as well as integer multiplies.

Since the Sandblaster processor provides an immediate form instruction, the INT unit extracts the immediate fields while maintaining the general purpose registers. The operands of the RR ops are always read from the GPRs. The integer unit also participates in the manipulation of special registers when executing the copy to/from special register (ctsr/cfsr) instructions.

In the case of load/store instructions, the 32-bit address generated is sent to storage subsystem for data delivery. Note that all memory access instructions are in the immediate form. Their signed integer fields are shifted based on the data size to be fetched from or stored to the storage space. Data sizes supported are 8, 16, 32, and 64-bits.

6.2.2.3. Vector Processing Unit (VPU)

The single instruction multiple data (SIMD) vector processing unit (VPU) is shown on the right side of Figures 6.1 and 6.2. The VPU works on multiple 16-bit and 40-bit data simultaneously. The number of concurrently manipulated data (vector) elements is implementation dependent. The Sandblaster base architecture implements a four element vector, i.e. four 16×16 multiply and 40-bit accumulate (four independent MAC ($a \times b + c$) operations).

The VPU contains the vector register file (VPR) that serves the same purpose as the GPRs except that it operates on vector/SIMD types. A vector, encoded either as a fractional or integer data type, is read from the vector file and is stored into architecturally visible registers (VRABC). From there the data proceeds through the multiply (MPY) blocks that perform parallel concurrent multiplication of the Vector data. The results are placed in the architecturally visible storage PABC. The ADD unit then can perform additional arithmetic operations and place the results in an accumulator (ACC) register. From there they can proceed through a parallel reduction unit where the results are summed in parallel. Since DSP arithmetic is nonassociative, the parallel reduction unit corrects the results to provide serial semantics. This is critical in many applications.

6.2.2.4. Instruction Cache Unit (ICU)

The instruction cache unit (ICU) stores instructions to be fetched for each thread unit. A cache memory works on the principle of locality. Locality can refer to spatial, temporal, or sequential locality. *Spatial locality* refers to the probability that a computer program is likely to access the same or neighboring memory locations during the period of execution. *Temporal locality* refers to the property that if a program accesses a sequence of references to n locations at a particular time, there is a high probability that subsequent references will be made into the sequence in the near future. *Sequential locality* refers to the property that if a reference has been made to a particular location s, then it is likely that a reference will be made to location $s+1$.

A cache is a fast memory that stores frequently used references from memory. It has characteristics of size, access time, and block or line size. A *mapping control* determines how data is stored in the cache and moved from memory into the cache. Processor references (or accesses) that are found in the cache are called *cache hits*. References not found in the cache are called *cache misses*. A cache typically fetches a *line* of memory from higher level memories. The size of the line is generally designed to be consistent with the expected spatial locality of programs.

A cache may be organized to fetch-on-demand or to prefetch data. Most processors, including the Sandblaster SB9600 product, use fetch-on-demand whereby, when a cache miss occurs, the cache controller will evict a resident line and replace it with the line referenced by the processor. In prefetch caches, the cache mapping controller tries to predict which lines will be required and move them into the cache before a processor references them.

Within the cache, there are three basic types of mapping organization. A *fully associative* mapping compares all bits of a processor reference request against all entries in a cache directory. A directory hit is found if the address is in the directory. Otherwise, a directory miss occurs. The address is then sent to the cache data array and the requested line is sent to the processor. A *direct-mapped* cache only stores the lower order line address bits. Since multiple addresses may map to the same location in the cache directory, the upper line address bits must be compared with the tag in the directory. If the tag matches the value stored in the directory, the data is sent to the processor. A *set-associative* cache operates like a direct-mapped cache except that multiple choices for the address may be present. Each of the line addresses corresponds to a location in a sub-cache. The upper bits of the address contained within the address are compared with each tag in the directory. If a match is found it signifies which set the data should be accessed from.

A cache can only accommodate a small number of index locations. Many addresses end up aliased to a small number of directory locations. When an index is valid but no tags match (a cache miss), one or more of the tags must be evicted (along with the data in the cache) and the correct data fetched and stored in the cache.

Many replacement policies are available to decide which data should be evicted. A *least recently used* (LRU) replacement policy attempts to exploit temporal locality by always removing the oldest nonaccessed location in the cache. Other replacement policies exist including *random* and *first in-first out* (FIFO) replacement. A feasible implementation of the LRU algorithm has demonstrated that for n resources, it requires n^2 1-bit registers to maintain the LRU state. Further enhancements have been shown in the literature that may reduce this to $n(n-1)/2$ [2].

In our implementation, the thread identifier register (TID) is used to select whether the left or right bank will be evicted. This effectively reduces the complexity of the set selection. In a four-way set associative cache, only one additional LRU bit is needed to select which of the two tags should be evicted. This gives a complexity of $n(n-2)/8$ for a general set-associative cache with set size n. Table 6.1 shows the bits required using a traditional cache as compared with this improved method.

Table 6.1 LRU complexity

Set associativity	Traditional cache	Standard enhancement [2]	Improved method
n	n^2	$n(n-1)/2$	$n(n-2)/8$
2	4	1	0
4	16	6	1
8	64	28	6
16	256	120	28
32	1024	496	120
64	4096	2016	496

6.2.3. Token Triggered Multithreading

Multithreading is a well-known technique for both hardware and software acceleration. The Delencor HEP was designed round about 1979 [13]. In this design, multiple instructions could be simultaneously active from multiple threads. The requirement was that each thread must complete the current instruction prior to issuing a subsequent instruction. When each context issues an instruction and contexts progress in sequence, this is sometimes termed barrel multithreading.

More recent embodiments of multithreaded processors make use of simultaneous multithreading (SMT) [14, 15]. In this approach, multiple thread units may issue multiple instructions each cycle. When combined with superscalar techniques such as out-of-order processing, the additional hardware required for SMT is not significant. Unfortunately, both superscalar and SMT techniques consume significant power. Other techniques may involve taking into account thread priorities but then care must be taken to ensure that deadlock does not occur.

As technology improves, processors are becoming capable of executing at very fast cycle times. Current state-of-the-art performance for 0.13 μm technologies can produce processors faster than 3 GHz. Unfortunately, such high-performance processors consume significant power. If power-performance curves are considered for both memory and logic within a technology, there is a region displaying approximately linear increase in power for linear increase in performance. Above a specific threshold, there is an exponential increase in power for a linear increase in performance. Even more significantly, memory and logic do not have the same threshold.

For 0.13-μm technology, the logic power-performance curve may be in the linear range until approximately 600 MHz. Unfortunately, memory power-performance curves are, at best, linear to about 300 MHz. This presents a dilemma as to whether to optimize for performance or for power. Fortunately, multithreading alleviates the power-performance trade-off. The Sandblaster implementation of multithreading allows the processor cycle time to be decoupled from the memory access time. This allows both logic and memory to operate in the linear region, thereby significantly reducing power dissipation. The decoupled execution does not induce pipeline stalls due to the unique pipeline design.

In a multithreaded processor, all threads of execution operate simultaneously. An important point is that multiple copies (e.g. banks and/or modules) of memory are available for each thread to access. The Sandblaster architecture supports multiple concurrent program execution by the use of hardware thread units (contexts). The architecture supports up to eight concurrent hardware contexts. The architecture also supports multiple operations being issued from each context. Unlike simultaneous multithreading, the Sandblaster processor uses a unique form of multithreading called token triggered threading (T^3), see Figure 6.3.

Token triggered threading is a unique form of multithreading where each hardware context is allowed to execute an instruction simultaneously, but only one context may issue an instruction on a cycle boundary. This constraint is also imposed on round-robin threading. What distinguishes T^3 threading is that in each clock cycle a token indicates the subsequent context that is to execute. Tokens may be sequential (e.g. round robin), even/odd, or based on

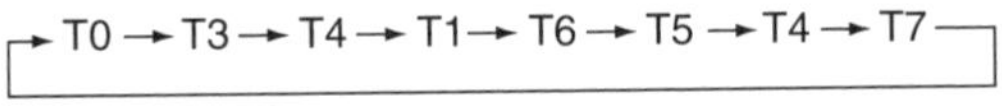

T0 → T3 → T4 → T1 → T6 → T5 → T4 → T7 ↩

Figure 6.3 Token triggered multithreading, even/odd sequence

```
L0: lvu %vr0, %r3, 8
|| vmulreds %ac0,%vr0,%vr0,%ac0
|| loop %lc0,L0
```

Figure 6.4 Sum of squares inner loop

other communications patterns. Compared with SMT, T^3 threading has much less hardware complexity and power dissipation, since the method for selecting threads is simplified, only a single compound instruction issues each clock cycle, and dependency checking and bypass hardware are not needed.

6.2.4. Compound Instructions

The Sandblaster architecture is a compound instruction set architecture. Historically, DSPs have used compound instruction set architectures to conserve instruction space encoding bits. In contrast, VLIW architectures contain full orthogonality, but only encode a single operation per instruction field, such that a single VLIW is composed of multiple instruction fields. This has the disadvantage of requiring many instruction bits to be fetched per cycle, as well as significant write ports for register files. These effects contribute heavily to power dissipation.

In the Sandblaster architecture, specific fields within the instruction format may issue multiple sub-operations including data parallel vector operations. Most classical DSP instruction-set architectures are compound. Restrictions may apply if a particular operation is chosen; however, each field controls a different execution unit. In contrast, a VLIW instruction set architecture may allow complete orthogonality of specification and then, either in hardware or through no operation instructions (NOPs), fills in any unused issue slots.

To illustrate the compound nature of our vector architecture, Figure 6.4 shows a single compound instruction with three compound operations. The program computes a sum of squares function common to many signal processing kernels.

The first compound operation, **lvu**, loads the vector register **vr0** with four 16-bit elements and updates the address pointer **r3** to the next element. The **vmulreds** operation reads four fixed point (fractional) 16-bit elements from **vr0**, multiplies each element by itself, saturates each product, adds all four saturated products plus an accumulator register, **ac0**, with saturation after each addition, and stores the result back in **ac0**. The loop operation decrements the loop count register **lc0**, compares it to 0, and branches to address **L0** if the result is not 0.

The code shown in Figure 6.4 is all encoded in a single 64-bit compound instruction. Each compound operation, including vector operations, is specified with at most 21 bits. Like most DSP architectures, arbitrary operations are not specifiable within the same instruction. The same instruction as Figure 6.4 may require more than 512 bits to encode on a VLIW machine. Furthermore, since the pipeline in a VLIW machine typically produces architecturally visible side effects (i.e. it is not transparent), it may take a deeply software-pipelined loop to obtain single-cycle throughput, thereby exploding the instruction storage requirements. To distinguish our approach further from VLIW and exposed pipeline architectures, each instruction is completely interlocked and architecturally defined to complete with no visible pipeline effects; this is critical for interrupt processing.

6.2.5. *Processing Datatypes*

The architecture supports a range of datatypes appropriate for convergence devices. This includes 8-, 16-, and 32-bit integers and 16 40-bit fixed point (fractional) types. Long 64-bit integers are supported with software emulation. Both single and double precision floating point is supported in software as it is not critical to convergence devices. Hardware implementations of floating point additionally consume significant power. This allows arbitrary C code to be compiled to the Sandblaster platform.

6.2.6. *Interrupts*

A challenge of visible pipeline machines (e.g. most DSPs and VLIW processors) is interrupt response latency. Visible memory pipeline effects in highly parallel inner loops (e.g. a load instruction followed by another load instruction) are not typically interruptible because the processor state cannot be restored. This requires programmers to break apart loops so that worst-case timings and maximum system latencies are acceptable. This convolutes the source code and may even require source code changes between processor generations.

In the Sandblaster architecture, interrupt processing is fully supported. Because of the interlocked pipeline, any thread can be interrupted on any instruction boundary – even inside highly parallel loops. In addition to typical interrupt processing, the Sandblaster architecture supports a high-speed, cross-thread interrupt mechanism which allows one thread to interrupt another in a single processing cycle. To deliver a cross-thread interrupt, the delivering thread writes a bit in a register that notifies the destination thread of the interrupt request. When the interrupt is delivered to the destination thread, its enable bit is automatically cleared. This prevents simultaneous reception of multiple interrupts. When the interrupt is serviced, the interrupt controller automatically re-enables interrupts for the thread.

6.2.7. *Synchronization*

Synchronization is accomplished through the use of load locked instructions (lwlock). The locked memory address is computed by summing the value of the input register (ra) and the shifted sign extended immediate field. The locked storage address is recorded into the locked address register (LAR). The target integer register (rt) is set to the value of the four bytes at the locked storage address. Concurrent programs may then check whether the storage location is locked prior to writing to the location.

The locked storage address must be four-byte aligned. Otherwise, the result of the lwlock instruction is unpredictable. Any detectible store to the locked storage address will cause the lock flag to be cleared. This includes the load/store instructions for the vector unit that are allowed to be aligned only to two-byte boundaries.

Using this mechanism, higher-level programming synchronization mechanisms can be constructed, such as semaphores and monitors.

6.2.8. *Java Support*

As noted earlier, Java-based services are being increasingly deployed in wireless networks and will increase in the future [10]. Java, which is similar to C++, is designed for general-purpose, object-oriented programming [16]. The appeal of the Java language is its 'write once, run anywhere' philosophy [17]. This is accomplished by providing a Java virtual machine (JVM) interpreter and run-time support for each platform [18].

JVM translation designers have used both software and hardware methods to execute Java bytecode. The advantage of software execution is flexibility; the advantage of hardware execution is performance. The Delft-Java architecture, designed in 1996, introduced the concept of dynamic translation of Java code into a multithreaded RISC-based machine with Vector SIMD DSP operations [14, 19, 20]. Another of the authors also explored dynamic translation [21]. The important property of Java bytecode that facilitated this translation is the statically determinable type state [16]. The Sandbridge approach embodies a unique combination of both hardware and software support for Java execution.

6.3. Processor Software Tools

Programmer productivity is one of the major concerns in complex DSP applications. Because most classical DSPs are programmed in assembly language, it takes a very large software effort to program an application. For modern speech coders, as in Ref. [22] for example, it may take up to 9 months or more before the application performance is known. Then, an intensive period of design verification ensues. If efficient compilers and simulation tools for DSPs were available, significant gains in software productivity could be achieved [44]. In reality, the availability of such tools will be an essential determinant of the rate of adoption new generation DSP architectures.

6.3.1. Compilers

The ideal DSP design process is illustrated in Figure 6.5. By codesigning the DSP compiler alongside the architecture, based on the intended application domain, trade-offs may be made

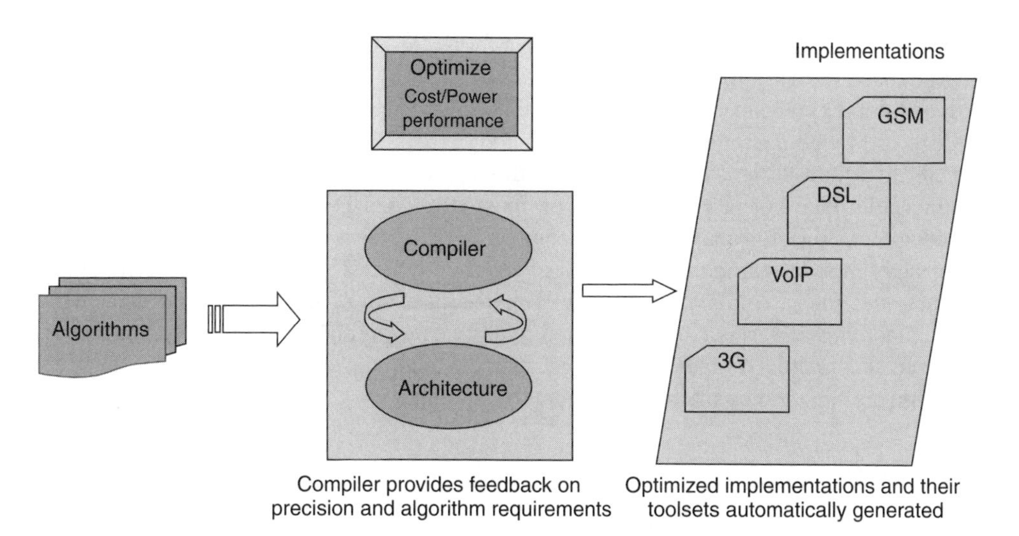

Figure 6.5 DSP compiler/architecture codesign

between the architecture and the compiler subject to the application performance, power and price constraints. Even so, there are a number of basic issues that must be addressed in designing a DSP compiler.

6.3.1.1. Basic Issues with Compilers

First, there is a fundamental mismatch between DSP datatypes and C language constructs. A basic data type in DSPs is a saturating fractional fixed-point representation; C language constructs, however, define integer modulo arithmetic. This forces the programmer to explicitly program saturation operations. The compiler must then deconstruct these idioms to recognize the underlying fixed point operations.

A second problem for compilers is that previous DSP architectures were not designed with compilability as a goal. To maintain minimal code size, multiple operations were issued from the same compound instruction. Unfortunately, to reduce instruction storage, it was common to use 16-bit encoding for all instructions. Often, three operations could be issued from the same 16-bit instruction. While this was good for code density, orthogonality suffered. Many special purpose registers were required and severe restrictions on operation combinations were imposed.

Early attempts to remove these restrictions used VLIW instruction set architectures with nearly full orthogonality. To issue four multiply accumulates (MACs) minimally requires four instructions (with additional load instructions to sustain throughput). This generality was required to give the compiler technology an opportunity to catch up with assembly language programmers.

6.3.1.2. High Level Languages

Because DSP C compilers have difficulty generating efficient code, language extensions have been introduced to high level languages [23, 24]. Typical additions may include special type support for 16-bit data types (Q15 formats), saturation types, multiple memory spaces, and SIMD parallel execution support. These additions often imply a special compiler, and the code may not be emulated easily on multiple platforms. As a result, special language constructs have not been successful.

In addition to language extensions, other high-level languages have been used. BOPS [3] produced a Matlab compiler which offers exciting possibilities, since Matlab is widely used in DSP algorithm design. Difficulties with this approach include Matlab's inherent 64-bit floating point type not being supported on most DSPs. On DSPs which do support 32-bit floating point, precision analysis is still required.

For algorithm design, tensor algebra has been used [25, 26]. Attempts have been made to automate this into a compilation system [27]. The problem of this approach is that highly skilled algorithm designers are still required to describe the initial algorithm in tensor algebra. However, this approach holds promise because the communications and parallelism of the algorithm are captured by the tensor algebra description.

6.3.1.3. Libraries

To reduce the programming burden of traditional DSPs, large libraries are typically built up over time. Often more than 1000 functions are provided, including FIR filters, FFTs, convolutions, DCTs, and other computationally intensive kernels. The software burden to generate libraries is high but they can be reused for many applications. With this approach, control code can be programmed in C and the computationally intensive signal processing functions are called through these libraries.

6.3.1.4. Intrinsics

Often, when programming in a high-level language such as C, a programmer would like to take advantage of a specific instruction available in an architecture but there is no mechanism for describing that instruction in C. For this case intrinsics were developed. In their rudimentary form, an intrinsic is an asm statement such as found in gcc [28].†

An intrinsic function has the appearance of a function call in C source code, but is replaced during pre-processing by a programmer-specified sequence of lower-level instructions. The replacement specification is called the intrinsic substitution or simply the intrinsic. An intrinsic function is defined if an intrinsic substitution specifies its replacement. The lower-level instructions resulting from the substitution are called intrinsic instructions [29]. Intrinsics are used to collapse what may be more than ten lines of C code into a single DSP instruction.

Early intrinsic efforts, like inlined asm statements, inhibited DSP compilers from optimizing code sequences [30]. A DSP C compiler could not distinguish the semantics and side effects of the assembly language constructs. Other solutions which attempted to convey side-effect free instructions have been proposed. These solutions all introduced architectural dependent modifications to the original C source.

Intrinsics which eliminated these barriers have been explored [31]. This technique represented the operation in the intermediate representation of the compiler. With the semantics of each intrinsic being well known to the intermediate format, optimizations with the intrinsic functions were easily enabled yielding speed-ups of a factor of more than 6.

The main disadvantage of intrinsics is that this approach moves the assembly language programming burden to the compiler writers. More importantly, each new application may still need a new intrinsic library, further constraining limited software resources.

6.3.1.5. Optimizations

In addition to classic compiler optimizations [32], there are some advanced optimizations which have proven significant for DSP applications. Software pipelining [33] in combination with aggressive in-lining has proven effective in extracting the parallelism inherent in DSP applications. Interestingly, some DSP applications are not data dependent. In these cases, profile directed optimizations are very effective at improving performance [34]. These techniques, when used with VLIW scheduling [35], have proven effective in DSP compilation. However, they can still can be more than two times less efficient than assembly language programmers.

† The asm statement in gcc is a method of signalling to the compiler that the text encompassed by the asm statement should pass through the compiler as a direct assembly language instruction.

6.3.1.6. The Sandbridge Compiler

Provision of effective design tools is essential for any new DSP architectural solution – thus Sandbridge has developed a supercomputer-class vectorizing compiler to accompany its SDR processor. A unique aspect of this compiler is that DSP operations are automatically generated using a technique called semantic analysis.

In semantic analysis, a sophisticated compiler searches for the meaning of a sequence of C language constructs. A programmer writes C code in an architecture independent manner – such as for a microcontroller – focusing primarily on the function to be implemented; if DSP operations are required, the programmer implements them using standard modulo C arithmetic†. The compiler analyzes the C code, automatically extracts the DSP operations and generates optimized DSP code without the excess operations required to specify DSP arithmetic in C code. This technique has a significant software productivity gain over intrinsic functions and does not force the compiler writers to become DSP assembly language programmers.

The Sandblaster architecture uses SIMD instructions to implement vector operations. The compiler vectorizes C code to exploit the data level parallelism inherent in signal processing applications and then generates the appropriate vector instructions. The compiler also handles the difficult problem of outer loop vectorization.

A final difficult consideration is vectorizing saturating arithmetic. Because saturating arithmetic is nonassociative, the order in which the computations are computed is significant. Because the Sandbridge compiler was designed in conjunction with the processor, special hardware support allows the compiler to vectorize nonassociative loops safely.

Figure 6.6 shows the results of various compilers on out-of-the-box ETSI C code for the AMR encoder. The *y*-axis shows the number of frequency (MHz) required to compute frames of speech in real-time. The AMR C code is completely unmodified and no special include files are used. Without using any techniques such as intrinsics or special typedefs, the Sandbridge compiler is able to achieve real-time operation on the Sandblaster core at hand-coded assembly language performance levels. Note that it is completely compiled from high-level language. Since other solutions are not able to automatically generate DSP operations, intrinsic libraries must be used. With intrinsic libraries the results for most DSPs are near the Sandbridge results; however, they only apply to the ETSI algorithms, whereas the Sandbridge compiler can be applied to arbitrary C code.

6.3.2. Performance Simulation

Efficient compilation is just one aspect of software productivity. Prior to having hardware algorithms, designers required access to fast simulation technology in order to evaluate at an early stage the likely performance of the DSP implementation.

A *simulator* is an interpreter of a machine language where the representation of programs resides in memory but is not directly executed by host hardware. Historically, three types of architectural simulator have been identified. An *interpreter* consists of a program executing on a computer where each machine language instruction is executed on a model of a target

† Modulo arithmetic is integer arithmetic defined over a set where the values are added or subtracted subject to a modulus. An example is $5+3$ mod 5 gives 3. Saturating arithmetic, in contrast, limits the maximum value to the largest value in the set. Using the same example, $5+3$ would limit the result to 5.

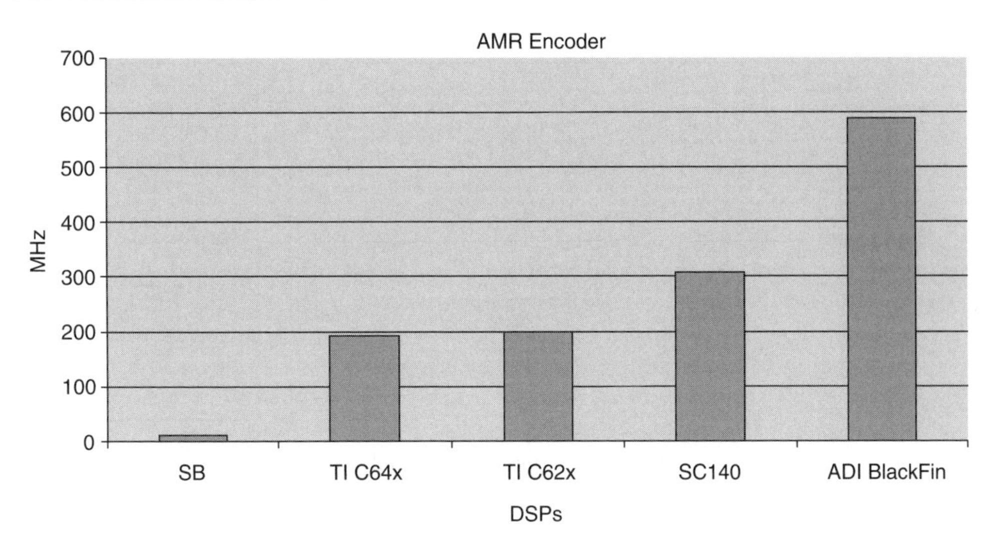

Figure 6.6 Out-of-the-box AMR ETSI encoder C code results. Results based on out-of-the-box C code. C64x IDE Version 2.0.0 compiled without intrinsics using -k -q -pm -op2 -o3 -d 'WMOPS = 0' -ml0 -mv6400 flags with results averaged over 425 frames of ETSI supplied test vectors. C62x IDE Version 2.0.0 compiled without intrinsics using -k -q -pm -op2 -o3 -d 'WMOPS = 0' -ml0 -mv6200 flags with results averaged over 425 frames of ETSI supplied test vectors. Starcore SC140 IDE version Code Warrior for StarCore version 1.5, Relevant Optimization Flags (encoder only): scc -g -ge -be -mb -sc -O3 -Og Other: no intrinsic used. Results based on execution of five frames. ADI Blackfin IDE Version 2.0 and Compiler version 6.1.5 compiled without intrinsics using -O1 -ipa -DWMOPS = 0 –BLACKFIN with results averaged over five frames of ETSI supplied test vectors for the encoder only portion

architecture running on the host computer. Because interpreted simulators tend to execute slowly, both statically and dynamically compiled simulators have been developed.

6.3.2.1. Interpreted Execution

Instructions-set simulators commonly used for application code development are cycle-count accurate in nature. They use an architecture description of the underlying processor and provide close-to-accurate cycle counts, but typically do not model external memories, peripherals, or asynchronous interrupts. However, the information provided by them is generally sufficient to develop the prototype application. Figure 6.7 shows an interpreted simulation system.

Executable code is generated for a target platform. During the execution phase, a software interpreter running on the host interprets (simulates) the target platform executable. The simulator models the target architecture, may mimic the implementation pipeline, and has data structures to reflect the machine resources such as registers. The simulator contains a main driver loop, which performs the *fetch, decode, data read, execute* and *write back* operations for each instruction in the target executable code.

An interpreted simulator has performance limitations. Actions such as instruction fetch, decode, and operand fetch are repeated for every execution of the target instruction. The

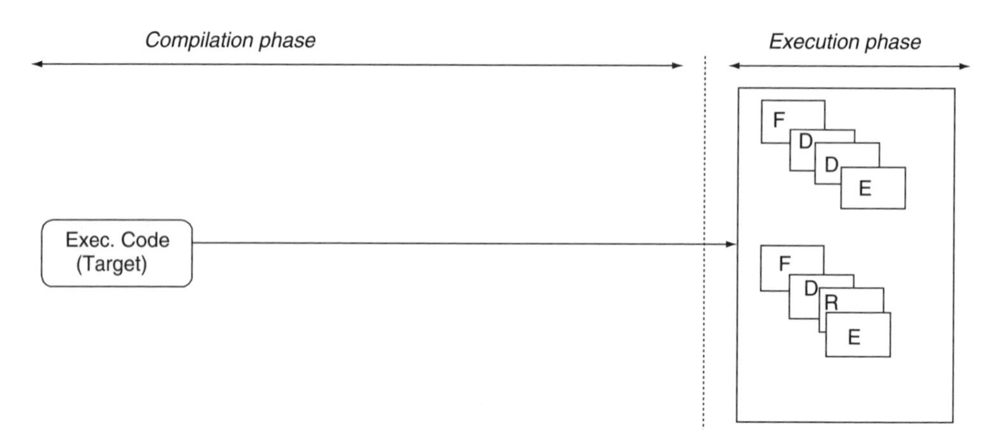

Figure 6.7 Interpreted simulation

instruction decode is implemented with a number of conditional statements within the main driver loop of the simulator. This adds significant overhead, especially considering that all combinations of opcodes and operands must be distinguished. In addition, the execution of the target instruction requires the update of several data structures that mimic the target resources, such as registers, in the simulator.

6.3.2.2. Statically Compiled Simulation

Figure 6.8 shows a statically compiled simulation system. In this technique, the simulator takes advantage of *a priori* knowledge of the target executable and performs some of the activities at compile time instead of execution time.

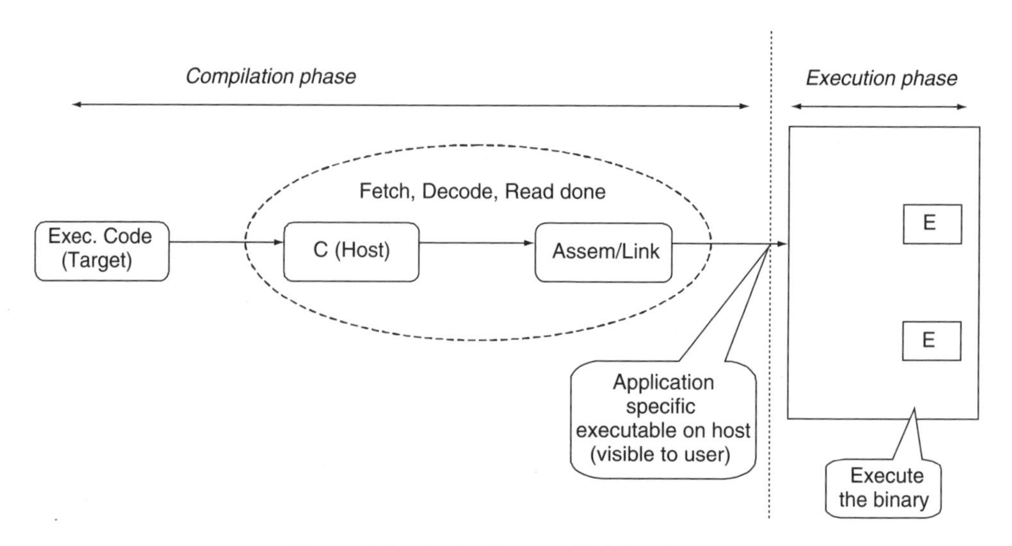

Figure 6.8 Statically compiled simulation

Using this approach, a simulation compiler generates host code for instruction fetch, decode and operand reads at compile time. As an end product, it generates an application specific host binary in which only the execute phase of the target processor is unresolved at compile time. This binary is expected to execute faster, as repetitive actions have been taken care of at compile time.

While this approach addresses some of the issues with interpretive simulators, there are further limitations. First, the simulation compilers typically generate C code, which is then converted to object code using the standard *compile→assemble→link* path. Depending on the size of the generated C code, the file I/O needed to scan and parse the program could well reduce the potential benefits to be gained by taking the compiled simulation approach. The approach is also limited by the idiosyncrasies of the host compiler, such as the number of labels allowed in a source file, size of switch statements, etc. Some of these could be addressed by directly generating object code, however, the overhead of writing the application specific executable file to the disc and then re-reading it during the execution phase still exists. In addition, depending on the underlying host, the application-specific executable (which is visible to the user) may not be portable to another host due to different libraries, instruction sets, etc.

6.3.2.3. Dynamically Compiled Simulation

Figure 6.9 shows the dynamically compiled simulation approach. In this approach, target instructions are translated into equivalent host instructions (executable code) at the beginning of execution time. The host instructions are then executed at the end of the translation phase.

This approach eliminates the overhead of repetitive target instruction fetch, decode and operand read in the interpretive simulation model. By directly generating host executable code, it eliminates the overhead of the compile, assemble, and link path and the associated file I/O that is present in the statically compiled simulation approach. This approach also ensures that the target executable file remains portable, as it is the only executable file visible to the user and the responsibility of converting it to host binary has been transferred to the simulator.

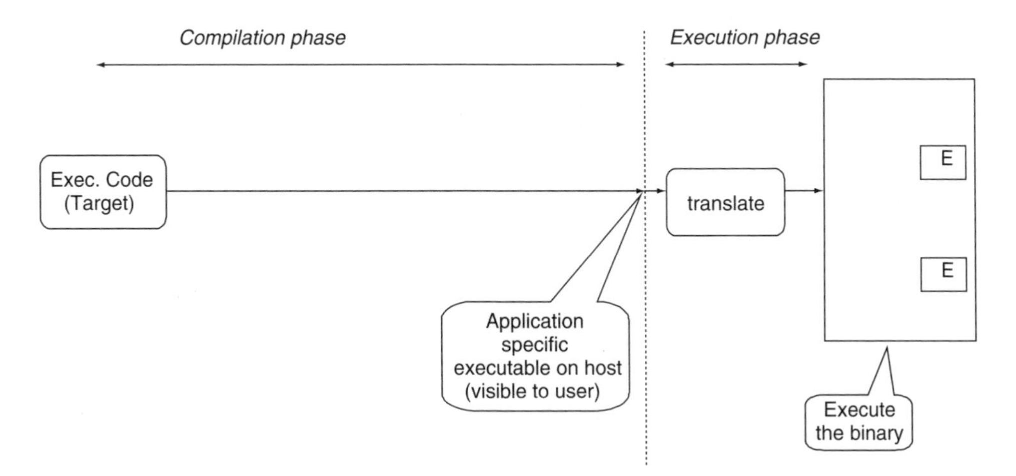

Figure 6.9 Dynamically compiled simulation

6.3.2.4. Tool Chain Generation

The Sandblaster tool chain generation is shown in Figure 6.10. The platform is programmed in a high-level language such as C, C++, or Java. The program is then translated using an internally developed supercomputer class vectorizing, parallelizing compiler. The tools are driven by a parameterized resource model of the architecture that may be programmatically generated for a variety of implementations and organizations.

The source input to the tools, called the Sandbridge architecture description language (SaDL), is a collection of Python† source files that guide the generation and optimization of the input program and simulator. The compiler is retargetable in the sense that it is able to handle multiple possible implementations specified in SaDL and to produce an object file for each implementation. The platform also supports many standard libraries (e.g. libc, math, etc.) that may be referenced by the C program. The compiler generates an object file optimized for the Sandblaster architecture.

The tools are then capable of producing dynamic and static simulators. A binary translator/compiler is invoked on the host simulation platform. The inputs to the translator are the object file produced by the Sandblaster compiler and the SaDL description of the processor. From these inputs, it is possible to produce a statically compiled simulation file. If the host computer is an x86 platform, the translator may directly produce x86 optimized code. If the

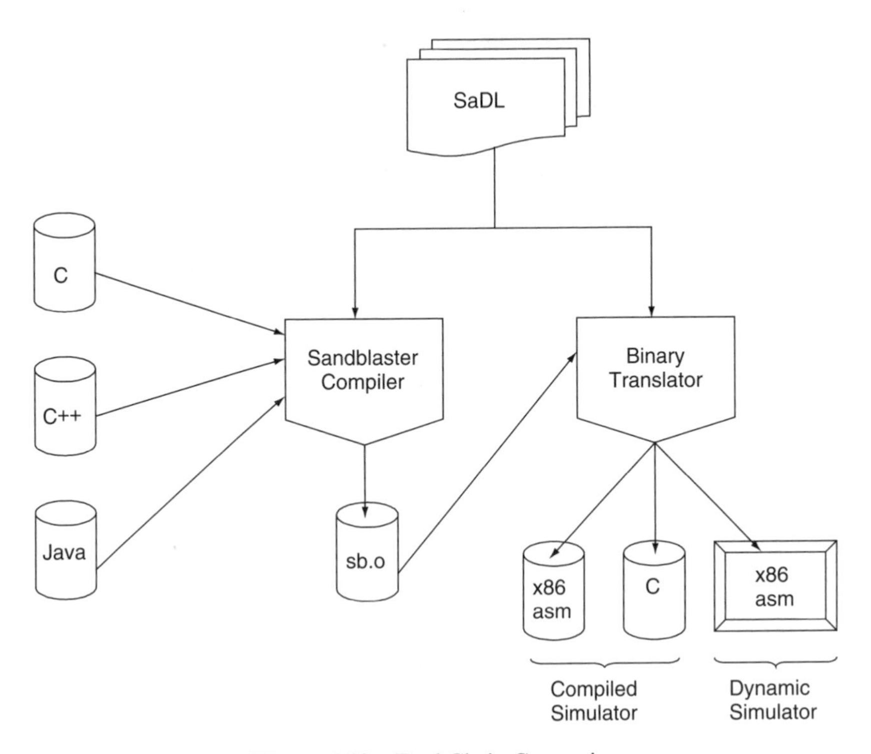

Figure 6.10 Tool Chain Generation

† Python is a high level programming language that is interpreted, interactive and object oriented.

host computer is a non-x86 platform, the binary translator produces a C file that may subsequently be processed using a native compiler (e.g. gcc).

For the dynamically compiled simulator, the object file is translated into x86 assembly code during the start of the simulation. In single-threaded execution, the entire program is translated and executed, removing the requirement for fetch–decode–read operations for all instructions. Dynamically compiled multithreaded simulation is more complicated and is described below.

Dynamically compiled, single-threaded, simulation translation is done at the beginning of the execution phase. Regions of target executable code are created. For each compound instruction in the region, equivalent host executable code is generated. Within each instruction, sophisticated analysis and optimizations are performed to reorder the host instructions to satisfy constraints. When changes of control are present, the code is modified to the proper address. The resulting translated code is then executed.

6.3.2.5. SaDL Description

The SaDL description is based on the philosophy of abstracting out the Sandblaster architecture and implementation specific details into a single database. The information stored in the architectural description file can be used by various parts of the tool chain. The goal is to keep a single copy of the information and replicate it automatically as and when needed.

The key part of the architectural description language is a set of Python files which abstract the common information. These files keep information about each opcode on the processor. The description of an opcode contains a number of attributes – for instance the opcode name, the opcode number, the format, and the input registers. In addition, it contains the appropriate host code to be generated for the particular opcode. These description files are then processed by a generator automatically to produce the C code and documentation. The produced C code is used by our just-in-time simulator and other tools.

Figure 6.11 shows an example of an opcode entry in our architecture description language. It contains the opcode name, number, format and the input resources. It also has calls to the functions (jx86_exec statement) that are called to implement the operation on the host platform. In addition, it contains both the mathematical description (doc_stmt) and the English description (doc_long) to document the opcode.

```
shr = opcode("shr", opcode=0xa4, format=(Rt, Ra, Rb),
      resources=binop_resources(),
      jx86_exec=jx86_shu_body("shrl")+jx86_intop_wback(),
      doc_full="Shift Right",
      doc_stmt=[
            EOp(EGP("rt"), "<-", EOp(EGP("ra"), ">>", EGP("rb")))
      ],
      doc_long="The target integer register, rt, is set to the
            value of the input register ra right shifted by the value
            of rb; 0s are shifted in to the high bits."
      )
```

Figure 6.11 Example SaDL Python descriptions

6.3.2.6. Dynamically Compiled Multithreaded Simulation

Dynamically compiled multithreaded simulation is more complex than the single-thread case because multiple program counters must also be accounted for. Rather than translating the entire object file as one monolithic block of code with embedded instruction transitions, and then executing it, in the multithreaded case we begin by translating each compound instruction on an instruction-by-instruction basis. A separate piece of code manages the multiple PCs, selects the thread to execute, and performs calls to the translated instruction(s) to perform the actual operation. The fetch cycle for each thread must be taken into account based on the scheduling policy defined in the SaDL implementation parameters. Properly speaking, the thread scheduling policy need not be considered for logical correctness; however, it facilitates program debugging. Although fetching with multiple program counters has an effect on simulation performance, compiled dynamic multithreaded simulation is still significantly faster than interpreted simulation.

When the simulator encounters a particular opcode during simulation, it calls the appropriate C function (generated from the processed architectural description files) for that opcode, makes a syntactic and semantic check, and generates the host code to be executed on the host processor. By using this approach, all the modifications to the architecture description are limited to a small set of files. This minimizes errors and maximizes productivity.

Figure 6.12 shows the post-compilation single threaded simulation performance of the same ETSI AMR encoder for a number of DSP processors. All programs were executed on the same 1-GHz laptop Pentium computer.

The Sandbridge tools are capable of simulating nearly 25 million instructions per second. This is more than two orders of magnitude faster than the nearest competitor and allows real-time execution of GSM speech coding on a x86 simulation model. To further elaborate,

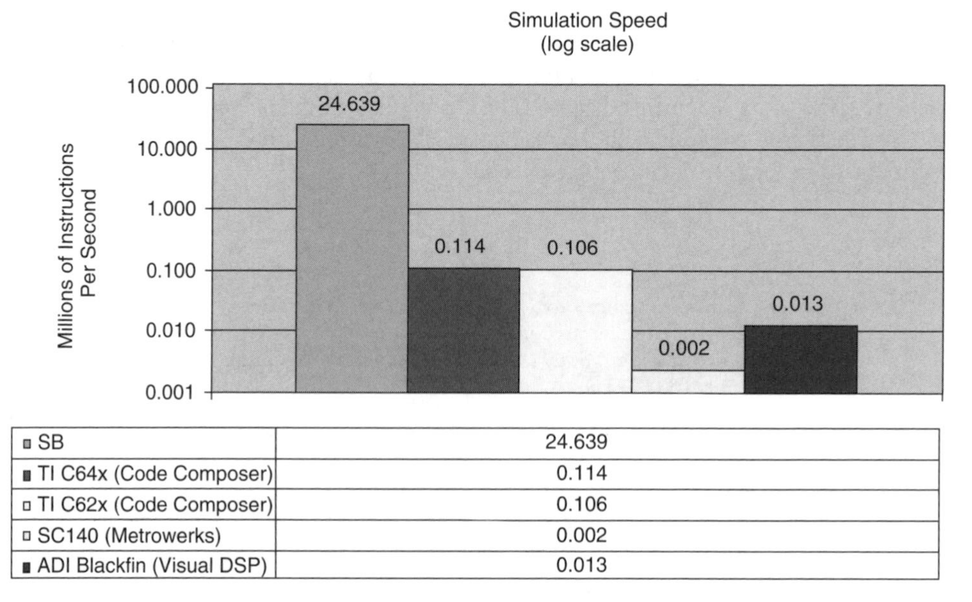

SB	24.639
TI C64x (Code Composer)	0.114
TI C62x (Code Composer)	0.106
SC140 (Metrowerks)	0.002
ADI Blackfin (Visual DSP)	0.013

Figure 6.12 Single-threaded simulation speed of ETSI AMR encoder on 1-GHz laptop

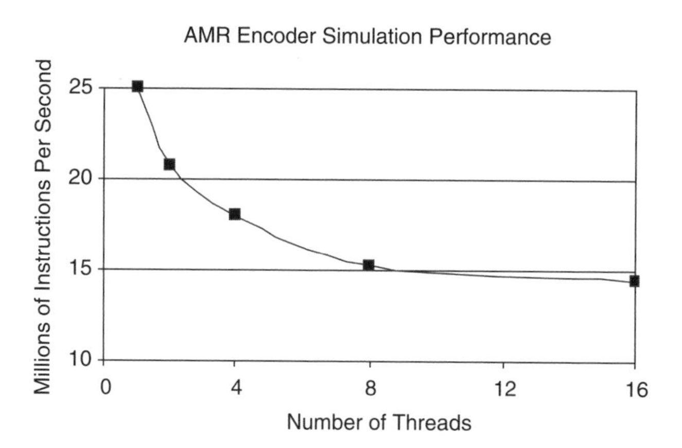

Figure 6.13 Multithreaded simulation speed on ETSI AMR encoder on a 1-GHz laptop

while some DSPs cannot even execute the out-of-box code in real time on their native processor, Sandbridge achieves multiple real-time channels on a simulation model of our processor. This is achieved by using our own compilation technology to accelerate the simulation.

Figure 6.13 shows the multithreaded simulation performance of the ETSI AMR speech encoder on out-of-the-box C code. The results show that the simulation performance degrades to 15 MIPS for eight threads and then very slightly more for additional threads. The degradation is due to the overhead of simulating multiple instruction counters. Since there are eight hardware threads simulated, there is only the overhead of scheduling contributing to additional degradation.

6.3.3. The Integrated Development Environment (IDE)

The Sandbridge Technologies Integrated Development Environment (IDE) provides an easy-to-use graphical user interface to all the software tools. The IDE is based on the Open source Netbeans integrated development environment [36]. This approach enables it to take advantage of any publicly available Netbeans modules that a user may prefer. The IDE is the graphical front-end to the C compiler, assembler, simulator and the debugger. The IDE provides the ability to create, edit, build, execute and debug an application. In addition, it provides the ability to mount a file system, access CVS,[†] access the web and communicate with the Sandblaster hardware board. A snapshot of a typical user session is displayed in Figure 6.14.

6.3.4. The Real-Time Operating System (RTOS)

The programming interface for the multithreaded processor is generic ANSI C code. The compiler is capable of automatically generating multithreaded code for parallel DSP loops. In keeping with an easy-to-use programming philosophy, additional access to multithreading

[†] Concurrent versioning system (CVS) is an open-source, network transparent version system that serves as a design repository for large distributed teams.

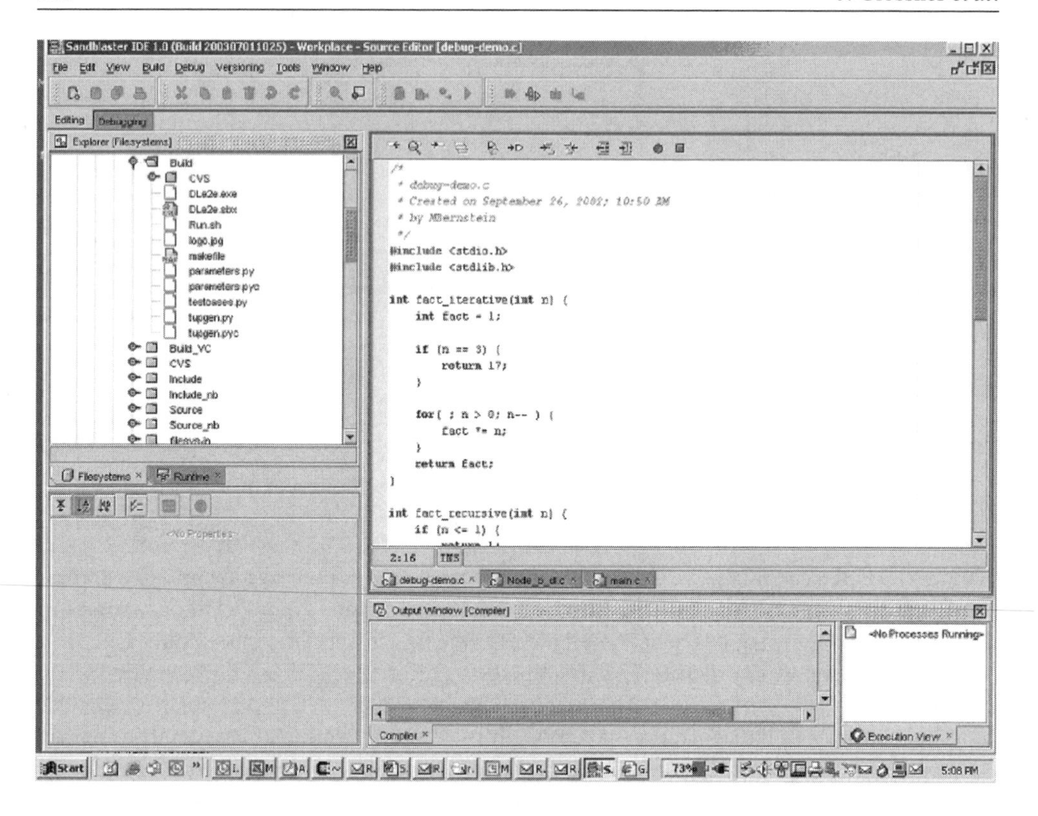

Figure 6.14 A typical Sandblaster IDE graphical user interface

is provided through the open standards of either Java threads or POSIX pthreads[†]. Since nearly all general purpose platforms support these standards, it is simple to port programs to the Sandbridge platform. An API is also supported to allow access to the underlying thread scheduler and for fast porting of third party RTOSs.

A multithreaded application has a number of advantages over a single-threaded application. As an example, one thread may be waiting for data from a peripheral while the other can perform some important computation. Thus, the latency of access from the peripheral is hidden.

A thread can be considered a lightweight process. The overhead of creating and deleting a thread is minimal compared to that of a process. A thread possesses an independent flow of control and maintains its own stack pointer, registers, run-time data, scheduling characteristics, and a set of pending and blocked signals. However, threads can share process ID, address space, working directories, file descriptors, etc. This enables them to access each other's memory locations, access files, and so on. Multiple threads can belong to a single process.

The first implementation of the Sandblaster processor provides access to a fixed number of threads. These threads are called hardware threads. However, an application (written in C) can contain a larger number of POSIX software threads limited only by the available memory. These are scheduled on the hardware threads by the Sandblaster RTOS kernel. The kernel provides the ability to create, destroy, and join the threads. The underlying operating system supports the ability to prioritize and schedule software threads on hardware contexts, to access

[†] Pthreads is an open POSIX standard for incorporating multithreaded programming into C code.

the peripherals such as ADCs and DACs, and to access various memories, etc. The RTOS has been kept lightweight to minimize the overhead.

6.4. A 3G System SDR Implementation

Implementing a third-generation wireless technology (3G) design is up to ten times more complex than previous 2/2.5G designs [37, 38]. Furthermore, modern communications devices will increasingly be required to connect to networks with multiple protocols. A factorial combination of choices presents intractable chip implementations. To facilitate successful system-on-chip (SoC) designs requires new approaches to communications systems implementation such as that described in the previous sections of this chapter.

6.4.1. The Conventional Multiple Processor Approach

Available communications systems have been developed in hardware due to high computational processing requirements. DSPs in these systems have been limited to speech coding and orchestrating the custom hardware blocks. In high-performance 3G systems over 2 million logic gates may be required to implement physical layer processing. The design typically contains an ARM processor, a DSP, and many hardware blocks for specific communications processing (e.g. WCDMA, cdma2000, IS-95, GSM, Bluetooth, etc.) – and this is just for the modem. Multimedia terminals typically include an additional ARM, an additional DSP, and additional hardware accelerators for multimedia functions. Choosing the proper combination of accelerators poses both implementation challenges and time-to-market constraints.

A complex 3G system may also take many months to implement. After logic design is complete, any errors in the design may cause a delay of up to 9 months in correcting and refabricating the device. This labour intensive process is counter productive to fast handset development cycles. An SDR design can take a completely new approach to communications system design.

6.4.2. The SDR Convergence Processor Approach

Rather than designing custom blocks for every function in the transmission system, a small and power efficient core can be highly optimized and replicated to provide a platform for broadband communications – an SDR processor capable of executing operations appropriate to broadband communications. This approach scales well with semiconductor generations and allows flexibility in configuring the system for future specifications and any field modifications that may be necessary.

The Sandbridge communications design approach begins with Matlab models of both the basestation and the terminal. Within Matlab we implement all conformance testing necessary to ensure that our algorithms perform properly. We then code in high-level ANSI C and use our sophisticated tool chain to produce production executable code immediately.

6.4.3. The UMTS WCDMA FDD Requirement

The major blocks for the transmitter and receiver for the UMTS WCDMA FDD-mode communication system [39] are shown in Figure 6.15. This target system was chosen because it is computationally intensive with tight constraints on latency.

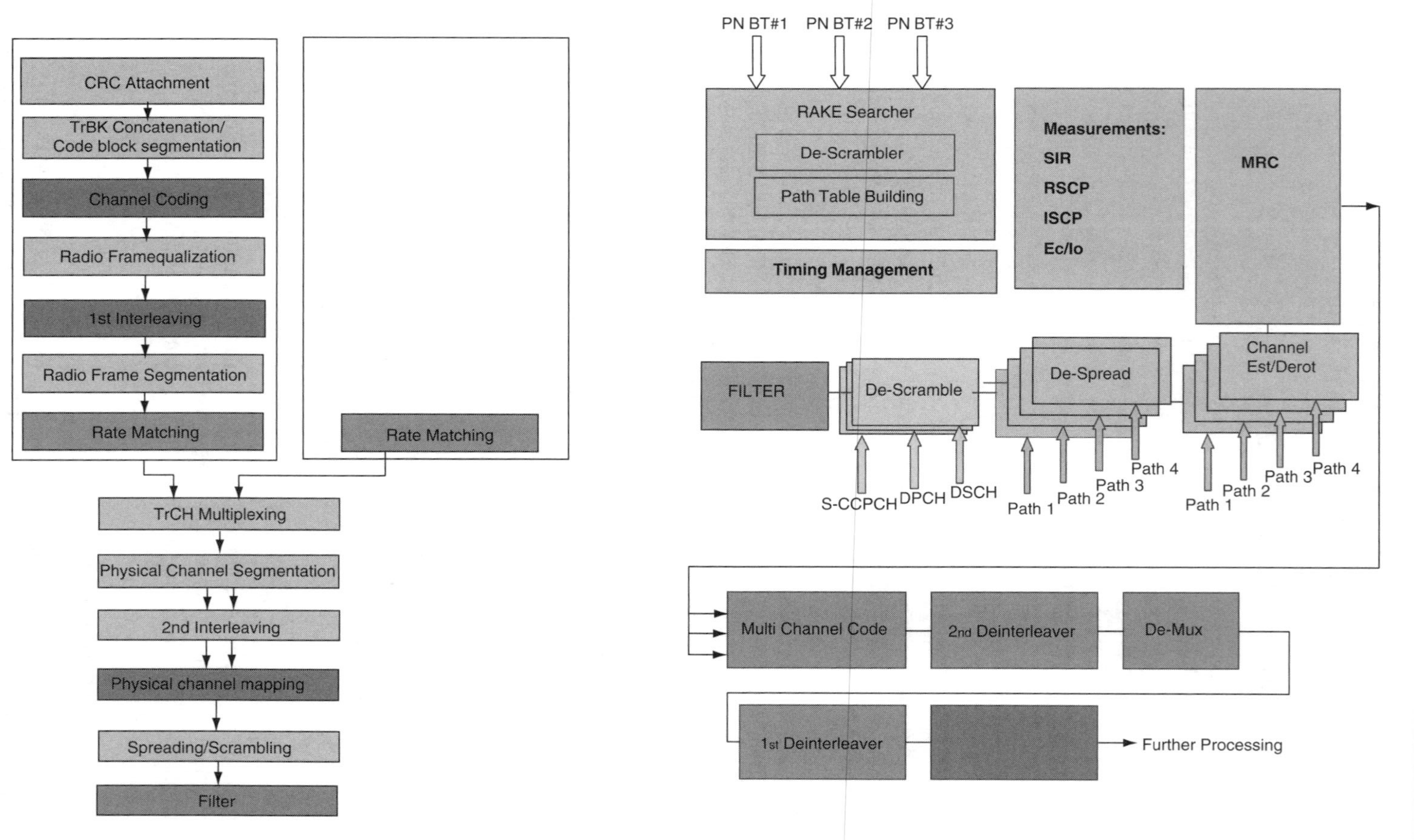

Figure 6.15 WCDMA transmission system

For the receiver, the incoming I and Q signals are filtered using a finite input response (FIR) representation of a root raised cosine filter. This filter is a matched filter in that both the transmitter and receiver use the same filter. The filter is ideally implemented on a DSP. As bit widths continue to increase, often consuming 10 to 14 bits in GSM and advanced communications systems, DSPs with appropriate data types may offer more efficient processing than custom silicon. After synchronization and multipath search, the strongest paths are descrambled, de-spread, equalized, and finally combined in the maximal ratio combining (MRC) block. The output of the MRC block is a soft representation of the transmitted symbols. The soft bits are then de-multiplexed, de-interleaved, and channel decoded. On the receiver side there is also the measurement block responsible for measuring and reporting to the base station the communication channel characteristics, as well as the received power at the terminal antenna. The power and communication channel characteristic measurements are necessary to keep the cell continuously functioning at maximum capacity.

Also shown in Figure 6.15 is the transmitter. In terms of computational requirements, it is significantly less complicated than the receive-chain processing. Additionally, each step of the processing chain is described by the WCDMA standard. After the cyclic redundant check (CRC) and transport block segmentation, the data is turbo or convolutional encoded, interleaved, assembled into radio frames, and then rate matched. The transport channels are parsed into physical channels, interleaved again, and mapped into transmit channels, spread, scrambled, and shaped before being sent to the DAC.

An important part of the WCDMA radio is generation of the RF front-end controls. This includes automatic frequency control (AFC), automatic gain control (AGC), and controls for the frequency synthesizers. These controls have tight timing requirements. Software implementations must have multiple concurrent accesses to frame data structures to reduce timing latencies. A multithreaded processor is an important component in parallelizing tasks and therefore reducing latency.

In WCDMA, turbo decoding is required to reduce the error rate. Because of the heavy computational requirements, this function is usually implemented in hardware. A high throughput WCDMA turbo decoder may require more than 5 billion operations per second. Implementing this function without special purpose accelerators requires high parallelism and innovative algorithms. We developed a new algorithm that reduces the latency and allows full 2 Mbits/s turbo decoding to be completely performed in software with no special purpose accelerators. The algorithm is described in Ref. [12].

6.4.4. *3G SDR Implementation and Performance*

Sandbridge Technologies has completely functional silicon for a single core multithreaded processor. The chip supports eight hardware thread units and executes many baseband processing algorithms in real time. In addition, a complete SDR product, which includes the SB9600 baseband processor as well as C code for the UMTS WCDMA FDD mode physical layer standard, is being developed. Using its internally developed compiler, real-time performance on a 768 kbits/s transmit chain and a 2 Mbits/s receive chain has been achieved, which includes all the blocks shown in Figure 6.15. As shown in Figure 6.16, the SB9600 contains four Sandblaster cores and provides processing capacity for full 2 Mbits/s WCDMA FDD-mode including chip, bit, and symbol rate processing

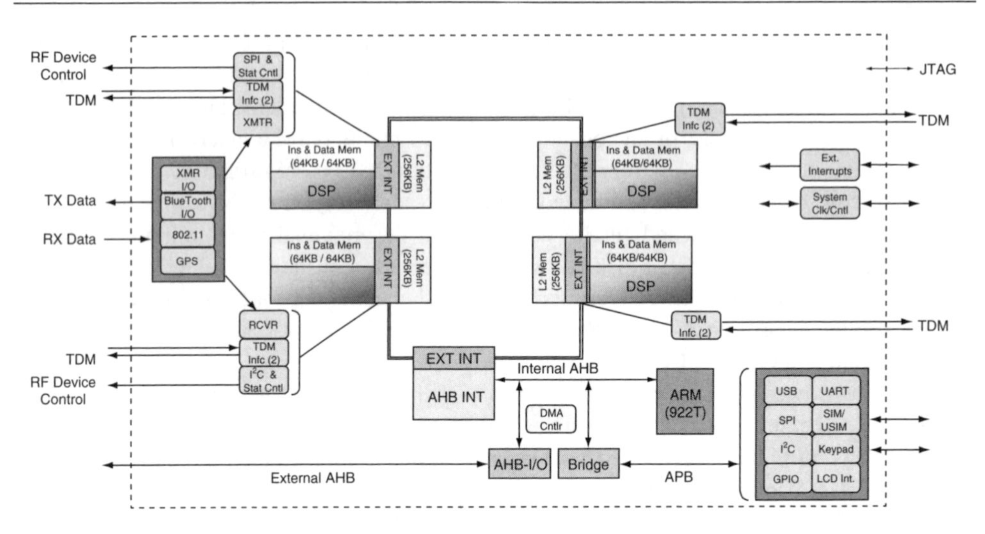

Figure 6.16 SDR SB9600 baseband processor

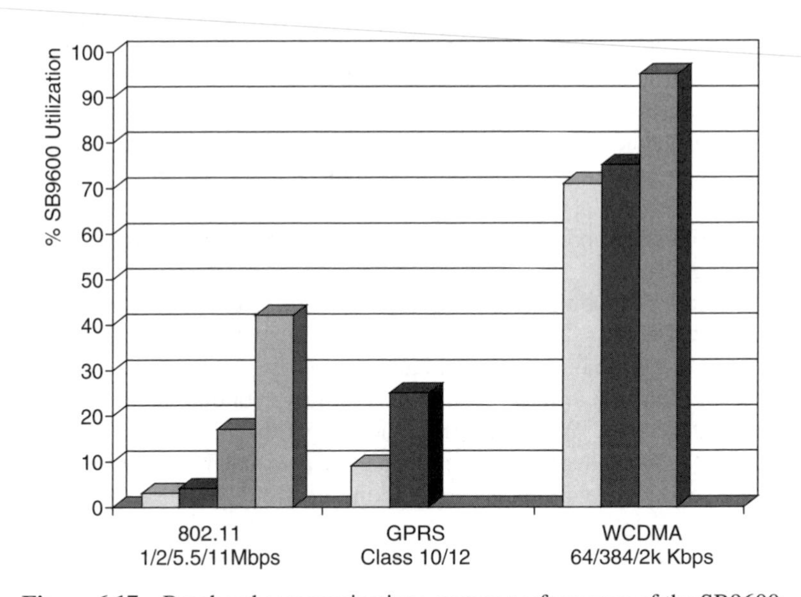

Figure 6.17 Baseband communications system performance of the SB9600

The measured performance requirements for IEEE802.11b, GPRS, and WCDMA as a function of SB9600 utilization for a number of different transmission rates are shown in Figure 6.17.

6.5. Conclusions

Sandbridge Technologies has introduced a completely new and scalable design methodology for implementing multiple transmission systems on a single SDR chip. Using a unique multithreaded architecture specifically designed to reduce power consumption, efficient broadband communications operations are executed on a programmable platform. The processor uses completely interlocked instruction execution providing software compatibility among all processors. Because of the interlocked execution, interrupt latency is very short. An interrupt may occur on any instruction boundary including loads and stores; this is critical for real-time systems.

The processor is combined with a highly optimizing compiler with the ability to analyze programs and generate DSP instructions. This obviates the need for assembly language programming and significantly accelerates time-to-market for new transmission systems.

To validate our approach, we designed our own 2 Mbits/s WCDMA, IEEE802.11b, GSM/GPRS, and GPS physical layers. First, we designed a MATLAB implementation to ensure conformance to the 3GPP specifications. We then implemented the algorithms in fixed point C code and compiled them to our platform using our internally developed tools. The executables were then simulated on our cycle accurate simulator thereby ensuring complete logical operation.

In addition to the software design, we also build RF cards for each communications system. With a complete system, we execute RF to IF to baseband and reverse uplink processing in our lab. Our measurements confirm that our communications designs, including 2 Mbits/s WCDMA, will execute within field conformance requirements in real time completely in software on the SB9600 platform.

References

[1] *http://www.sdrforum.org*

[2] G.A. Blaauw and F.P. Brooks Jr., *Computer Architecture: Concepts and Evolution*, Addison-Wesley, Reading, MA, 1997.

[3] D. Strube, 'High performance DSP technology brings new features to digital systems', *Electronic Product Design*, 1999 (October), pp. 23–26.

[4] G.G. Pechanek, C.J. Glossner, W.F. Lawless, D.H. McCabe, C.H.L. Moller and S.J.Walsh, 'A machine organization and architecture for highly parallel, scalable, single chip DSPs', In *Proceedings of the 1995 DSPx Technical Program Conference and Exhibition*, San Jose, California, 42–50.

[5] B. Case, 'Philips hopes to displace DSPs with VLIW', *Microprocessor Report*, 1997 (December), 12–15.

[6] O. Wolf and J. Bier, 'StarCore launches first architecture', *Microprocessor Report*, 1998 (14), **12**.

[7] J. Fridman and Z. Greenfield, 'The TigerSHARC DSP architecture', *IEEE Micro*, 2000, **20**, 66–76.

[8] J. Turley and H. Hakkarainen, 'TI's New "C6x DSP Screams at 1,600 MIPS"', *Microprocessor Report*, 1997, **11**.

[9] B. Ackland *et al.*, 'A single-chip 1.6 billion 16-b MAC/s multiprocessor DSP', *Proceedings of the Custom Integrated Circuits Conference*, 1999, pp. 537–40.

[10] J. Yoshida, 'Java chip vendors set for cellular skirmish', *EE Times*, 2001, 30th January.

[11] J. Glossner, E. Hokenek and M. Moudgill, 'Multithreaded processor for software defined radio', *Proceedings of the 2002 Software Defined Radio Technical Conference*, November 11–12, 2002, San Diego, California, Volume 1, 2002, pp. 195–99.

[12] J. Glossner, D. Iancu, J. Lu, E. Hokenek and M. Moudgill, 'A Software defined communications baseband design', *IEEE Communications Magazine*, 2003, **41**(1), 120–28.

[13] See *http://www-ee.eng.hawaii.edu/~nava/HEP/introduction.html*

[14] J. Glossner, M. Schulte and S. Vassiliadis, 'A Java-enabled DSP', in *Embedded Processor Design Challenges, Systems, Architectures, Modeling, and Simulation (SAMOS)*, Lecture Notes in Computer Science 2268, E. Deprettere, J. Teich and S. Vassiliadis (Eds), Springer-Verlag, Berlin, 2002, pp. 307–25.

[15] D.M. Tullsen, S.J. Eggers and H.M. Levy, 'Simultaneous multithreading: maximising on-chip parallelism,' in *22nd Annual International Symposium on Computer Architecture*, June 1995, pp. 392–403.

[16] J. Gosling, 'Java intermediate bytecodes', *ACM SIGNPLAN Workshop on Intermediate Representation (IR95)*, January 1995, pp. 111–18.

[17] J. Gosling and H. McGilton, *The Java language environment: a white paper*, Sun Microsystems Press, 1995.

[18] T. Lindholm and F. Yellin, 'Inside the Java virtual machine', *Unix Review*, 1997, **15**(1), 31–39.

[19] J. Glossner and S. Vassiliadis, 'The Delft-Java engine: An introduction', *Lecture Notes in Computer Science*. Third International Euro-Par Conference (Euro-Par '97), Passau, Germany, August 1997, pp. 776–80.

[20] J. Glossner and S. Vassiliadis, 'Delft-Java dynamic translation', *Proceedings of the 25th EUROMICRO Conference (EUROMICRO '99)*, Milan, Italy, September 1999.

[21] K. Ebcioglu, E. Altman and E. Hokenek, 'A Java ILP machine based on fast dynamic compilation', *IEEE MASCOTS International Workshop on Security and Efficiency Aspects of Java*, Eilat, Israel, January 1997.

[22] European Telecommunications Standards Institute, Digital Cellular Telecommunications System, ANSI-C code for the GSM Enhanced Full Rate (EFR) speech codec (GSM 96.53), March 1997, ETS 300 724.

[23] K.W. Leary and W. Waddington, 'DSP/C: A standard high level language for DSP and numeric processing', *Proceedings of the International Conference on Acoustics, Speech and Signal Processing*, IEEE, 1990, pp. 1065–68.

[24] B. Krepp, 'DSP-oriented extensions to ANSI C', *Proceedings of the International Conference on Signal Processing Applications and Technology* (ICSPAT '97), DSP Associates, 1997, pp. 658–64.

[25] J. Granata, M. Conner and R. Tolimieri, 'The tensor product: A mathematical programming language for FFTs and other fast DSP operations,' *IEEE Signal Processing Magazine*, 1992 (January), 40–48.

[26] C.J. Glossner, G.G. Pechanek, S. Vassiliadis and J. Landon, 'High-performance parallel FFT algorithms on M.f.a.s.t. using tensor algebra' *Proceedings of the Signal Processing Applications Conference at DSPx'96*, March 11–14, 1996, San Jose Convention Center, San Jose, California, pp. 529–36.

[27] N.P. Pitsianis, 'A Kronecker compiler for fast transform algorithms', *8th SIAM Conference on Parallel Processing for Scientific Computing*, March 1997.

[28] R. Stallman, 'Using and porting GNU CC', *Free Software Foundation*, June 1996, version 2.7.2.1.

[29] D. Batten, S. Jinturkar, J. Glossner, M. Schulte and P. D'Arcy, 'A new approach to DSP intrinsic functions', *Proceedings of the Hawaii International Conference on System Sciences*, Hawaii, January 2000.

[30] D. Chen, W. Zhao and H. Ru, 'Design and Implementation Issues of Intrinsic Functions for Embedded DSP Processors', in *Proceedings of the ACM SGIPLAN International Conference on Signal Processing Applications and Technology (ICSPAT '97)*, September 1997, pp. 505–09.

[31] D. Batten, S. Jinturkar, J. Glossner, M. Schulte, R. Peri and P. D'Arcy, 'Interaction between optimizations and a new type of DSP intrinsic function', *Proceedings of the International Conference on Signal Processing Applications and Technology (ICSPAT '99)*, Orlando, Florida, November 1999.

[32] A. Aho, R. Sethi and J. Ullman, *Compilers: Principles, Techniques and Tools*, Addison-Wesley Publishing Company, CA, 1986.

[33] M. Lam, 'Software pipelining: An effective scheduling technique for VLIW machines', In *Proceedings of the SIGPLAN '88 Conference on Programming Language Design and Implementation*, Atlanta, GA, June 1988.

[34] S. Jinturkar, J. Thilo, J. Glossner, P. D'Arcy and S. Vassiliadis, 'Profile directed compilation in DSP applications', *Proceedings of the International Conference on Signal Processing Applications and Technology (ICSPAT'98)*, September 1998.

[35] W. Hwu, 'Super block: An effective technique for VLIW and superscalar compilation', *Journal of Supercomputing*, **7**, 229–48.

[36] *www.netbeans.org*, 'Netbeans IDE'.

[37] J. Glossner, E. Hokenek and M. Moudgill, 'Wireless SDR solutions: the challenge and promise of next generation handsets', *Communications Design Conference November 2002*, San Jose, California. Available at *http://www.commdesignconference.com/proceedings.htm*

[38] J. Glossner, M. Schulte and S. Vassiliadis, 'Towards a Java-enabled 2 Mbps wireless handheld device', in *Proceedings of the Systems, Architectures, Modeling, and Simulation (SAMOS) Conference*, Samos, Greece, July 14–16, 2001.

[39] 3GPP, 3rd Generation Partnership Project, *Technical Specifications V3.8.0 (2001–09)*.

[40] M. Saghir, P. Chow and C.G. Lee, 'Towards better DSP architecture and compilers', *Proceedings of the International Conference on Signal Processing Applications and Technology*, October 1994, pp. 658–64.

[41] B. Nichols, D. Buttlar and J. Proulx-Farrell, *Pthreads Programming: A POSIX Standard for Better Multiprocessing*, first edition O'Reilly & Associates, Sebastopol, CA, 1996.

[42] T. Boudreau, J. Glick, S. Greene, J. Woehr and V. Spurlin, *NetBeans: The Definitive Guide*, first edition, O'Reilly & Associates, Sebastopol, CA, 2002.

[43] H. Holma and A. Toskala, *WCDMA for UMTS Radio Access for Third Generation Mobile Communications*, John Wiley & Sons, Inc., New York, 2001.

[44] S. Jinturkar, J. Glossner, E. Hokenek and M. Moudgill, 'Programming the Sandbridge multithreaded processor', Accepted for publication at the 2003 Global Signal Processing Expo (GSPx) and International Signal Processing Conference (ISPC), March 31–April 3, 2003, Dallas, Texas.

Part III

Basestation Technologies

THE REAL PERFORMANCE CONSTRAINTS AND TRADE-OFFS OF A NEW CELLULAR SYSTEM ARE UNDERSTOOD AS IT IS DEPLOYED. LESSONS MAY BE LEARNED AND REFINEMENTS INCORPORATED, BUT IF THIS CAN BE DONE BY BASESTATIONS BEING RECONFIGURABLE IT IS MUCH MORE COST EFFECTIVE. THIS IS BUT ONE OF THE DRIVERS FOR SDR BASESTATION SOLUTIONS.

7

Cost Effective Software Radio for CDMA Systems

Alan Gatherer, Sundararajan Sriram, Filip Moerman, Chaitali Sengupta and Kathy Brown

Texas Instruments, Dallas, Texas

7.1. Introduction

7.1.1. The Phases of Baseband Development

The development of baseband radio equipment traditionally passes through three distinct phases. In phase zero, a completely flexible solution is developed using DSPs and FPGAs. This early modem is designed to test out the algorithms in trials. Already at this stage of development the software-to-hardware partition is being developed. Often in the next stage, phase one, the modem is roughly a copy of the phase-zero modem architecture with the FPGA replaced by ASIC and the major bugs removed. Developers do not go to a completely hardwired solution because of potential changes in the standard being implemented, as well as upgrades for better performance, or even channel density, that occur as a greater understanding of the modem's operation is gained from field experience. Also, some operations are already carried out efficiently in the DSP and conversion to an ASIC solution is seen to have little value. The third phase is generally a cost reduction of the second phase and may well involve a change in architecture. A more detailed description of the design cycles for 2G and 3G modems is given in [1].

7.1.2. Software Radio and Technology Intercept

From the software defined radio (SDR) disciple's point of view, the modem in phase zero would already be a software-only solution developed on a generic platform. However, in practice, standards bodies tend to aim new technology at a level realizable by state-of-the-art ASIC technology. Therefore all-software real-time solutions are rarely available to intersect a new standard at its initial development. The best the SDR disciple can therefore hope for is an evolution rather than a revolution. Even this evolution will remain in check due to the

Software Defined Radio: Baseband Technologies for 3G Handsets and Basestations. Edited by W. Tuttlebee
 ISBN: 0-470-86770-1

development of newer and more complex standards that the developing modems have to support. How 'soft' the radio is can hence be seen as a balance between newly emerging standards and newly emerging SDR technology.

7.1.3. Technology Intercept versus Time

Figure 7.1 shows how this balance, or technology intercept, plays out conceptually in time. Note that as each new standard emerges the MIPs requirement jumps above what SDR technology is capable of and aligns itself closer to a mixed DSP/ASIC solution. Whether SDR can catch up depends on the length of time to the next standards evolution. The reader must be in no doubt though that the increasing complexity of standards and the shrinking time to market for these new standards means that modem manufacturers will endeavour to produce the softest radio that they can achieve cost effectively.

The cellular evolution from 2G to 3G is a great example of this. Just as SDR is beginning to make inroads into the chip rate part of CDMA and WCDMA for basestations, the standard is evolving to HSDPA and EVDV, and the incorporation of adaptive antenna arrays are being considered. All three of these more complex requirements move the solution further away from the capability of today's SDR approaches.

7.1.4. Evolution from Today's Basestation Technology

In this chapter we describe an evolutionary step towards SDR for CDMA based standards. In our approach we take the solution which is dominant in today's basestations systems (C64x DSP with ASIC for chip rate processing) and evolve it in two ways.

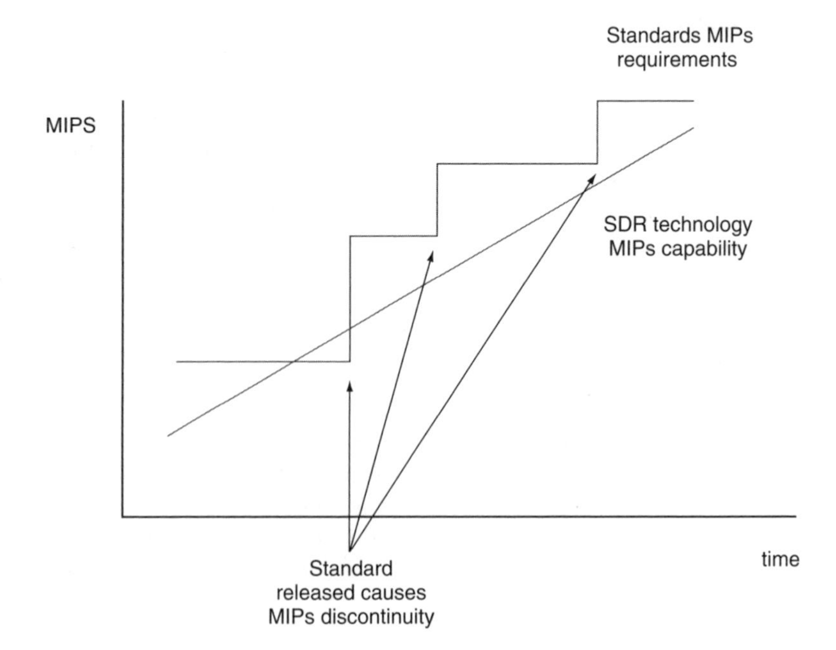

Figure 7.1 MIPs requirements versus SDR technology

- First, we add to the DSP processing capability, a step made possible by the increasing operating frequency and efficiency of the DSP; this results in a drift of functionality from ASIC to DSP.
- Second, we redefine the remaining ASIC functionality to maximize its flexibility so that it becomes a generic 'function call accelerator' for the DSP. Thus, the ASIC becomes software definable by the DSP.

The goal is to make the resulting architecture completely software defined within the scope of the problem at hand, in this case CDMA. We call this 'domain focused SDR'. The resulting solution has enough processing power to also perform algorithms outside of the immediate CDMA-based application domain, such as GSM and EDGE. Thus the result is truly SDR within the domain of wireless basestation modems.

Most if not all of the attempts to bring CDMA closer to SDR in the basestation modem have focused on specialized processing of some kind. One can divide these efforts into 'tightly coupled' and 'loosely coupled' extensions to the software capability.

- By tightly coupled we mean extensions to the instruction set of a CPU or hardware accelerators that perform specialized data manipulation on the data coming into the CPU. This manipulation would take many more cycles without the extension. Such tight coupling generally produces a speed up of factors between 2 and 10 and the scope of this improvement is generally limited by the ability of the CPU to supply the extension with data. An example of this technique would be the Galois Field multiplier added to TI's C64x DSP.
- By loosely coupled we mean a second slave processor, with a specialized data path, that either receives data independently of the CPU or has data routed between it and the DSP memory under control of the CPU using, for instance, DMA. In this case the coprocessor is not limited by the bandwidth of the CPU but it has the extra complexity of its own control structure.

It is possible that a loosely coupled coprocessor (LCCP) could be another CPU with a tightly coupled coprocessor (TCCP). A LCCP will often achieve greater gains than a TCCP, and unlike a TCCP, it does not burden the CPU and can run in parallel. Synchronization with the CPU will tend to be controlled by interrupts and message passing. We believe that the LCCP will be the dominant force in the evolution of SDR implementations for WCDMA and cdma2000.

7.1.5. Chapter Structure

In what follows we will focus, for illustration, on a WCDMA implementation for basestations and describe how a LCCP can provide a more SDR solution while both reducing the cost and increasing the performance of previous generations. In Section 0 we describe the software/hardware trade-off that occurs today in many WCDMA base station modems, and in Section 7.3 we proceed to show how different levels of 'real timeliness' influence decisions on the partitioning of processing. Section 7.4 describes an evolution of the chip rate processing towards an efficient SDR solution. Sections 7.5 and 7.6 discuss in more detail what needs to be done to achieve the potential of this architecture, highlighting design constraints and trade-offs.

Throughout this chapter we use the recently announced TI cellular infrastructure (TCI) modem platform to illustrate the decisions that are made when developing a modem. The TCI modem is a domain-focused SDR using DSPs and LCCP chip rate accelerators. This is described in Section 4. We focus on the receive portion of the modem as this poses the most computationally intensive requirements.

7.2. Today's Software/Hardware Trade-off in WCDMA Modems

The functionality of a WCDMA modem receiver is shown in Figure 7.2. The figure is shown divided into three categories; modem functionality commonly done in software today, modem functionality commonly done in hardware today, and modem functionality that is in the 'grey area' of being done either way, depending on the designer.

7.2.1. Hardware Implemented Elements

The complexity of the CDMA code search process requires in the order of 10 to 100s of Giga complex-adds-per-second, whilst the preamble detection requirement is even higher. For this reason there is little debate about the use of some sort of hardware accelerator for implementation of these functions.

Despreading of the control and data channels is much lower in complexity but does, however, require the maintenance of a large number of streams of data. For instance, each control channel requires three separate channels (early, on-time and late) in order to maintain time tracking. Further, this number is multiplied by the number of rake fingers, perhaps eight on average. Therefore support of 128 users requires the maintenance of about 3200 despread control and data streams – a not inconsiderable processing load.

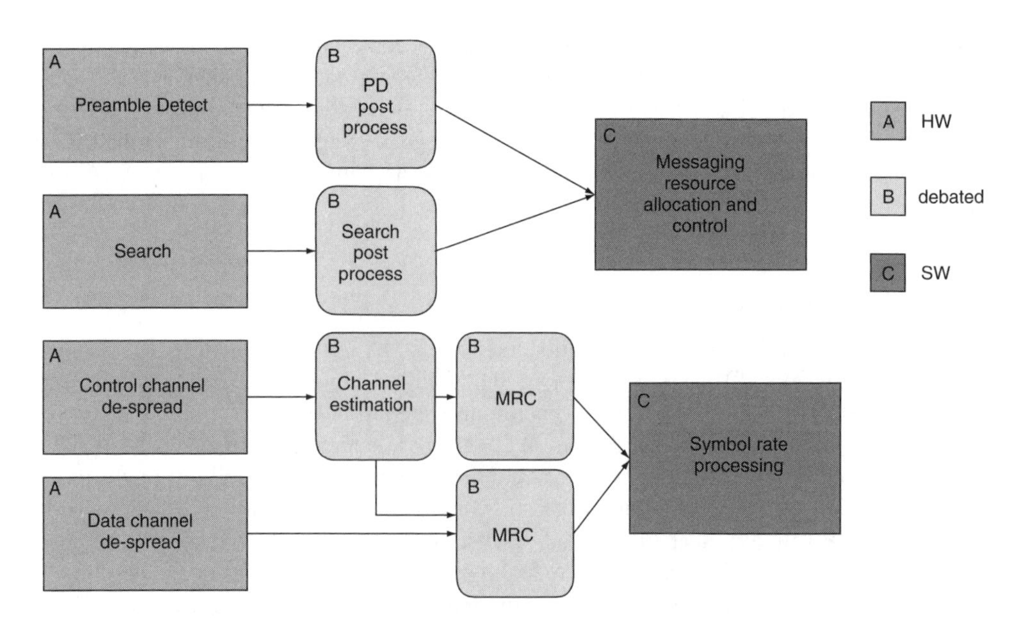

Figure 7.2 Simplified CDMA modem block diagram

7.2.2. Software Implemented Elements

Symbol rate processing (turbo and Viterbi decoding, interleaving, rate matching, etc.), shown on the right hand side of Figure 7.2, is commonly implemented in software; the operating frequency requirements are very reasonable at a few tens of MIPs per channel if the turbo and Viterbi operations are accelerated. Acceleration of the turbo and Viterbi operations is commonly done because the functionality is well defined and because the processing load (MIPS) of doing these functions in software is significant. Similarly the control and maintenance functions, such as finger allocation and RACH message set up, are well suited to software implementation.

7.2.3. The 'Debatable' Functionality

The remaining functionality (marked as 'debated' in Figure 7.2) is best done in software in our opinion and this is how it is performed on the TCI platform. There is a strong advantage to putting this functionality in software as many of the functions are still under development and/or may change depending on the modem location (pico cell or macro cell for instance).

Channel estimation, frequency estimation, adaptive array processing, search and preamble detect post processing are all good examples of functions that are regarded as 'secret sauce' by modem manufacturers, and which can change with time as better algorithms are devised and more is learned about the cellular environment. However, they may be found in hardware or software, depending on the modem design approach adopted by the manufacturer. The MIPS load of these functions is reasonable but their functionality is closely linked to the de-spread operation, so the de-spread accelerators have to be designed carefully for optimal performance.

The above discussion has considered partitioning of the modem function purely in terms of the MIPS requirements. However, smart partitions also have the advantage of being easy to program. In order to understand how to partition a modem design intelligently, one must also consider the temporal domains of the design.

7.3. Temporal Processing Domains

7.3.1. Temporal Domains in WCDMA

Development of a SDR modem is simplified by considering the different temporal domains of processing. An example of a temporal domain is the sub-chip, rate-processing domain. In this domain the samples processed are coming in at a multiple of the 3.84 MHz chip rate. Processing in this domain includes anything in front of the correlation operation and includes interpolation, buffering and subsampling. It is also important to note that in order to minimize buffering (and therefore cost), the processing loop tends to work on not more than a few tens of chips of data. Table 7.1 illustrates a temporal domain partition of the WCDMA modem.

7.3.2. Partitioning Implications

In an SDR system, mixing temporal domains on the same processor can certainly be done but this does require extra care because functions will have different real time constraints. For this reason it is recommended that the temporal domain implementations be separated as much as possible.

Table 7.1 Temporal domains

Temporal domain	Output Buffering requirement	Example functions
Sub chip rate	10s of chips	Interpolation
Chip rate	10s of symbols	Correlation, dispreading, energy accumulation
Chip rate assist	Slot or a few slots	MRC, channel estimation
Symbol rate	A frame	Imterleaving, TFCI decoding

In the case of the WCDMA modem in Figure 7.2 the 'A' blocks can be grouped into sub-chip and chip domains, the 'B' groups into chip-rate assist domains and the 'C' block into symbol-rate domains. Already the symbol-rate functionality is often put on a separate DSP for this reason. In the TCI platform we perform the sub-chip-rate and chip-rate processing using accelerators, though the control of these accelerators is performed in software on a DSP. The DSP therefore performs sub-chip and chip-rate processing in software using accelerators. The DSP controls this processing at the rate of the chip-rate assist domain and can therefore easily mix this processing with other chip-rate assist domain processing such as MRC and channel estimation. An exception to this 'rule-of-thumb' is given in Section 7.6, where a tight latency loop has to work its way through the modem.

7.4. The Evolution of Chip-Rate Processing

Given the partitioning discussion of the last two sections, we now consider how WCDMA modem technology may change over the coming few years.

7.4.1. The Key Challenge

In a CDMA receiver, the spread spectrum signal is sampled after down conversion and is then processed digitally. The sample rate used is typically 4–8 times the chip rate, and the bit widths of the sampled signal are typically 16-bit complex. As a result the digital front end responsible for processing this signal (the 'chip-rate processor') works at a relatively high computation rate, as described in Section 7.2. The chip-rate processor is responsible for performing different kinds of correlation operation that, in effect, convert the chip-rate data into a much lower symbol-rate data stream. The symbol-rate stream is at a low enough rate that modern programmable digital signal processors are able to process this stream and complete all the symbol rate functions (interleaving/de-interleaving, CRC, channel de-multiplexing, rate matching, channel decoding, etc.).

From the above discussion it is clear that one of the key challenges for a true software defined radio is the high chip-rate processing required of a CDMA transceiver front-end. With current silicon technology, we are at a stage where an all software solution for chip-rate processing is definitely feasible, but is not very cost effective from a power, silicon-area, and board-area perspective. We believe that at the 0.1-μm technology node a coprocessor based HW-SW solution is the most viable. However, as DSP processor speeds continue to rise, and the instruction-set architecture continues to evolve, it is only a matter of time before the chip-rate processing migrates completely to software.

7.4.2. Evolutionary Approaches

The following steps describe an evolution of chip-rate processing from a complete hardware solution, to a hybrid hardware–software solution, and finally to a fully software defined solution.

7.4.2.1. The Hardwired ASIC Approach

This has been the traditional approach for implementing chip-rate processing functions, simply because for a long time this was the only viable approach to meet the high computation rate requirement. However, such an approach makes it difficult to deploy computational resources flexibly under widely varying traffic conditions, and to adapt to varying channel conditions. This approach is also becoming more and more expensive due to skyrocketing mask costs, and test and validation costs. Finally, such an approach becomes infeasible for supporting multi-standard platforms.

7.4.2.2. Coprocessor Controlled by a DSP

This approach is that which has been adopted in the TCI platform. The TCI110 coprocessor coupled with the TCI100 DSP achieves the dual objectives of low cost per channel and high flexibility by employing a number of architectural innovations.

The low cost per channel objective is achieved through a high channel density – the chip-set is designed to handle up to 64 voice user channels. The flexibility objective is achieved through a judicious partitioning of the receiver baseband functionality between the TCI110 and the TCI100 DSP. The TCI110 is a highly configurable and parallel task-based coprocessor that handles all the compute-intensive correlation functions that are required in a CDMA receiver; the correlation results are fed to the TCI100 DSP for further processing.

Software running on the TCI100 DSP establishes various tasks on the TCI110 through a memory-mapped interface. A 'task' can be thought of as a hardware function call made to the TCI110; once set up, a task runs continuously until it is modified or overwritten. A task on the TCI110 could be a single rake finger, for example, or a search operation for one user.

7.4.2.3. Reconfigurable Architectures

Many innovative FPGA-like architectures have been proposed recently for handling the intensive chip-rate processing functions in a CDMA transceiver. The idea here is to be able to configure the hardware itself, thereby achieving flexibility. It is still too early to predict how far this approach will go. A key question to be answered is if the tools for programming reconfigurable processors can realistically evolve to a maturity similar to the programming tools available for conventional processors. Another question to be answered is whether or not the power consumption and silicon area overhead of this approach is justifiable in the long run. Our opinion is that reconfigurable processors are only a stop-gap solution in the march towards an all-soft SDR.

7.4.2.4. Massively Parallel Processors

Some companies are attacking the SDR problem by providing massively parallel single-chip processors containing hundreds of CPUs. The same two issues of programming tools and silicon costs as in the case of reconfigurable architectures also arise for this approach. From our experience 50–70% of the silicon area requirement in chip-rate processing is memory. This favors small and fast processing units connected to a high performance memory subsystem, and goes somewhat counter to a massively parallel approach, where the emphasis is on achieving a high MOPS count by employing a large number of processors with relatively small amounts of memory.

7.4.2.5. Digital Signal Processors with Instruction Set Extensions

We believe that, long term, the approach to take in order to get to an SDR is the evolution of the DSP to incorporate appropriate instruction set extensions.

DSP processor speeds are rising rapidly to rival the speeds of general purpose processors. At the same time, the memory subsystem has also evolved in order to be able to feed the high-speed CPU core with high-rate, real-time data. In addition it will be possible to extend the instruction-set of these DSP processors with specialized functions that accelerate wireless communications functions. For example, specialized operations for de-spreading or matched filtering can be added to the general purpose instruction mix. This will enable efficient software implementation of compute-intensive functions such as chip-rate processing, using tools and a software architecture with which programmers are already very familiar. Unfortunately, this solution is not yet available and we will have to wait for its emergence. There is no cost-effective, all-software, solution to the WCDMA modem today, due to the extreme computational load of the front-end chip-rate processing. Companies that have tried this approach have ended up with solutions that are an order of magnitude more expensive than an optimized solution.

7.4.3. Today's Solution

We believe that the second approach described above – the DSP-controlled coprocessor – provides the domain-focused SDR solution that is needed today, at a price close to that of the hard wired approach, with today's technology. Given this architectural choice, there are still important issues to consider. We explore two of these, interface and latency management, in the next two sections.

7.5. The Importance of Efficient Interfaces to Minimize Overhead

The TCI110 is an example of a LCCP for chip rate acceleration of DSP software. A badly designed accelerator may have much processing power but may require many DSP MIPs to support control and data transfer requirements, thereby raising the overall cost of the whole modem. In this section we consider the general requirements for a LCCP to operate without generating excessive DSP overhead.

7.5.1. Coprocessor Data Exchange

The LCCP exchanges data not directly with the DSP's CPU, but with the DSP memory. The DSP memory acts as a buffer, allowing a low interaction level between the CPU and the LCCP and thereby enabling a high efficiency of both the CPU and the LCCP data path. Therefore, an efficient data exchange between the LCCP and the DSP memory should not rely on the intervention of the CPU; it should fully exploit the capabilities of DMA. Let us examine the two types of data exchange between the LCCP and the DSP memory.

The LCCP needs samples to perform correlations; it also typically operates on a fixed number of sample streams. A sample stream could be associated with an antenna, for example, and each sample stream transports samples at a fixed rate. There is no issue in setting up DMA channels that will periodically feed the samples into the LCCP. The LCCP generates correlation results which are produced by a large number of correlation tasks. These correlation tasks dump different, and often small, numbers of correlation results at different instants of time. Moreover, the number of correlation results may vary over time.

7.5.2. Interface Requirements

These requirements give rise to the following three observations:

- It is not possible statically to configure DMA channels that will accommodate the transfer of correlation results – the DMA channels need to be configured dynamically.
- We should not rely on the CPU for the dynamic configuration of the DMA channels.
- The dynamic configuration of the DMA channels adds overhead to each transfer. Thus transfers should be as big as possible so that the transfer time largely exceeds the overhead time.

We thus conclude that the LCCP should be able to dynamically configure a DMA channel and that it should do so whenever it has generated a critical mass of correlation results or whenever a critical deadline is met. It should allow the DSP to specify both critical masses and critical deadlines. Also note that correlation results from different correlation tasks may be used to reach the critical mass.

7.5.3. Programmable Interface Configuration

These issues are all accommodated within the TCI platform by means of a sophisticated host interface on the TCI110, which allows transfer of groups of data from the TCI110 to the TCI100 DSP without requiring CPU intervention. It also allows the CPU to define the conditions under which it will be interrupted once the appropriate data is available. Each task written to the TCI110 by the CPU on the TCI100 contains information to define all of the flexibility in transfer. This allows the TCI platform to run with minimum overhead due to interrupt processing and data movement, giving it a significant cost advantage.

The DSP controls the operation of the accelerator using task definitions. Each task, once set up, performs a simple operation continuously, dumping data into an interface buffer. An example is a single finger despread operation that takes data real time and produces symbols. It is then up to the DSP to retrieve this processed data and process it further. The parallel operation of the two processing elements (DSP and accelerator) requires a third element to coordinate the transfer of data. This is conveniently provided by the DMA on the DSP. The DMA on the C6x TI DSP can perform a variety of functions based on an internal list. The order of operations is therefore as follows:

1. The DSP software requests a task be performed on the correlation accelerator with the results being dumped into a memory location in the DSP. It also may or may not request notification when the data is available.
2. A hardware abstraction layer on the DSP processes the request sending all the relevant information to the accelerator. The DSP may now perform other actions. The Host Interface on the accelerator is now primed to operate in unison with the DSP DMA to provide the correct transfer of data. It can perform this without intervention by the DSP.
3. The task is processed by the data path controller on the accelerator. The data sent by the DSP software contains information about when the task is to start or stop, so the task may lie dormant until the correct time. When the task is activated, it starts generating data which the host interface will eventually transfer back to the DSP by writing to the DMA internal list, the instructions given to it by the hardware abstraction layer.
4. If required the DSP is notified by interrupt that data is available. The process for this new data is activated by the hardware abstraction layer using a software interrupt. Notice that the correlation accelerator knows what time it is (i.e. it has a clock synchronized to the input data that is aware of slot and frame boundaries) but the DSP is only aware of the existence of real time. This allows the correlations to occur in real time as the data arrives and is rapidly discarded (to minimize the input buffer size). But the DSP need only meet deadlines, usually at the slot rate or above, and has no need to be synchronized at the chip level. For a 720 MHz DSP like the TI C64x there are 48000 cycles every slot so the timing requirements on the DSP are quite loose and a software solution becomes manageable. The DSP controls the chip rate accelerator to the chip level by sending clock information on when tasks need to be activated. New tasks are created fairly infrequently so control of the accelerator can coexist with the other functions on the DSP without onerous overhead due to interruption of longer processes.

7.6. Real Time Processing on a Programmable Device

7.6.1. Sub-slot Latency Demands

Sections 7.2 and 7.3 above have described the various domains of real-timeness encountered in a WCDMA modem – varying from sub-slot (120–130 μs in 3GPP FDD), slot (667 μs in 3GPP FDD), to frame (10 ms) and multiple-frame processing. In this section we focus upon the tightest (sub-slot) latency requirements and explore the impact of these on how soft the chip-rate solution can be. For this purpose we consider the downlink (DL) power control processing in 3GPP FDD as an example.

7.6.2. Downlink Power Control Processing

Figure 7.3 shows the processing time available for DL power control processing (reception of transmit power control (TPC) bits in uplink (UL) DPCCH and applying the changed power to downlink (DL) DPCCH pilots).

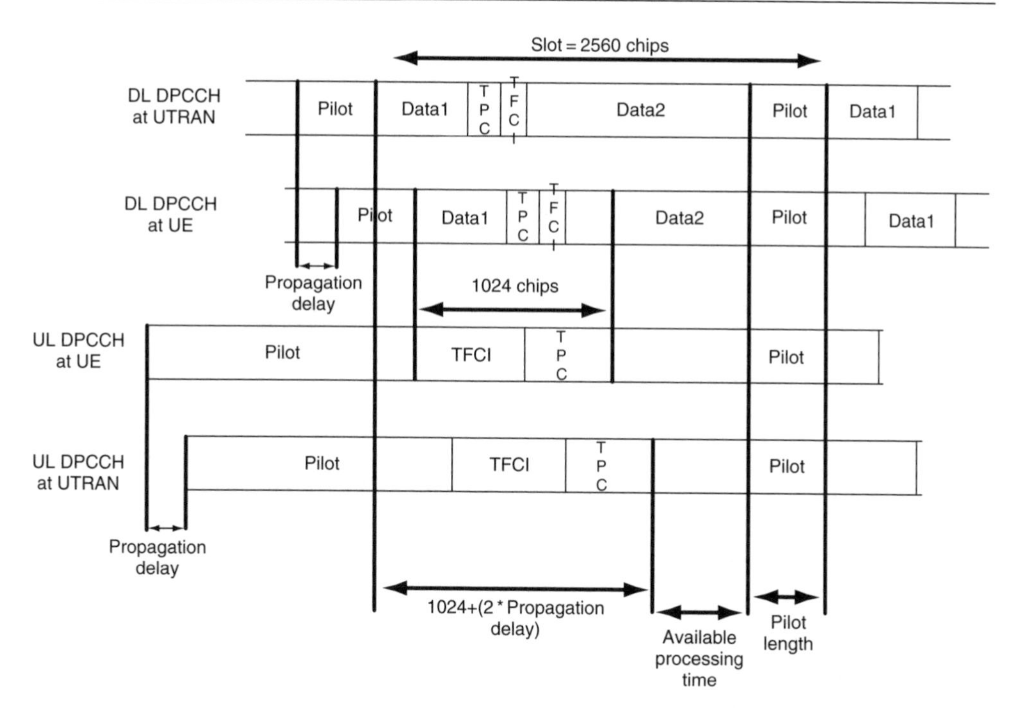

Figure 7.3 DL power control latency requirement from 3GPP (TS 25.214) specifications

7.6.2.1. The Processing Time Budget

As seen in the figure, the available processing time is a function of two factors:

- the DL DPCCH pilot length, and
- whether power correction is applied to the DL DPCCH pilots immediately following the UL TPC field or one slot later.

Let us take a typical case at spreading factor 128 (such as would be used for a voice call) and evaluate the available processing time budget.

If the DL Pilot length is N_p (bits) and the one-way propagation delay is N_d (bits) then:

$$\text{Available processing time} = (2560 - (1024 + 2 \times N_d)) - N_p$$

For the case where:

- the DL Pilot length is $N_p = 2\text{symbols} \times 128 = 256$ chips, and
- assuming 20-km cell radius, the one-way propagation delay is $N_d = 256$ chips

then we find the processing time budget is given as

$$\text{Available processing time} = (2560 - (1024 + 2 \times 256) - 256) = 768 \text{ chips.}$$

In the best case another whole slot (i.e. total of 3328 chips) is available, if power correction is applied to the DL DPCCH pilots one slot later.

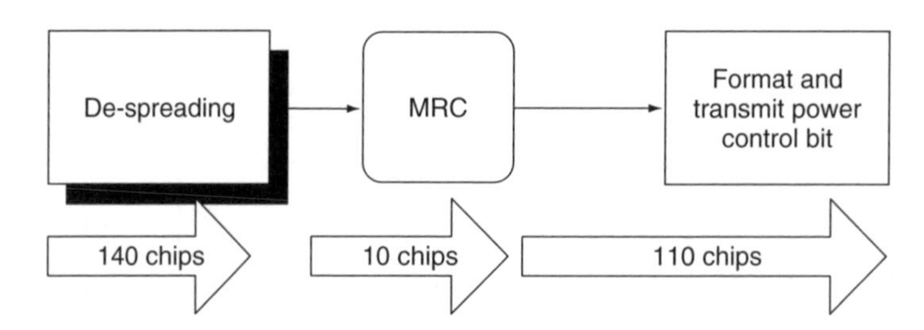

Figure 7.4 Example latency in a tight latency loop

7.6.2.2. Processing Latency for a Single UE

Let us now consider an implementation that is as soft as possible, thus providing flexibility of algorithm (such as channel estimation) implementation. The despreading, due to its highest operating frequency requirement, is performed in a receiver LCCP followed by the demodulation of the bits and interpretation of the TPC command in DSP software. The latency chain for DL power control processing will now have three components, as shown in Figure 7.4.

- Latency through the chip-rate despreading Rx LCCP (this is, say, 140 chips assuming some buffering of chips prior to the LCCP and then processing in vector-mode);
- latency through the chip-rate DSP software (assume this to be ten chips);
- latency due to communication with the transmitter LCCP followed by processing within the transmitter LCCP (assume this to be 110 chips).

This adds up to a total of 260 chips for a single user equipment (UE), which is well within the available time budget.

7.6.2.3. Processing Latency for a Multiple UE Implementation

We now consider the case of 64 UEs on the same implementation under the following assumptions:

- In the worst possible case, all 64 UEs are perfectly aligned in time and so we need to process all of them within the same time budget.
- The LCCPs are designed to exploit maximum parallelism across UEs which are completely independent from each other.
- The processing inside the DSP is serialized across UEs as it is all on one CPU.
- The data transfer from the LCCP to the DSP is also serialized across UEs but a UE is processed in the DSP as soon as its data arrives in the DSP.

In these circumstances, for the last UE to be processed in the DSP software, the total latency will be $140+64\times10+110=890$ chips, which exceeds the available time budget of 768 chips and implies the need to delay the application of the TPC command by an extra slot, at the cost of some performance loss.

7.6.3. Overcoming the Software Bottleneck

In the above example, the latency on the DSP is ten chips every slot for 64 users; this corresponds to 150 MHz on a 600 MHz device. Thus, it is not a case of requiring more average MHz from the processor. Neither is the data transfer from the LCCP to the DSP a bottleneck in the latency chain. Rather, the bottleneck is the serial processing in software itself and the *peak* MHz requirement, as opposed to the average. Other than putting the entire loop processing in a dedicated ASIC, and thereby losing all flexibility of algorithm implementation, there exist two options for solving this problem while keeping the software in the loop.

7.6.3.1. Multiple Cores

The first option is to employ multiple cores, each supporting a sub-set of the users. The practical problem with this approach is having to distribute the required processing efficiently among the cores so as to keep each of the cores busy.

7.6.3.2. Faster Processing

The second solution is to employ a faster processor and to complement low latency requirement functions with higher latency requirement functions in the background. This would enable all latency requirements to be met, provide flexibility due to software processing, as well as efficiently using the higher available MHz.

7.6.3.3. Temporal Domain Combination

The latter approach points towards combining the low latency chip-rate temporal domain functions (such as power control loops, and channel estimation which occur on the control channel) with the higher latency chip-rate temporal functions (such as demodulation of the data channels which are not part of the feedback loop). In the TCI platform, the chip-rate data and control channels are processed in the same DSP allowing the high and low latency channels to average out.

This example shows that the decision on how soft a modem can be and how the modem functionality is partitioned across accelerators and DSPs is not simply dictated by the operating frequency requirements of the various functions but also by latency constraints, which imply a high peak to average operating frequency ratio.

7.7. The TCI Platform

All the conclusions of the discussion thus far are brought together in the brief description in this section of the TI cellular infrastructure (TCI) receive architecture [2, 3].

The top level receive architecture is shown in Figure 7.5. The DSP dynamically controls all of the ‘A’ and ‘B’ processing blocks illustrated earlier in Figure 7.2. Therefore the user has just one single DSP device to program and does not have to worry about the synchronization issues that arise with a multiple-CPU architecture.

The chip-rate accelerator is a slave device performing all of the high MIPS, but straightforward correlation operations. A software API is provided which allows this resource to be accessed in the form of a simple ‘function call’.

The DSP also controls all the post processing of the correlated data. Both low latency and high latency chip-rate assist domain data is processed to allow for load balancing of the MIPS. The symbol rate domain processing is commonly done on another DSP, though symbol rate domain and slot rate domain can be placed on the same DSP.

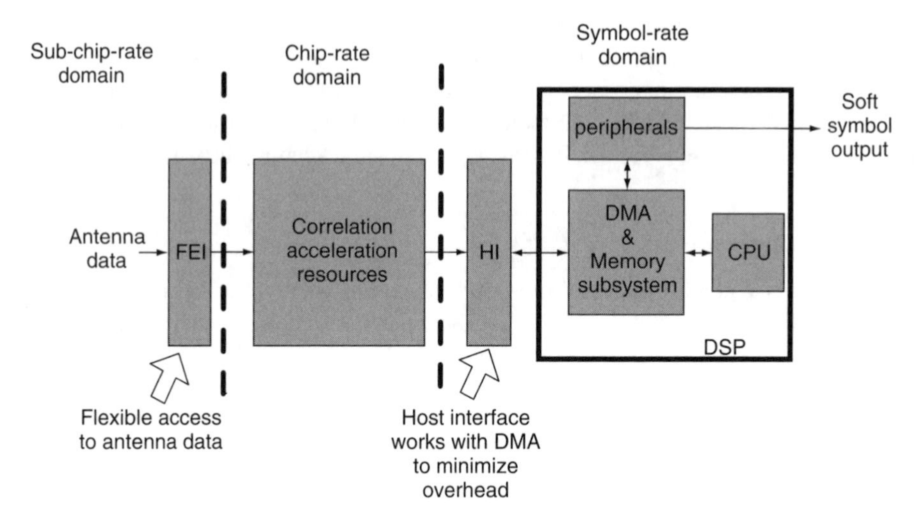

Figure 7.5 Basic block diagram of the TCI platform

7.8. Summary

In the early stages of modem development, fully soft modems could not be achieved without acceleration. In the TCI platform we use accelerators for chip-rate processing that allow DSP software to control the function of the modem. The modem is completely flexible for a CDMA system. We call this 'domain focused SDR'.

We have considered the temporal domains of processing and have shown that there are four within the WCDMA modem. We endeavour to keep these domains on separate processors to simplify operating system and scheduling overhead; however we also find that for tight latency loops it is preferable to mix low and higher latency processes, in order to allow the higher latency process to use the MIPS that would otherwise be wasted if the processor was dominated by the peak MIPS load dictated by the low latency process.

References

[1] A. Gatherer *et al.*, 'DSP based Architectures for Mobile Communications: Past, Present and Future', *IEEE Comm. Mag.*, January 2000.

[2] D. Hocevar *et al.*, 'Programmable DSPs for 3G Base Station Modems', chapter 4 in 'The Application of Programmable DSPs in Mobile communications', Gatherer and Auslander, Wiley 2002.

[3] S. Sriram *et al.*, 'A UMTS Baseband Receiver Chip for Infrastructure Applications', Hotchips 2003.

8

DSP for Basestations – The TigerSHARC

Michael J. Lopez, Rasekh Rifaat and Qian Zhang

Analog Devices, Inc., Norwood, MA

Cellular basestations are extremely complex systems that possess a different set of software-defined radio requirements than mobile terminals. A baseband board inside a basestation may have to process a hundred or more data streams and handle additional tasks such as multi-antenna beamforming, mobility management, and interaction with the network infrastructure. This can lead to extremely high computational loads and data throughput. It also creates many difficult system design problems, including the issue of scheduling all of these tasks, probably among a number of different processors. A processing platform, therefore, must not only be able to handle the high number of computations, but also have sufficient memory and I/O bandwidth, and work seamlessly in a multiprocessor environment. Even size and power dissipation, while somewhat less constrained than in terminals, are still limited by board area, power supply, and thermal issues.

In the bigger picture, the long life and high development cost of basestations means that a given baseband architecture may need to support multiple generations along the cellular roadmap. For example, a service provider may transition from GSM to 3G, add advanced features such as multi-user detection, or switch to an OFDM-based system. This illustrates the attractiveness of a software-defined radio solution, as long as performance will not be sacrificed. In this chapter, we describe the powerful but flexible approach that the Tiger-SHARC processor from Analog Devices, Inc., uses to meet these challenges.

8.1. Introduction and Philosophy

The TigerSHARC digital signal processor provides a high-performance, fully programmable solution for software-defined radio. It extends the DSP model with special communications-oriented instructions, large internal memory, and high data throughput to satisfy the demands of a wireless basestation. For example, the ADSP-TS201 processor, one member of the

Software Defined Radio: Baseband Technologies for 3G Handsets and Basestations. Edited by W. Tuttlebee
 ISBN: 0-470-86770-1

TigerSHARC family, has the resources to implement efficiently the entire 3G baseband physical-layer protocol stack, including path search, despreading, and decoding. Even so, it continues to share the advantages typically associated with a DSP, such as low power, ease of programming, and deterministic cycle counts for real-time processing.

The TigerSHARC processor follows the software-defined radio philosophy in that it achieves its performance without the use of application-specific hardware accelerator blocks. Instead, the ADSP-TS201 processor handles even the most computationally intensive communications tasks with a purely instruction-based approach. This is accomplished in two ways:

- by enhancing the parallelism of traditional instructions, and
- by defining a suite of specialized instructions.

It is important to say that although these new instructions are communications oriented, they are defined generically, so as not to be tied to a particular standard or parameter set. Programmers can then reuse these instructions for many applications. Indeed, many of them have already been employed in ways that were unforeseen at design time. However, because these instructions were developed within the framework of a DSP, they could be implemented in silicon in a very efficient manner.

The flexibility of this instruction-based solution can manifest itself at a number of levels:

- First of all, the TigerSHARC processor can be used for a variety of different standards and applications. Basestation providers can then design a common platform for multiple air interfaces, such as GSM and UMTS.
- Secondly, even within a particular standard, the same baseband processing board can be dynamically reconfigured to meet various traffic conditions and therefore make the best use of its resources. For example, a system with mostly voice streams tends to require a large amount of low-level processing due to the pure number of streams, while high bit-rate data traffic tends to require more resources toward channel decoding. A TigerSHARC-based system can effectively balance the processing load back and forth, since both types of processing are done on the same hardware.
- Finally, even within a single subroutine, a programmer can employ different bit precisions, fixed- and floating-point computation, and many other offered options, because the TigerSHARC processor does not use hardware-selected modes, only instruction-level flags.

The TigerSHARC processor was designed with high-bandwidth, multiprocessor systems in mind. The ADSP-TS201 processor has over 40 Gbps of interprocessor communications bandwidth, 24 Mbits of embedded DRAM, and 14 independent channels of direct memory access (DMA). The TigerSHARC architecture contains many features to promote a seamless multiprocessing environment across either of two separate interfaces.

The TigerSHARC processor is designed as a 'balanced' platform that allows all the good features to be used concurrently. The processor architecture is designed such that the internal buses can feed adequate data to the processing units so that they can run at full strength, while at the same time, the external I/O units can move high speed data into or out of on-chip memory in the background. One can design such a system in C/C++ (either generic or optimized with intrinsics[†]) or assembly code for ease of development and upgrades. Finally,

[†] 'Intrinsics' refers to platform-dependent enhancements to a high-level programming language.

as baseband processing requirements continue to increase, the TigerSHARC architecture will continue to develop along a roadmap of code-compatible parts.

In the remainder of this chapter, we describe the TigerSHARC architecture in more detail, and then discuss how its features address various software-defined radio issues.

8.2. Device Architecture

8.2.1. Highlights

The TigerSHARC processor is an instruction-set architecture DSP, but with many specific enhancements to achieve the performance demanded by basestation applications. In this way, it rivals the throughput and computational performance of more hardware-defined solutions while overcoming their main limitations. The ADSP-TS201 processor runs at 600 MHz and can outperform DSPs and microprocessors running at much higher clock rates, especially on communications-specific algorithms. The main functional blocks are shown in Figure 8.1 and described below.

8.2.1.1. Architectural Features

The TigerSHARC processor has a static superscalar DSP architecture (that is, with parallelism determined at compile time) that is organized in a SIMD (single instruction, multiple data) fashion on several levels. At the highest level, there are two main compute blocks that can be operated on in parallel using a SIMD instruction. These compute blocks are labeled X and Y in Figure 8.1. Within a compute block, the ADSP-TS201 processor has four computational units, of which up to two may be utilized in a given cycle per compute block. These computational units provide different functionalities: arithmetic logic unit (ALU) for arithmetic operations, shifter for bitwise manipulations, multiplier, and a communications logic unit (CLU) that implements special communications-oriented instructions. Finally, within a given instruction, the processor operates on packed data in a parallel SIMD fashion.

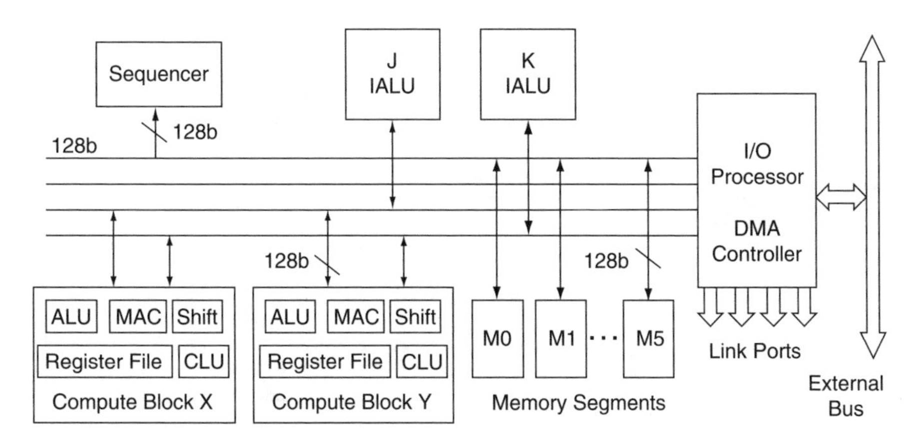

Figure 8.1 ADSP-TS201 processor block diagram

8.2.1.2. Computational Performance

Table 8.1 provides examples of the computational performance available in a single cycle on the TigerSHARC processor. For an ADSP-TS201 processor running at 600 MHz, this translates into 4.8 giga 16-bit fixed-point MACs and 4.8 giga 16-bit fixed-point additions per second. Note how the fixed-point performance numbers increase inversely with data size because of the instruction level parallelism. Its floating-point capabilities include 1.2 giga 32-bit floating-point multiplications and 1.2 giga 32-bit floating-point additions per second. As well, there is great flexibility in regards to which arithmetic operations may be combined in a single cycle. For example, note that multiplications and additions can execute simultaneously. The TigerSHARC's extensive instruction set[†] supports bit FIFOs, bitfield manipulations, complex multiply-accumulates, etc. Table 8.2 summarizes a set of general purpose benchmarks.

TigerSHARC execution proceeds similarly to a RISC (load/store) architecture in that data is first fetched from memory into the compute block register files and is then operated upon from there. Up to four instructions may be issued in one cycle via a single instruction line, allowing load/store operations to run in parallel with compute instructions and other operations. Each of the compute blocks has a bank of 32 general purpose 32-bit registers, along with several special purpose registers that facilitate single-cycle execute latency for a few key instructions.

Supporting the compute blocks are the integer ALUs J and K, which are primarily used for address calculation in parallel with the main compute units. They support many standard ALU addressing modes along with useful features including circular buffering, bit reversal, and data alignment.

The TigerSHARC processor, like most high performance CPUs, is a pipelined processor that incorporates special features to minimize the most harmful side effects caused by this pipelining. First of all, the pipeline is closed, meaning that whenever a pipeline stage tries to access data from a source that is being updated by another instruction later in the pipeline, the processor stalls the pipeline until the results are ready. This has the advantage that, in contrast to the open pipeline selected in other DSPs, it is impossible to write incorrect code that violates pipeline rules on the TigerSHARC processor. Additionally, precise interrupts/exceptions can be implemented that allow interrupts to remain on even during the most critical of DSP loops. The sequencer also includes a branch target buffer that predicts branches so that it is possible to perform zero overhead looping, procedure calls and interrupts.

Table 8.1 Computational throughput examples

Operation	Throughput in outputs/cycle		
	32-bit inputs	16-bit inputs	8-bit inputs
Fixed-point additions	4	8	16
Fixed-point multiply accumulates	2	8	N/A
Floating-point additions	2	N/A	N/A
Floating-point multiplications	2	N/A	N/A

[†] Example instructions are discussed later in the chapter and listed in Table 8.3.

Table 8.2 General purpose algorithm benchmarks on the ADSP-TS201 processor

Benchmark	Speed	Clock cycles
32-bit algorithm, 1.2 billion MACs/s peak performance		
1024 point complex FFT (Radix2)	16.77 μs	10,061
50-tap FIR on 1024 input	45.83 μs	27,500
16-bit fixed-point algorithm, 4.8 billion MACs/s peak performance		
256 point complex FFT (Radix 2)	1.83 μs	1,100
50-tap FIR on 1024 input	12.0 μs	7,200
32-bit floating-point algorithm		
[8 × 8] [8 × 8] matrix multiply (complex)	2.33 μs	1,399
I/O DMA transfer rate		
External port	800 Mbytes/s	N/A
Link ports (each, per direction)	600 Mbytes/s	N/A

In addition to its raw computing power, the processor supports extensive interprocessor communications using two distinct systems:

- the external port, and
- the link ports.

The external port is a standard 32/64-bit host port with the ability to connect to eight Tiger-SHARC processors in a single bus without any additional glue logic. The external port also has built-in support for SRAM, SDRAM, and a host processor. The four on-board link ports specialize in providing point-to-point transactions between TigerSHARC processors, between a TigerSHARC and an FPGA or ASIC, or even between TigerSHARCs on different boards.

The DMA (direct memory access) provides a very thorough and powerful mechanism to move data around a system without the intervention of the core. Additional features include DMA chaining (i.e. starting another DMA automatically when the first is done) and two dimensional DMA (allowing for the reading and writing of arrays).

8.2.2. SDR Features of the Device

The TigerSHARC processor is a general purpose DSP architecture possessing a number of key features that make it well suited for baseband processing in basestations, as described below.

8.2.2.1. Large Memory

The ADSP-TS201 processor includes 24 Mbits of embedded DRAM, a large amount by typical DSP standards but highly desirable from the standpoint of a basestation baseband processing system. Examples where a large amount of memory is needed include the antenna buffers, scrambling code buffers, decoding buffers, interleaving, rate matching, and de-spreading. In these cases and others, having a large on-chip memory minimizes off-chip data transfers and gives the system greater flexibility in the types of algorithm that can be implemented. For instance, one can often alter an algorithm to trade-off compute cycles for memory usage in order to comply with the needs of a particular system.

8.2.2.2. High Bandwidth

In a basestation design, moving sampled channel data into an uplink baseband processing board is usually a tough challenge. For example, in a typical UMTS-FDD system, this may result in an incoming data rate of 1.47 Gbps (3.84 Mchips/second × 2 times oversampling × 16 bits/sample × 12 antennas). This is well within the capability of the ADSP-TS201 processor, which can accept 4.8 Gbps (600 MHz × 8 bits/cycle) over each of its four link ports. It is worthy of notice that interfacing a link port with a data source, typically an FPGA, is simple and straightforward, and a free reference design is available. Hence, one can easily build a data distribution network by interconnecting multiple ADSP-TS201 processors and FPGAs with link ports.

There is also extremely high bandwidth between the memory subsystem and the compute blocks of the processor. The two compute blocks can access up to 256 bits/cycle from the main memory configured as all read or half read/half write. This feature is crucial in order to perform high-throughput processing such as CDMA despreading within the DSP.

8.2.2.3. Interprocessor Communications

The TigerSHARC processor can communicate with other DSPs in several glueless fashions. Up to eight TigerSHARC processors can communicate over the external port, with the memory and registers of each processor in the cluster mapped to a unified address space. Each TigerSHARC processor has arbitration logic that also allows it to share the common bus with memory, peripherals, and a host processor for seamless inter-DSP communication. In addition to the external port, there are four link ports for point-to-point communication. These link ports can connect TigerSHARC processors together or interface to FPGAs or other devices.

8.2.2.4. Communications Logic Unit

Each compute block on the ADSP-TS201 processor contains a computational unit that implements special communications-oriented instructions. These instructions are designed to speed up key signal processing algorithms such as channel decoding and computing correlations while keeping maximum software programmability. This functionality is described in more detail in Section 8.3.

8.2.2.5. Development Tools

The ADSP-TS201S processor is supported with the CROSSCORE suite of software and hardware development tools, including Analog Devices emulators and the VisualDSP++ development environment.

The VisualDSP++ project management environment lets programmers develop and debug an application. This environment includes an easy-to-use assembler, an archiver (library builder), a linker, a loader, a cycle-accurate instruction-level multiprocessor simulator, a C/C++ compiler, and a C/C++ runtime library that includes DSP and mathematical functions. The compiler has been developed for efficient translation of C/C++ code to DSP assembly. For programmers who wish to explore the full potential of the TigerSHARC processor, the assembler offers an easy-to-use assembly language that is based on an algebraic syntax.

The VisualDSP++ debugger has a number of important features. Data visualization is enhanced by a plotting package that offers a significant level of flexibility. The pipeline viewer enables cycle-accurate tracing of program execution at instruction level. The cache viewer provides tracking of memory I/O status. Statistical profiling enables the programmer nonintrusively to poll the processor as it is running the program.

The expert linker can be used to manipulate the placement of code and data on the embedded system visually. It allows the programmer to view memory utilization in a color-coded graphical form, easily move code and data to different areas of the DSP or external memory with the drag of the mouse, or examine run-time stack and heap usage.

Analog Devices DSP emulators use the IEEE 1149.1 JTAG test access port of the ADSP-TS201S processor to monitor and control the target board processor during emulation. The emulator provides full speed emulation, allowing inspection and modification of memory, registers, and processor stacks. Nonintrusive in-circuit emulation is assured by the use of the processor's JTAG interface – the emulator does not affect target system loading or timing. The emulator supports simultaneous debugging of multiple TigerSHARC processors.

8.3. Special Instructions for Communications Signal Processing

In addition to the high-performance arithmetic operations described above, the TigerSHARC processor is equipped with a plethora of special purpose instructions to meet the challenges of signal processing for communications. These instructions can be tailored to implement sophisticated and computationally intensive algorithms, such as CDMA finger despreading, the Viterbi algorithm, and turbo decoders. Within the ADSP-TS201 processor, they are implemented in a special computational unit known as the communications logic unit (CLU). In the following, we describe several such instructions within the context of specific signal chain components, as well as touching upon a few additional uses. The instructions under discussion are summarized in Table 8.3.

8.3.1. Chip-Rate Processing in 3G Cellular Baseband

The ADSP-TS201 processor contains many features that enable high performance chip-rate processing for CDMA based third generation wireless communications. (The cdma2000, UMTS-FDD, UMTS-TDD, and TD-SCDMA systems all fit under this umbrella.) In fact, the TigerSHARC processor was the first commercial DSP designed to perform many of these tasks, formerly the traditional domain of ASICs or other more hardware-oriented solutions. The key functionality required is the ability to correlate soft channel data against hard-bit chip sequences on a large scale. The ADSP-TS201 processor accomplishes this through two classes of instructions, DESPREAD and XCORRS, which are described through the following examples.

Table 8.3 Specialized instructions in the ADSP-TS201 processor. The given performance is split between two compute blocks X and Y, so the performance per compute block is half of the value listed. Each instruction also has several options that may be accessed by flags or special registers. Note that for the instructions marked as *, one factor in each CMAC has full precision while the other factor is a 2-bit complex number.

Instruction	Purpose	Calculations per cycle		
		8-bit inputs	16-bit inputs	32-bit inputs
DESPREAD	Multiply accumulate with 2-bit complex chip sequence	N/A	16 CMACs*	N/A
XCORRS	Perform parallel DESPREADs for shifts of chip sequence	N/A	256 CMACs*	64 CMACs*
ACS	Perform one trellis stage update of Viterbi/turbo decode	16 states updated	8 states updated	N/A
TMAX	Perform special arithmetic operation (see text)	16	8	N/A
VMAX	Perform parallel maximizations	16	8	N/A
PERMUTE	Rearrange data, including movement and negation	16 values	16 values	N/A

8.3.1.1. Finger Despreading

The despread operation is used in CDMA systems to convert chip-rate data into a lower-rate symbol stream. In a typical despreading operation, a set of chip samples is multiplied against a 2-bit complex code sequence, with the results accumulated to form the resulting symbol value. In the TigerSHARC processor, this can be implemented with consecutive executions of the DESPREAD instruction. As shown in Figure 8.2, each DESPREAD instruction correlates eight samples of 16-bit complex input data[†]. The input data reside in the core registers that are accessible to other computational units. The complex codes are stored in special purpose CLU registers. The correlation result is held in an accumulator in the CLU until the desired length of correlation is completed. All of this can be done with single-cycle latency, because the core can fetch new input data in parallel with executing the DESPREAD. Furthermore, the core can pull the intermediate correlation result from the CLU after the completion of each DESPREAD. This feature may be useful, for example, when one wishes to do frequency correction several times in between despreading operations. DESPREAD,

[†] We refer to complex data with B-bit I and Q components as $2B$-bit complex data.

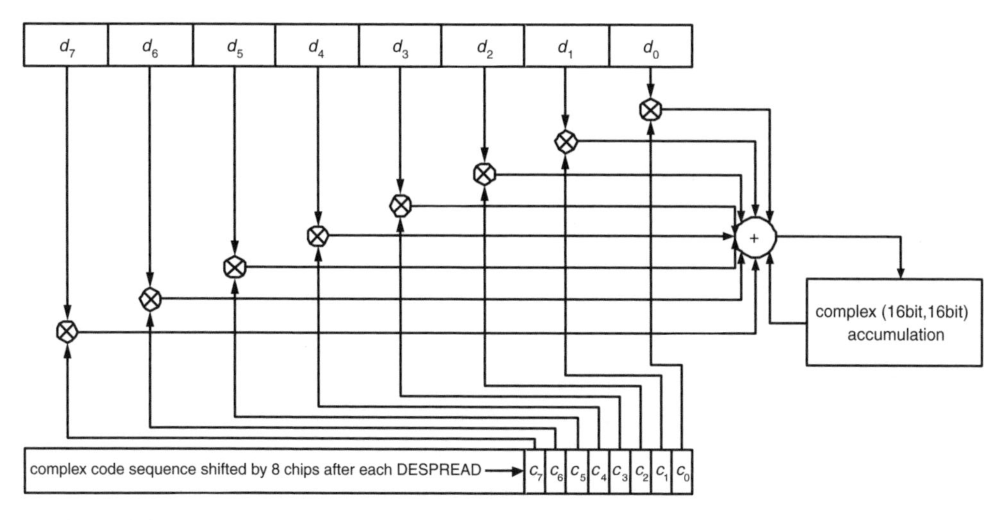

Figure 8.2 The DESPREAD instruction. d_k is the kth data sample, while c_k is the kth code

together with the companion data fetch and code movement instructions, provides a simple yet flexible and efficient way to implement potentially sophisticated despreading algorithms.

8.3.1.2. Path Search

Power delay profile estimation is another computationally intensive chip-rate processing function. Here the basic chip-rate operation is the correlation between soft antenna data and a known pilot code sequence in a time window whose size typically ranges in the order of hundreds of chips. Since this correlation, already significant, must also be performed across many shifts of the input data, it clearly presents a very heavy burden on a processor.

To facilitate efficient implementation of this operation, the ADSP-TS201 processor introduces the XCORRS instruction. From the processing efficiency point of view, the basic XCORRS operation can be thought of as 16 parallel DESPREAD instructions. In a single execution, the XCORRS instruction despreads eight samples of 16-bit complex inputs with 16 separate code sequences, each of them a time-delayed version of a longer code sequence. If we consider each 16-bit by 2-bit product as a complex multiply, then this amounts to 128 complex MACs per cycle in each compute block. An alternate version of XCORRS, selectable by a software flag, performs eight parallel despreads on four samples of 32-bit complex data. Just like DESPREAD, XCORRS is simple to use, as shown in an example in Figure 8.3.

Here the antenna data (16-bit complex numbers) are stored in quad registers R0 to R3 while the complex pilot code sequence is in quad registers THR0 to THR3. The 16 delays are obtained by shifting the contents in THR3:0. The despreading results are accumulated in 16 registers TR15:0. Because of the high bus bandwidth, antenna data can be fetched into registers at the same time as an XCORRS instruction is performed. This allows uninterrupted execution of an arbitrary number of XCORRS.

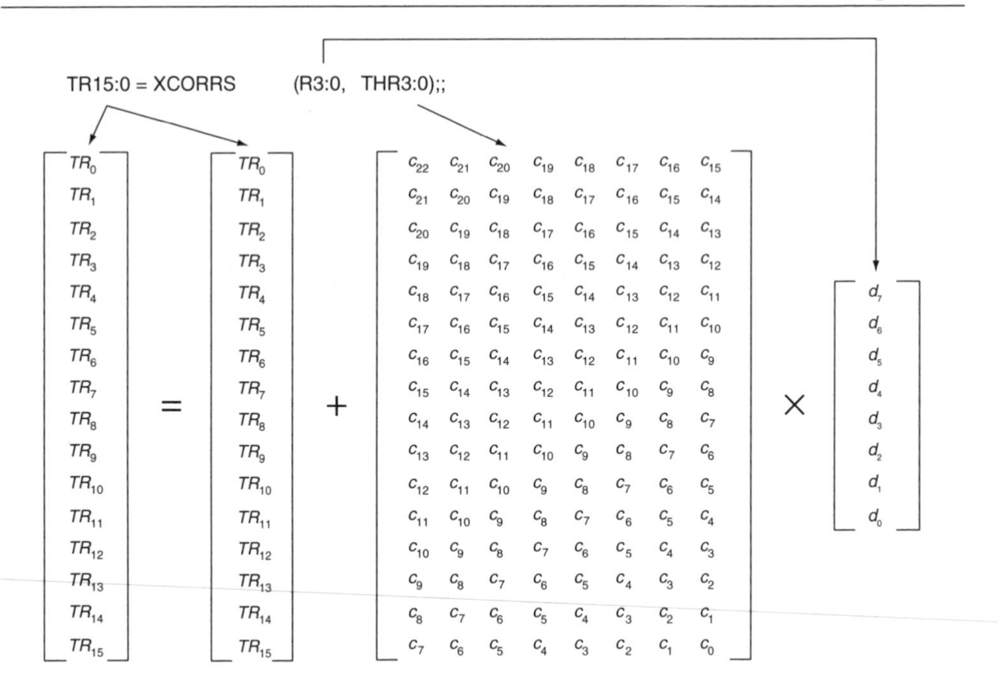

Figure 8.3 XCORRS instruction operation

In addition to programming simplicity, XCORRS comes with a variety of options that allow a programmer to control the behavior of the correlation process. For example, one can specify the effective portion of the code sequence in THR3:0 and thus control the length of the correlation. Also, in the same way as DESPREAD, the core can extract intermediate correlation results while executing XCORRS.

Similarly, the XCORRS instruction is efficient at computing the power delay profile estimation of random access channels, which provide the initial acquisition of cellular users to the basestation. In this case, the length of the search depends on the maximum propagation delay from a user to the basestation, making the processing complexity proportional to the cell size. Once again, a software solution such as the ADSP-TS201 processor provides the scalability that is lacking in any hardware solution.

The ADSP-TS201 processor can use its parallel computational units to perform other operations at the same time as XCORRS and data fetching. For the power delay profile estimation application mentioned above, for example, this parallel processing may include frequency correction, scaling, and pilot pattern removal.

8.3.1.3. Other Chip-rate Processing Applications

XCORRS allows a basestation to combine interpolation with despreading at very little computational overhead. For example, antenna data may be sampled at twice the chip rate, while the finger location may have a resolution of four- or eight-times the chip rate. Traditionally, the antenna data is appropriately interpolated to reflect the finer timing resolution of the finger. The ADSP-TS201 processor provides the alternative of doing interpolation at symbol level after despreading, thus reducing the computational requirement and suppressing the distortion that may otherwise arise in chip-rate interpolation. To do post-despread interpolation, the despread of the even samples of antenna data can be done in compute block X, while the despread of the odd samples can be done in compute block Y at the same time, using a SIMD XCORRS instruction. A total of 32 despread results will be generated. After a symbol is despread in this way, the XCORRS outputs of each block are passed through an interpolation filter to yield the desired timing phase. This feature is particularly useful when a delay-locked loop (DLL) is applied. In this case, the 32 despread results are available for interpolation to produce the early, on-time and late fingers.

Another application of XCORRS is spreading. In this case, XCORRS serves to convolve a sequence of soft complex numbers with a 2-bit complex signature sequence. If the soft complex numbers are pre-combined with antenna weights, the XCORRS outputs can be ready for D/A conversion without the need for intermediate computations. The 2-bit signature sequence represents the pseudo-random code that is used to spread the data symbols. The ADSP-TS201 processor can produce 32 chips at 16-bit precision or 16 chips at 32-bit precision in one clock cycle.

Besides the special instructions for communications signal processing, the ADSP-TS201 processor provides support to chip-rate processing in many other ways. We mention two examples here. First, the TigerSHARC processor is efficient at computing Hadamard transforms, which are heavily used in UMTS-FDD for processing random access channels and in the decoding of Reed–Muller codes. The wide registers and a special simultaneous addition-and-subtraction instruction (which, for example, can compute four additions and four subtractions at 16-bit precision in each compute block in a single cycle) promise a low cost implementation of this algorithm. In the other application, the ADSP-TS201 processor can generate long scrambling codes (UMTS-FDD) on the fly. This function involves linear feedback shift register manipulations and has traditionally been considered too computationally complex to implement in a DSP. However, the ADSP-TS201 processor can take advantage of its large internal memory, wide bus bandwidth, and wide register size to run a block-oriented scrambling code generator that works on many bits simultaneously. The outcome is a highly efficient scrambling code generator with negligible processing cost as compared with other chip-rate processing functions, such as the finger despreader.

8.3.2. Channel Decoding and Equalization

Unlike CDMA chip-rate processing, channel decoding has been a traditional domain of DSPs. However, each succeeding generation of wireless networks presents new challenges as data rates get larger, capacities increase, and coding schemes become more intricate. In this subsection, we explain how special instructions for TigerSHARC help to meet requirements for what is typically the most computationally complex symbol-rate functionality, channel decoding. Through a suite of carefully designed instructions, the ADSP-TS201 processor can sustain high bit rates (in the Mbps region for a single processor) for a large variety of state-of-the-art coding schemes. It is also flexible enough to address codes for potential future standards, like low-density parity-check codes, or other functions related to decoding, such as multipath equalization.

8.3.2.1. Decoding of Convolutional and Turbo Codes

The standard decoders for the most popular channel codes, convolutional and turbo codes, involve operations on trellises such as the one shown in Figure 8.4. Because the decoder must

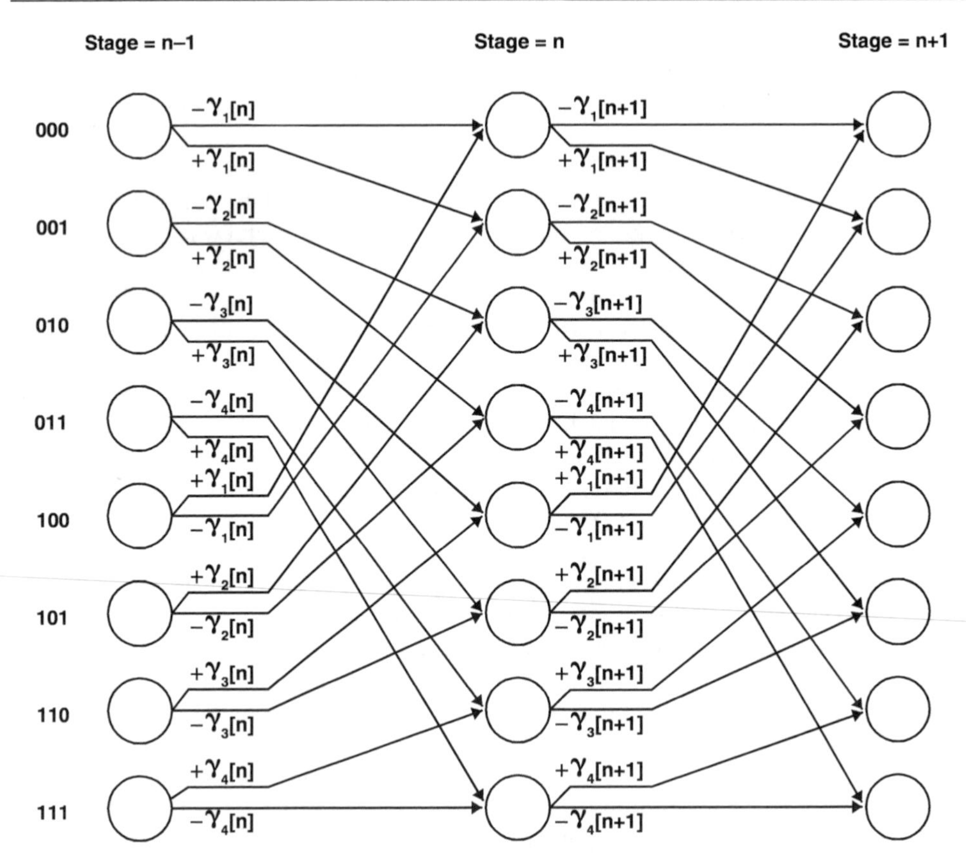

Figure 8.4 Example trellis for Viterbi or turbo decoding

perform computations and comparisons over each possible state transition at every stage, the total complexity of these decoders can be quite high. To keep up with this demand, the ADSP-TS201 processor has instructions that increase both the depth and breadth of computation that can be done on a single cycle. In this way, it can achieve high computational throughput while retaining the flexibility of a software approach.

Consider once again the trellis in Figure 8.4. At stage n, the accumulated metrics for each state are updated based on possible transitions from previous states. For example, in Viterbi decoding of convolutional codes, the metric for state '000' is computed as

$$\alpha_{000}[n] = \max(\alpha_{000}[n-1] - \gamma_1[n], \alpha_{100}[n-1] + \gamma_1[n]),$$

where $\gamma_k[n]$ are the transition metrics based on the realized channel output. A suboptimal 'MaxLogMAP' turbo decoder can use the same trellis update equation. However, the 'LogMAP' turbo decoder provides better performance [1] by replacing 'max' with a different operation which we define as 'tmax':

$$\text{tmax}(a, b) = \max(a, b) + \ln(1 + e^{-|a-b|}).$$

The ADSP-TS201 processor contains an add–compare–select (ACS) instruction that updates several states simultaneously in this way, using either the 'max' or 'tmax' form. For example, with 8-bit transition metrics and 16-bit accumulation, one compute block can update eight states in a single instruction. A second option for 16-bit transition metrics and 32-bit accumulation updates four states in a single instruction. Binary trellises with larger numbers of states can be dealt with by breaking them into subtrellises. Therefore, updating one complete stage of a 64-state (i.e. constraint-length 7) convolutional code, using both compute blocks, can require as few as four cycles. Similarly, one could update one stage of both the forward and backward iterations of an eight-state turbo decoder in a single cycle. The ACS instruction also saves a record of the selected transitions for use in a traceback routine.

This instruction was designed with flexibility in mind, in keeping with the TigerSHARC philosophy. To achieve the required level of parallelism with single-cycle latency, the hardware makes use of a special register set and the particular symmetries among the transition metrics shown in Figure 8.4, which are typical of practical codes. Other than these relationships between transition metrics, however, there are no constraints on the number of states, coding rates, generator polynomials, or block size. The instruction-level implementation allows the programmer to apply add–compare–selects in many ways, or in combination with any other instructions, using assembly language or C/C++ intrinsics.

Supporting ACS and other instructions are several varieties of PERMUTE instructions that can juggle data across registers and parts of registers in various ways.

8.3.2.2. Decoding of Low-density Parity-check Codes

The ADSP-TS201 processor also defines a number of related lower-level building blocks that may offer somewhat lower throughput than the ACS instruction, but open up even more applications. For example, a standalone TMAX instruction is used to compute log likelihood probabilities in the final steps of turbo decoding. Perhaps even more interesting is that TMAX can also be used as a key operation in decoding low-density parity-check codes, a different kind of high-performance channel code that does not make use of trellises [2]. Low-density parity-check codes, in some ways, have even better properties than turbo codes and are an active subject of both research and standardization [3]. Even more, it turns out that the TMAX operation is a basic operation in the iterative decoding of an entire family of coding schemes [4]. Making it a basic instruction in TigerSHARC gives system designers the power to address all of these codes without the need for additional hardware.

8.3.2.3. Channel Equalization for GSM/EDGE

The ADSP-TS201 processor also has tools for efficiently computing trellis operations when the trellis does not have all of the structure assumed by the ACS instruction. For example, the Viterbi equalizer (or a related lower-complexity algorithm) is a key part of the GSM and EDGE signal chains, but its trellis will likely not have the proper symmetry constraints on γ to use ACS. A different example would be a trellis decoder with nonbinary transitions. For these cases and others, the TigerSHARC processor has a separate maximization instruction called VMAX. This is defined for various bit precisions and levels of parallelism, and, like ACS, saves the trellis history decisions. Although not technically part of the CLU, the

VMAX instruction falls under the same category as a generically defined, instruction-level resource that can significantly improve cycle counts.

These specialized instructions together give the TigerSHARC processor a powerful capability for many kinds of processing. This comes without the use of modes, hardware-defined acceleration blocks, or tying up DMA channels. Still more, it allows the programmer the freedom to trade off complexity between symbol-rate, chip-rate, or other processing, depending on the needs of a particular application or network configuration. This versatility, ease of programming, and computational power make an appealing combination for software-defined radio.

8.4. System Design

The possibility of an all-software, baseband-processing solution has been discussed for many years, but practical implementations have been slow to develop until now for several technical reasons. First of all, the amounts of memory and complex multiplications available have not been sufficient for key portions of the signal chain. We have seen in this textbook how modern processors, including the TigerSHARC, are reaching the point of providing these resources. The second major impediment to implementation has been the daunting task of distributing and scheduling the processing of such a complex system for a large number of users. It is important to take a look at the software structure necessary to operate such a system efficiently, while at the same time meeting the requirements for latency that are critical to the system.

8.4.1. CDMA System Load Balancing

An interesting phenomenon occurs in CDMA systems when both chip-rate and symbol-rate processing occur in software on the same DSP. Depending on the type of traffic, such as voice or high-rate data, the dominant computational routines will vary. However, because the total system processing requirement remains approximately constant, the software-defined radio can dynamically shift the processing load to meet any of these scenarios. This represents an advantage over a hard-partitioned system, which must statically allocate its resources to meet worst-case requirements at each level of processing. Below, we describe this situation in more detail.

Chip-rate processing in CDMA systems is essentially the same, regardless of channel data rate. The effect on computation is that there is a fixed amount of this type of processing that must occur for each channel on the system. If there is a large number of channels, then there is a large amount of processing. However, a system operating at capacity may have a few high data rate channels or many low data rate (voice) channels, resulting in an uneven amount of processing.

Conversely, the processing power needed for symbol-rate processing on a given channel is extremely dependent upon its data rate. Channels with higher data rates require more sophisticated error correction schemes that take up more processing power. A typical system with only a few high data-rate users can easily take more symbol-rate processing power than one with hundreds of voice channels.

Now consider the possible mixes of high-to-low data rate channels that a basestation system can encounter while still operating within capacity. If the system is only serving a few high data-rate channels, then the processing requirement for chip rate is very low, while the processing requirement for the symbol rate is extremely high. Next, consider the other extreme where there are many low data-rate (voice) users sharing the same bandwidth. The processing requirement for the chip rate is now significantly higher because of the number of channels, while the symbol-rate processing requirement is much lower because of the much less complex channel decoding scheme. This point is illustrated in Figure 8.5, which shows the TigerSHARC processing load remaining constant as various combinations of high and low data rate channels are mixed.

In a more traditional approach, where an FPGA or ASIC is used for chip-rate processing, there is a split between the hardware needed for chip-rate and symbol-rate processing. The result, shown in Figure 8.6, illustrates the extra processing requirement needed in order to support different mixes of channels. By having both the chip-rate and symbol-rate processing in software, the amount of hardware needed is significantly reduced to the point that an ASIC or FPGA can be eliminated with only a slight increase in the amount of DSP hardware needed for chip-rate processing because of the over capacity inherently built into systems based on ASICs and FPGAs.

More information about load balancing using the TigerSHARC processor may be found in Ref. [5].

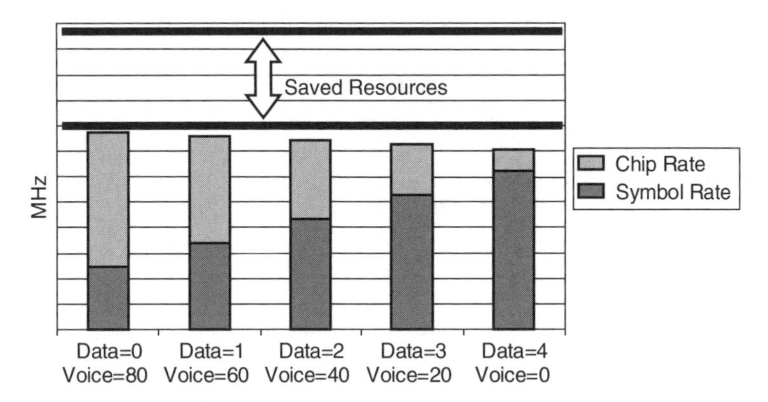

Figure 8.5 Load balancing illustration

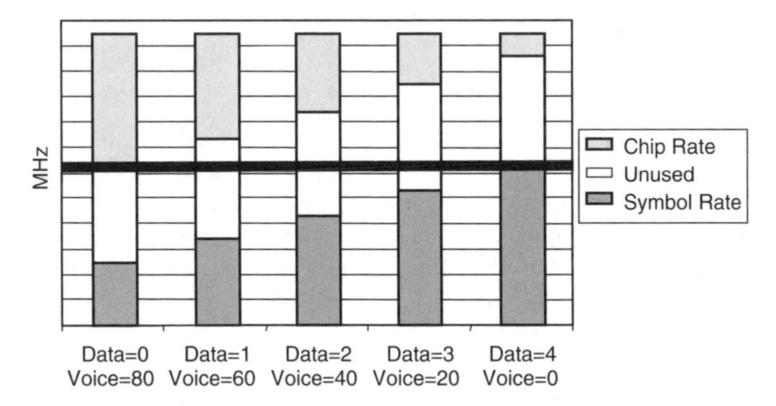

Figure 8.6 Hard-partitioned system requirements

8.4.2. Scheduling

With large multi-user systems, there are several key factors that affect the construction of the system scheduling software. The most important, by far, is the ability of the system to switch its processing back and forth between different users and tasks. For example, there may be a low priority algorithm running for one user, then an interrupt occurs or timer expires, and a high priority task must start up and execute for another user.

To illustrate a concrete example, consider a UMTS-FDD system, where the data from the different users arrives asynchronously. For such a system, it may be necessary to have two or three context switches during the chip-rate processing of each user in each slot, depending on the resolution of the processing. In the case of 128 users, this amounts to 576,000 context switches/second for chip-rate processing alone (128 users $\times$ 1500 slots/second $\times$ 3 context switches/slot). A context switch could cost between 50 and 100 cycles to save the state of the machine, resulting in 28.8 to 57.6 Mcycles/s being used up in context switches just for the chip-rate section alone. This is an extremely high number and would definitely be excessive.

There are, however, several techniques that can be used to mitigate the cost of these context switches. Two of these are:

(i) *Fixed-granularity functions for a given priority level*: In order to minimize context switches, a granularity level for processing can be created that can be shared by all users and, as such, does not require any intermediate context switching. For instance, in the UMTS-FDD example, consider performing all the chip-rate processing for 256 chips across all users in a single master loop that cannot be interrupted. This loop could be swapped in, executed quickly, and then swapped out, resulting in only one context swap every 256 chips regardless of the number of users. With this strategy, the chip-rate processing now requires as few as 0.75 Mcycles/s of context switching (3.84 Mchips/s/ 256 chips/context switch = 15,000 context switches/s).

(ii) *Cooperative multitasking with fine-grained functions*: Alternatively, one can keep the number of cycles for each function extremely low and only allow the operating system to switch tasks between the fine-grained functions (i.e. via a system call). This allows the system to check quickly if a higher priority function needs to run and then switch to it without the overhead of a preemptive multitasking systems overhead. However, in a system such as this one, latency must be taken into account from the start to ensure that the proper priority levels and granularity are selected.

8.4.3. Pre-despreading versus Frame Buffering

In the UMTS-FDD system, a very special design challenge arises that requires more attention than in other CDMA systems. Because of the dynamic data rate flexibility and the particular signaling methods used, the format of the user data is not known until the end of a 10 ms physical frame. This means that the basestation does not know what the spreading factor is for the data and therefore cannot recover the data completely from the input chips before the frame is finished. There are two approaches to this problem – pre-despreading and frame buffering.

In the pre-despreading case, a low spreading factor is used to provide an initial reduction of the data coming in from the antenna. The reduced data is then held in memory for each user until the end of the frame, when the frame format is known and the final despread is accomplished.

In the frame buffering case, all of the chips from all of the antennas are stored in memory. When a user's frame finishes and the frame format is known, the data is despread just once at the appropriate spreading factor.

The more advantageous system depends on the memory trade-off that the system designer is willing to make. For a low number of users, the pre-despreading approach uses less memory, but more computations for its calculation than frame buffering. However, if the number of users grows, then frame buffering becomes much more memory efficient, since the amount of memory needed for this method primarily depends on the number of antennas rather than the number of users.

8.4.3.1. Pre-despreading Requirements

A concrete example can be used to illustrate the demanding requirements of these two techniques. For pre-despreading, the amount of memory required in a typical UMTS system is:

3.84 MHz [chip rate] / 4 [Spreading Factor] * 128 [users] * 2 bytes [IQ symbol]
* 8 [fingers] / 100 Hz [frame rate] = 19.6608 Mbytes/frame

In addition to the memory capacity requirement, there is also a bandwidth requirement of:

3.84 MHz [chip rate] / 4 [Spreading Factor] * 128 [users] * 2 bytes [IQ symbol]
* 8 [fingers] * 2 [read/write operations] = 3.84 Gbytes/second

It may be possible to decrease the amount of memory by performing channel compensation and finger combine before the spreading factor is available, decreasing the requirement for memory capacity and bandwidth by a factor of 8 (fingers). However the required amount of computation would increase by:

3.84 MHz [chip rate] / 4 [Spreading Factor] * 128 [users] * 8 [fingers]
* 4 MAC [per complex multiply] = 3.93216 GigaMACs/second.

8.4.3.2. Frame Buffering Requirements

For the frame buffering case, the amount of memory needed for a system with 2× oversampling can be calculated as:

3.84 MHz [chip rate] / 100 Hz [frame rate] * 2 [oversampling] * 2 bytes [IQ sample]
* 1.3 [overlap margin] = 0.19968 Mbytes/antenna

For a typical six-sector system with main and diversity antennas, the memory buffer is:

0.19968 Mbytes/antenna * 12 antennas = 9.58464 Mbytes

The memory bandwidth needed for the frame buffering case can be computed as:

3.84 MHz [chip rate] * 128 [users] * 8 [fingers] * 2 [memory accesses] = 7864.32 Gbytes/second.

8.4.3.3. Design Options

This particular design challenge can be solved in either way on the TigerSHARC processor because of the extremely large internal memory of the chip. The ADSP-TS201 has 24 Mbits of onboard memory with a sustained internal memory bandwidth of:

256 bits/cycles * 600 MHz [cycles/second] / 8 [bits/byte] = 19.2 Gbytes/second,

which can more than sustain either of the above solutions on a single chip or over multiple chips.

In general, frame buffering is considered to be the better design choice if the memory is available, since it reduces both the total memory and cycle count needed to despread the user data for a large number of users. However, pre-despreading has a strong following from the 'ASIC/DSP' design community, since it follows a much more predictable behavior conforming to the 'fixed granularity functions' strategy described in Section 8.4.2. It can also be more advantageous if the number of users or antennas is low.

8.4.4. Processing Distribution Across Multiple DSPs

For a basestation built on the software-defined radio concept, one of the major system design choices is in the distribution of processing among the physical chips on a baseband board. The configuration can be dynamic or static, and can have varying amounts of flexibility depending on the processing power, memory, and interprocessor communications capability of the key units. Because the TigerSHARC processor addresses all of these areas, it enables systems to use a number of resource allocations strategies among DSPs, several of which are explained below.

8.4.4.1. Distribution by User

The simplest way to partition the processing load is to assign each user or group of users to a DSP. In this case, each DSP works as a stand-alone processor, computing the entire signal chain for its dedicated user(s). System design can proceed similarly to single processor systems. Compared with other types of system, such a system is easier to design, debug and maintain. It also easily scales to different numbers of users. However, this partitioning may not be memory efficient because it requires that antenna data be duplicated in each processor.

8.4.4.2. Distribution by Antenna Data

If the antenna data is too large to be stored in a single processor, it can be partitioned among several processors. In this case, these processors need to pass parameters and results between each other.

8.4.4.3. Distribution by Functionality

Another way to partition the processing load is to assign each type of processing to one or several dedicated DSPs. This may be desirable as one migrates from a system that has fixed partitioning among FPGA/ASIC/DSP to an all-DSP system. One can replace an FPGA or ASIC with one or more DSPs without changing any other part of the system. This allows a low-risk evolution path toward a DSP-based full software defined system.

8.4.4.4. Multilevel Processing Distribution

Finally, the processing load can be distributed at multiple levels. For example, the system may consist of several groups of DSPs. Processing load may be distributed to each group on a user-by-user basis. Within each group, the processing load may be further distributed by antenna data or functionality.

8.5. Advanced Receiver Techniques

As a communications standard becomes more entrenched, equipment makers look for ways to increase system performance beyond that of entry-level configurations. For current and soon-to-be-deployed systems such as 3G cellular, this is primarily expected to take the form of interference-reducing techniques such as multi-user detection [6] and smart antennas. With its large memory, high computational capabilities, and emphasis on software programmable solutions, the TigerSHARC architecture is well positioned for migration to these types of enhanced system. In this section, we consider how these algorithms may be implemented on the ADSP-TS201 processor.

8.5.1. Multi-User Detection (MUD)

We consider two types of MUD receivers: interference cancellation [7] and joint detection [8]. We discuss the former within the context of UMTS-FDD and the latter for UMTS-TDD/ TD-SCDMA. We will show that both receivers can be efficiently implemented on the ADSP-TS201 processor, although the underlying signal processing algorithms are quite different.

8.5.1.1. Interference Cancellation

The basic idea of interference cancellation is to estimate and subtract interference from the received waveform in order to detect a desired signal more easily. First, the receiver demodulates the channel output data, yielding rough symbol estimates for each interfering user. These estimates, though distorted, provide some information about the composition of the overall received signal. The estimated symbols are then passed to a modulator to regenerate the interference signal had these estimates been the actual transmitted symbols. Subtracting this presumed interference signal from the actual channel output data yields a cleaner signal that can be used to detect the desired user. Actual interference-canceller implementations may include a great many modifications of this procedure, often across multiple iterations, but the general idea of estimating and then subtracting potential interference exists throughout.

For UMTS-FDD we consider a parallel interference canceller, in which the symbols from all users are estimated, and consequently subtracted, simultaneously during a given iteration.

In such a receiver one may choose to cancel interference at either the chip level or symbol level. Although both methods are mathematically equivalent, implementation concerns favor chip-level cancellation. Symbol-level cancellation requires maintaining a table of cross correlations between symbols from each pair of users; in light of the asynchronous arrival time of the data streams, their time varying behavior, and the probable different data rates, this table can become too large and complex to manage in practical situations. This is particularly at issue for systems that use long scrambling codes. In this case, the cross correlation coefficients will differ from symbol to symbol within the duration of a frame, making their estimation prohibitively complex.

Implementing a practical interference canceller, in our case for a UMTS-FDD basestation, presents a number of challenges:

- First, the demodulation and regeneration operations require large amounts of computation, especially as they are performed at chip rate.
- Second, storing the regenerated signals requires large memory resources.
- Third, performing interference cancellation for different types of channels and probably at multiple stages, involves high complexity task management.
- Finally, if an interference canceller is used as an enhancement to a conventional rake receiver based modem, compatibility becomes a critical issue.

In the following, we will show that the ADSP-TS201 processor has all the necessary elements to resolve these issues, thus providing a good platform for implementing an interference canceller for UMTS-FDD.

To implement the demodulation and regeneration operations efficiently, we can build on the discussion of the ADSP-TS201 processor's special-purpose instructions from Section 8.3. The demodulation operation involves despreading the channel output data, for which both the DESPREAD and XCORRS (in the case of despreading with interpolation) instructions are highly adept. Signal regeneration is basically a spreading operation, but with the additional problem of emulating the pulse-shaping filter in the handset transmitter and any continuous-time receive filters in the basestation. In the ADSP-TS201 processor, there is an efficient way to combine this filtering with the spreading operation. First, note that to achieve any oversampling precision, it is sufficient to regenerate the data at the 2× rate: we decompose the interpolation filter, which may be at 4× or 8× oversampling, into its polyphase components and then use only those components corresponding to a particular user's delay. This well-known method from multirate filter theory [9] results in much shorter filters and eliminates unnecessary CMAC operations. More relevant for the current discussion, we can perform the spreading and filtering (with these shorter filters) with a single XCORRS operation and make use of the instruction's entire CMAC capability. The procedure involves premultiplying an estimated symbol by each coefficient of a particular polyphase filter component and then using the resulting vector and the spreading/scrambling code as inputs to XCORRS. The 16-bit or 32-bit version of the XCORRS instruction should provide enough precision for most practical applications. With the use of XCORRS flags, SIMD instructions, and instruction-line parallelism to push the multiplies to the background, XCORRS can regenerate data with large spreading factors at a theoretical throughput as high as 32 output samples per cycle.

Regenerated chip-rate signals can consume a significant amount of memory. The amount of memory depends on the unit size of the signal on which the interference canceller operates. Since UMTS-FDD signals from different users arrive asynchronously and continuously at the basestation, it is not possible to collect the entire signal of every channel and then do interference cancellation. The system designer must therefore select an appropriate unit size: too large a size will result in long latencies and high memory usage, while a very small unit size will lead to inefficiencies as the receiver tries to capture all the fingers of the asynchronous channels. A reasonable set of parameters is a one slot (2560 chips) unit size, drawn from an overall buffer length of 2- to 3-times this number. In this simple example, consider a sector with two diversity antennas and sampling at 8-times the chip rate of 3.84 MHz, with each regenerated sample represented by a 16-bit complex number. In this case, the total memory requirement could be 1.85 Mbits (2 antennas * 3 slots/antenna * 2560 chips/slot * 8 samples/chip * 16 bits/sample). The 24 Mbit internal memory of the ADSP-TS201 processor is comfortably sufficient to hold the regenerated interference signal.

The design of a practical UMTS-FDD interference canceller is complicated by a number of factors. First, there are multiple types of uplink physical channels. It is possible that each type of channel will be treated differently during interference cancellation. Next, an interference canceller needs to fit to the radio environment for which it is designed. Radio environments can vary enormously from indoor office to outdoor neighborhood, from urban city to rural country, from small pico cells to much larger macro cells, etc. Moreover, the design of an interference canceller can be quite different for different traffic patterns, e.g. voice-oriented versus data-oriented traffic. To make the matter worse, some of these factors may not be known until the interference canceller is deployed. Given the multiplicity and uncertainty of these design factors, it is unlikely that a universal interference canceller will work well for every type of cell. The implication is that a hardware based design is most likely too rigid to match to a real world environment. The only seemingly feasible solution is a full software design with a large enough pool of elements being assembled to fit a variety of channel types, propagation environments, and traffic conditions.

A full software interference canceller demands immense performance from the employed processor. Let us consider an example in which interference cancellation runs in two threads. In the first thread, a single-iteration parallel interference canceller is applied to the dedicated physical control channels (DPCCH) so as to achieve better channel estimation results. In the second thread, a multi-iteration parallel interference canceller is applied to the dedicated physical data channels (DPDCH). In this thread, the estimated symbols in a low-rate channel, e.g. voice channel, may be obtained directly from the demodulator, while the estimated symbols in a high-rate channel undergo the additional processing of channel decoding before regeneration and cancellation (Turbo MUD) [10]. With further consideration that channels are asynchronous and have different data rates, it becomes a highly complex task to manage all the channels and to execute the two threads in a timely manner. With its high computational power, large internal memory and high internal bus bandwidth, the ADSP-TS201 processor is an appropriate platform for implementing this task.

Finally, an ADSP-TS201-based design significantly reduces compatibility concerns when upgrading to MUD. With its multiprocessing features and support of both conventional and multi-user detection, the migration path to more and more complicated receivers can be accomplished through a series of new software routines. If necessary, additional ADSP-TS201 processors can be added to the baseband board to increase the overall computational power. In contrast, less flexible designs require a separate add-on card for the transition to MUD. The TigerSHARC processor can therefore reduce cost, risk, and processing constraints in the roadmap toward more advanced receivers.

8.5.1.2. Joint Detection

The joint detector (JD) [8] has been extensively studied for CDMA cellular systems that use short spreading codes. A JD exploits the known structure of the spreading codes and knowledge of the users' dispersive channels to mitigate both multiple access interference and inter-symbol interference. Typical JD approaches include zero forcing block linear equalizer (ZF-BLE), MMSE block linear equalizer (MMSE-BLE), and their decision feedback variants. In a UMTS-TDD or TD-SCDMA basestation, the use of JD is necessary for achieving high capacity or spectral efficiency. A pictorial diagram of a ZF-BLE for UMTS-TDD/TD-SCDMA is shown in Figure 8.7. Channel estimation of each user is performed on the midambles (training sequence). The midambles are specifically designed to allow an efficient FFT implementation of channel estimation. The matched filter for a user is the combination of the estimated channel and the employed pseudo random code. A cross correlation matrix is calculated in a similar way. This matrix represents the multiple access interference and inter-symbol interference. The Cholesky decomposition of this matrix yields two linear filters that will be used to suppress the interference. Our analysis is based on this JD receiver.

JD is known for its high computational complexity. With its efficient parallel processing design, the ADSP-TS201 processor can meet complex multiplication and accumulation (CMAC) operations requirements of these algorithms. However, what makes the ADSP-TS201 processor stand out above other DSPs are its floating point and chip-rate processing capability. These capabilities allow better numerical results and reduce processing complexity.

Matched filtering is a high computation complexity block in JD. A matched filter is the convolution of a pseudo random code of length Q and a channel impulse response of length W. In a conventional DSP, this matched filer executes $(Q+W-1)$ CMACs per output. In the ADSP-TS201 processor, the matched filter can be implemented as a rake receiver. First, the XCORRS instruction despreads the antenna data with the pseudo-random code at required delayed positions. Then, the channel impulse response filter sifts through the despread results to yield the desired output. Since the value of Q is small, less than 16, the execution of XCORRS requires a small number of cycles and can be easily multiplexed with the CMAC operations. Therefore, the total cost is W CMACs per output. For example, consider $Q=16$ and $W=16$. A conventional DSP requires 31 CMACs, while an ADSP-TS201 processor needs only 16 CMACs.

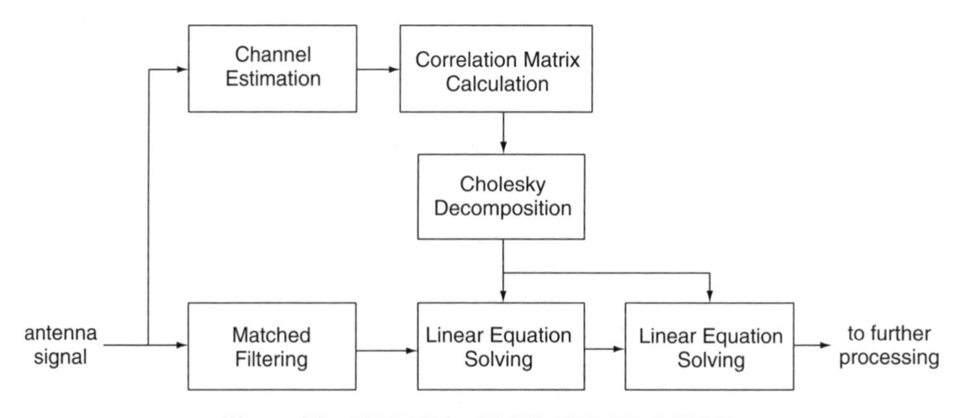

Figure 8.7 ZF-BLE for UMTS-TDD/TD-SCDMA

Another high complexity block is Cholesky decomposition [11]. We note that this block involves division operations. The floating point capability of the ADSP-TS201 processor is well suited for such processing. In particular, the RSQRTS instruction yields the square-rooted reciprocal with 8-bit precision (floating point). With two iterations of RSQRTS in a few more cycles, the result can be improved to 22-bit precision (floating point), which should be sufficient for JD.

8.5.2. *Smart Antennas*

Smart antenna systems use an array of antenna elements to localize users, boosting the gains of their signals and differentiating them from background noise and interference [12]. For cellular and many other systems, this represents the next step in a natural progression that has divided users into cells, sectors, and now spatially localized users within sectors. Additionally, smart antennas can be used in both the uplink and downlink directions, whereas multi-user detection is a strictly receiver based technique. In the following, however, we concentrate on receiver processing because it tends to be the more computationally complex of the two directions.

The TigerSHARC processor supports a variety of smart antenna configurations with software-defined baseband solutions. This has several advantages over implementing the functionality at passband, IF, or in an add-on baseband card. In the integrated baseband system, smart antenna processing becomes simply another set of software routines, which can be developed and shared across multiple processors just like any other piece of code. This can improve efficiency in terms of using the available computational resources and limiting unnecessary data movement, especially off-board data movement. It also presents an easier upgrade path (in terms of both hardware and software) as a system migrates to more and more complex systems. Finally, an integrated software solution enables new algorithms (such as post-despread beamforming combining) that can reduce computations in ways that are not possible in a hard-partitioned system.

On the low-complexity end of smart antennas, consider a switched beam receiver system. The receiver needs to synthesize a number of beams using the antenna data. Then, the strongest beam is selected for each user (and, perhaps, for each subcarrier or multipath component per user) and processing proceeds mostly as for a single antenna system. The major new tasks, then, are forming the beams and selecting beams for each user. The former is basically complex multiply-accumulates, which is a strength of TigerSHARC. Using 32-bit complex fixed-point arithmetic in both compute blocks, the TigerSHARC processor can synthesize four beams from four antennas in 8 cycles per time sample. For a UMTS-FDD system running at two times oversampling, this comes to about 62 MIPs (3.84 Mchips per second * 2-times oversampling * 8 cycles per sample), which is easily within the capability of a 600 MHz ADSP-TS201 processor. As explained in Section 8.3, the ADSP-TS201 processor can perform post-despread interpolation to achieve higher sampling resolution if desired, and has sufficient memory to support these types of on-chip combining. To select among the beams, one may simply repeat the path searcher or similar functionality for each beam.

From here, smart antenna processing can grow in complexity up to fully adaptive systems, where each user (or subcarrier or multipath component) may have a unique set of beamforming coefficients applied to the antenna data. In this case, it is not sufficient to generate a small number of beams for use by all; for N users, at least N beams are needed. An add-on card approach would have to generate all the beams at sample rate, then pass this information to the baseband card. The software-defined solution has the benefit of being able to rearrange functions to create a mathematically equivalent system than has many fewer computations. For example, in a CDMA system such as UMTS or cdma2000, the ADSP-TS201 processor can perform the beamforming after despreading and hence at a much lower data rate. In an OFDM system that multiplexes users across subcarriers, one can perform the beamforming after the FFT demodulation and therefore avoid duplicating the FFT for different users.

In a fully adaptive system, the computation of the beamforming weights alone can also be a significant task. An operator may choose methods such as minimum mean square error (MMSE) estimation, and could implement them with adaptive algorithms such as least mean square (LMS) or recursive least squares (RLS). In any case, this will likely involve matrix multiplies, inversions, and/or factorizations. The TigerSHARC processor's floating-point capability, including instructions that compute reciprocals and reciprocal square roots, can help greatly in these operations.

8.6. Summary

The ever increasing demand for capacity and quality of service calls for employment of advanced technologies in basestations, such as multi-user detection, smart antenna, multiple input–multiple output (MIMO), adaptive coding/modulation/scheduling, etc. These technologies exemplify not only massive processing and data throughput requirements, but also enormous complexity in coordinating numerous tasks. A full software solution will likely be the only practical answer to these challenges. The TigerSHARC family of processors, with its powerful and versatile processing capability, large on-chip memory, high speed point-to-point links and glueless multiprocessor clustering interface, provides an attractive platform for implementing such solutions.

To deal with fast evolution of the wireless communications standards and technologies, each generation of the TigerSHARC processor is designed to enable low risk and an economical roadmap of basestation design, be it from pico cell to macro cell, from traditional rake based system to more advanced MUD/smart antenna/MIMO-based systems, or from GSM to GPRS/EDGE to UMTS, to name just a few. As a powerful DSP, a TigerSHARC processor can serve as a simple upgrade to boost capacity, or as an add-on to enable new features, or as an upgrade to facilitate a new standard, etc. In any case, the TigerSHARC processor provides a stable and cost-effective migration course.

Acknowledgments

The authors would like to thank Andrew McCann for his advice and useful comments in the writing of this chapter.

References

[1] P. Robertson, E. Villebrun, and P. Hoeher, 'A comparison of optimal and sub-optimal MAP decoding algorithms operating in the log domain,' *Proceedings of the IEEE International Conference on Communications*, June, 1995, Vol. 2, pp. 1009–13.

[2] X.-Y. Hu, D.-M. Arnold, and A. Dholakia, 'Efficient implementations of the sum-product algorithm for decoding LDPC codes,' *Proceedings of the IEEE Global Telecommunications Conference (GLOBECOM)*, November, 2001, Vol. 2, p. 1036.

[3] T.J. Richardson, and R.L. Urbanke, 'The capacity of low-density parity-check codes under message-passing decoding,' *IEEE Trans. Inform. Theory*, 2001, **47**, 599–618.

[4] F.R. Kschischang, B.J. Frey, and H.-A. Loeliger, 'Factor graphs and the sum-product algorithm,' *IEEE Trans. Inform. Theory*, 2001, **47**, 498–519.

[5] K. Lange, G. Blanke, and R. Rifaat, 'A software solution for chip rate processing in CDMA wireless infrastructure,' *IEEE Communications Magazine*, 2002, **40** (Feb), 163–67.

[6] S. Verdu, *Multiuser Detection*, Cambridge University Press, Cambridge, UK, 1998.

[7] M.K. Varanasi, and B. Aazhang, 'Multistage detection in asynchronous code-division multiple-access communications,' *IEEE Trans. Commun.*, 1990, **38**, 509–19.

[8] A. Klein, G.K. Kaleh, and P.W. Baier, 'Zero forcing and minimum mean square error equalization for multiuser detection in code division multiple access channels,' *IEEE Trans. Veh. Technol.*, 1996, **45**, 276–87.

[9] P.P. Vaidyanathan, *Multirate Systems and Filter Banks*, Prentice Hall, Englewood Cliffs, NJ, 1993.

[10] M. Moher, 'An iterative multiuser decoder for near-capacity communications,' *IEEE Trans. Commun.*, 1998 **46**, 870–80.

[11] G.H. Golub, and C.F. Van Loan, *Matrix Computations*, Third Edition, The Johns Hopkins University Press, Baltimore, MD, 1996.

[12] H.L. Van Trees, *Optimum Array Processing*, John Wiley & Sons, Inc., New York, 2002.

9

Altera System Architecture Solutions for SDR

Paul Ekas

Altera Corporation, San Jose, CA

A key challenge in the implementation of SDR systems is the definition of the implementation architecture that enables system reconfigurability with the additional constraints of high processing performance, low power, and low cost. Field programmable gate arrays (FPGAs) or programmable logic devices (PLDs) are natural components of software defined radios (SDRs) because of their programmable reconfigurability and their efficient implementation of high performance processing. In this chapter we explore the specifics of what FPGAs are and how they can be used to advantage in a variety of different SDR design situations. Of importance are not only the devices themselves but also their support tools and the intellectual property (IP) available to program them. To confirm their practical usefulness today we will describe, as appropriate, the FPGA product line of Altera Corporation and project its future directions in this application area.

9.1. Setting the Scene

9.1.1. Cellular Baseband SDR Requirements

This chapter addresses, with FPGA solutions, the SDR requirements described in the early chapters of this book. The important general distinctions to keep in mind are that requirements vary between the cellular basestation and the cellular handsets and portable terminals. In each case the total metric of quality will include performance, low power and low cost through a high level of integration. However, the power and size budgets are larger for the higher-performance multichannel basestation than for the single-channel very low-power and low-cost budgets of the handset.

Both handset and basestation share the same needs for multiple functions over time within a single mode of operation and for multimode reconfigurability. Both must also have provisions for being updated in the field. However, the lifetime of a particular product solution may

Software Defined Radio: Baseband Technologies for 3G Handsets and Basestations. Edited by W. Tuttlebee
 ISBN: 0-470-86770-1

need to be longer in the basestation than for the handset where the higher volumes may justify more frequent product redesigns.

The cost of redesigns, however, remains important in both basestation and handset applications. Regardless of the care used in the initial design to allow for reconfigurability and updating, new requirements may force a redesign. Thus, the cost in time and nonrecurring engineering of a new design are part of the trade-off in deciding on an implementation strategy. Reuse of previous designs in the form of high-level intellectual property (IP) and a fully supported software tool set to allow full exploration of the design space is mandatory. Designability remains an important requirement even when most of the design has migrated into software through the use of FPGAs.

9.1.2. FPGAs are Not What They Were

Change has been rapid in the capabilities of FPGAs so it is important to have a current view of this broad category of devices. Tracing the evolution of FPGAs from their early history to present day implementations help make it possible to project the likely future of FPGA architectures and tools and their applicability to SDR.

9.1.2.1. Origins

Prior to the existence of FPGAs, the commercial success of the gate array confirmed that high quality integrated circuits of random logic could be produced with lower design costs in shorter amounts of time by simple interconnects on an array of logic gates. It was a natural step to replace their final-metal-layer interconnect with a switchable array of buses to allow the interconnect to be programmable. Thus the term field programmable gate array (FPGA). Most of the benefits of high-density circuits were realized with little nonrecurring engineering cost, plus the device could be readily altered it if it was incorrectly designed, the common debugging experience. Only later did this flexibility begin to be used to reconfigure the system for different uses and to bring additional overall benefits.

9.1.2.2. From Homogeneity to Heterogeneity

Initial FPGA devices used a homogeneous array of programmable logic elements with conventional input/output circuitry. Early simple tools working from logic equations were replaced with ones that allowed for higher-level descriptions and the use of functional IP. As chip densities continued to increase, the additional silicon die area was used for implementing other semiconductor components, including memories and more complex I/O and peripheral functions. The latest moves to more heterogeneous architectures include whole microcontrollers, with their own bus system and peripherals, and very-high-bandwidth interface technology. Application demands should continue to enhance and tune the types and composition of architectural elements included in FPGAs.

9.1.2.3. Semiconductor Technology – from Trailing to Leading Edge

Early FPGAs were implemented on trailing edge semiconductor technology yielding rather small and slow devices. That has completely changed since the late 1990s, when FPGAs moved to leading-edge process technologies with corresponding speed, power and cost benefits. Historically, process technology has been driven by two major types of device – memories and processors. In the evolution of memories, however, the semiconductor process technology became so specialized that it was no longer suitable as a driver for all other types of device. FPGAs have filled the void. FPGAs fulfill the key technical requirements for driving process technology – they utilize standard CMOS processes, they are regular structures, and they are high volume products. FPGA manufacturing is so closely aligned to standard process technologies that they can be produced on the same manufacturing lines as processors, graphics chips and custom ASICs. At the same time, FPGAs are very scalable architectures. Devices can be implemented from a low of thousands of logic elements to many millions as process technology evolves. The additional engineering costs to design the high-end devices are minimal within the cost structure of new device families. While fewer and fewer applications can justify the development of application specific semiconductor solutions, FPGAs are intrinsically evolving to fill the gap.

9.1.2.4. The Evolutionary Trend

All this points to a strong evolutionary roadmap for FPGAs which includes high levels of integration, high processing performance, low power, and low cost, all in a programmable component suitable for custom system architecture implementation. Utilizing FPGAs, SDR systems can yield the best combination of power, flexibility, and performance with a continuing roadmap that further differentiates the value of an FPGA based system.

9.2. SDR Design Choices

Design choices in a cellular baseband SDR system are always uniquely concerned with long-term system flexibility. SDR systems require a working system today that can be enhanced to meet different requirements tomorrow. A key challenge in SDR is designing an implementation architecture that can meet the reconfigurability requirements along with cost, power and performance demands. Two main semiconductor components can be used to implement the reconfigurability required in SDR, processor-based components and FPGA-based components, and combinations thereof. In general, it is believed by the present author that combinations of these two technologies will be utilized to implement optimal SDR system architectures.

9.2.1. SDR Conceptual Architecture – Dynamic Functions

As has been discussed elsewhere [1], system reconfiguration may be implemented utilizing three different techniques:

- using parameterized radio (and protocol) modules;
- exchange of (a) single component(s) within a module, and
- exchange of complete radio modules or protocol layers.

9.2.1.1. Module Parameterization

In the first case, that of parameterized radio modules, the design of the modules must take into account all the permutations necessary to implement the system. This is feasible and is required within a narrow range of operation such as GSM-based standards or UMTS-based standards. Typically, wireless communications standards are defined with a variety of operational modes that must be dynamically supported in real time to take advantage of, or to compensate for, operating conditions such as data throughput demands, and physical system impairments such as multipath and user loading. However, parameterized modules become impractical as a method of supporting reconfiguration across complex standards. In the case of GSM and UMTS, for example, the physical layer communications are so radically different technologically and so complex in and of themselves that commercial implementations invariably implement them separately in software and hardware, and merge them at a higher level of hardware and software integration. Going further, it is reasonably impossible to rely on parameterized modules as a method of supporting undefined future radio standards.

9.2.1.2. Component Exchange

In the second case, exchanging a single component is useful where particular algorithms are nonoverlapping in implementation but are similar in functionality. This can be seen, for example, in forward error correction with Viterbi and turbo decoding, where either algorithm can be chosen within the same standard but where the physical implementation is significantly different. Texas Instruments, as a case in point, has two hardware coprocessors on their TIC6416 device that implement these two algorithms. In fact, each of these hardware coprocessors on the TI device has been implemented as a parameterized function to support a variety of requirements across different wireless systems. However, even these parameterized coprocessors are only capable of limited variation and may not support new features as wireless system standards evolve, as is the case with HSDPA in the 3GPP WCDMA standard.

9.2.1.3. Module Exchange

In the third case, the exchange of complete radio modules or protocol layers is reasonable where most of the system operation is significantly different and is better described with more than one unique implementation. Again, GSM and UMTS showcase an example of this, as some components are effectively identical or simply permutations of the other, such as the network signaling, whereas the physical layer is entirely different. In another case, GSM to cdma2000 requires a complete replacement of both the physical and network-layer implementations as there are insufficient similarities between the two standards to attempt to utilize the same soft implementation across both.

9.2.1.4. Accommodating the Options

Utilizing a combination of processors and FPGAs enables complete coverage of these three reconfiguration approaches. The processor can dynamically switch between major sections of software when switching between standards, while the FPGA can be completely reconfigured to implement an architecture customized for that particular standard. This turns out optimally to meet the design demands of SDR with radio configurations that can be independently developed, tested, and loaded onto radios.

The key questions for system architects are what is the best trade-off in partitioning between processors and FPGAs for their system requirements, and what portions of the system should be parameterized modules versus full replacement modules. Fortunately, the functional partitioning of SDR systems has received considerable attention with an emphasis on the definitions of control and data interfaces. As these standard interfaces mature, SDR system designers can develop various implementation architectures to optimize their equipment for the end-user environment. The use of processors in combination with FPGAs will help facilitate a freedom of architecture implementation that enables an optimal partitioning between software and programmable hardware.

9.2.2. Mapping to an Implementation Architecture – System Partitioning

Key considerations regarding system implementation include ease of design and support, performance, power, cost and reconfigurability.

In the architecture design process, the first step is to identify what needs to be reconfigurable and what can be fixed function. There will be fixed function components, certainly, that are part of the physical and electrical implementation without being truly part of the radio. These can include specific interconnect and backplane technology, power supplies and various physical interface support. Outside of this, it will be assumed all of the other electronics need reconfigurable capability. Although the analog and RF circuitry may need reconfigurability, it will be ignored in this chapter as FPGAs do not provide an alternative solution for these types of component.

As mentioned earlier, all other digital logic components, apart from stand-alone memory and storage devices, can be implemented in one of two leading component categories, namely processors and FPGAs. Both of these technologies provide extremely mature components and tools that are widely used in the general electronics industry.

The next step is to identify the different types of operation that are required and the most suitable type of processing for those operations. This logically divides into four major operational groups that we will continue to use throughout the architectural design choice process.

9.2.2.1. System Control and Configuration

System control and configuration are two related tasks in that they maintain and control the state of the system. System control is the dynamic operation within a wireless standard whereas the dynamic configuration is the change between wireless standards. These tasks are control flow intensive and thus require complex software implementations with little concentration of computational load. In SDR, these are the applications written on top of CORBA [2] and the SCA [2]. In general, system control and configuration will be performed by control processors running large C- or similar-based programs which must be memory efficient and maintainable by high-level C language tools. The system control and configuration will likely reside on control processors.

9.2.2.2. Signal Processing Data Path and Control

Signal processing data path and control tasks are typically where the bulk of the processing load resides. Example system modules are the physical layer communications in TDMA, CDMA, and OFDM based systems along with encryption and some of the networking functions.

The concentrated computational load in signal processing makes it amenable to parallel and cascaded data path elements. Where data path elements are utilized, the related active real-time control functionality may also demand dedicated control logic integrated with the data path processing. In systems where the signal processing does not consume significant processing capabilities, typically TDMA and FM systems, the signal processing and its related control can be implemented in software.

Depending on the wireless communication standard being supported, the processing demands can vary significantly, resulting in software-only systems or software-plus-hardware based systems. These systems will be implemented in a combination of digital signal processors and FPGA based dedicated logic architectures.

9.2.2.3. Memory

Active program memories must be large because of the complexity of link and communication protocols; these requirements are compounded for multichannel operation in basestations. Likewise, data memories are large because of the long interval processing requirement for coding, interleaving, adaptive equalization, error correction and echo cancellation. Inevitably the multimode configurability increases active memory requirements, even if mode changes come infrequently from nonvolatile memories. System memories follow the usual hierarchy of small, close, fast and static through moderate-sized, off-chip, slower and dynamic until finally the largest, slowest and nonvolatile memories which reside off-chip. Memories will not be considered further in this chapter except in relation to the description of FPGA device support for internal and external memory structures.

9.2.2.4. Input/Output

System I/O is significantly different for the basestation and handset. Handsets require low data-rate I/O with common peripheral interfaces, whilst the basestation demands special very high-bandwidth, highly prescribed, interfaces. In both cases the signal data rates are after frequency down conversion and they are digital signals. System I/O will not be considered further in this chapter except in relation to the description of FPGA device support for I/O standards.

9.2.2.5. Control and Signal Processing Partitioning

A typical wireless system maintains an orthogonal data flow between the system control and signal processing data flows as shown in Figure 9.1.

9.2.3. Software/Hardware Partitioning

Even in the straightforward system partitioning of Figure 9.1 there remains what is often the central design choice regarding configurability – what processing is done in hardware directly and what through the execution of program code. There are three broad choices for the processing hardware and we here summarize their relative strengths and weaknesses, considering the three major criteria of performance, power, and flexibility for SDR.

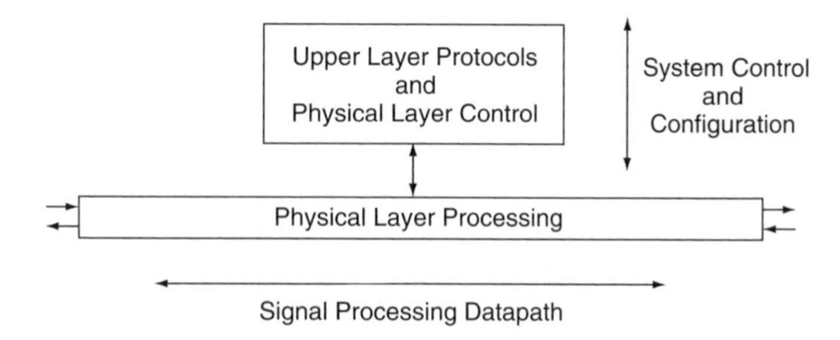

Figure 9.1 Orthogonal dataflows of system control and signal processing

9.2.3.1. Programmable Processors – Microcontrollers, DSPs and Soft Processors

The commercial success of general-purpose microcontrollers and programmable signal processors assures that a wide choice of hardware–software trade-off options are available to the baseband SDR designer. Key among these choices will be software availability, performance, and suitability to the given task. Among the tasks in SDR, the control layer requires support for CORBA and potentially Java. Generally, the control layer will be implemented on a microcontroller. DSP processors are far more suitable for the physical layer signal processing requirements in SDR. For these reasons, SDR systems will likely include a control processor and a DSP processor. This is not surprising, given that existing mobile phones all include both a control processor, usually the ARM, and a DSP processor.

Another alternative for the control processor is an FPGA-based soft processor [3, 4]. A soft processor differs from a traditional processor in that it is implemented directly onto the programmable FPGA fabric. Here is how soft processors compare with traditional hard processors:

Advantages

- Soft processors are far more flexible than hard processors. Soft processors can be enhanced with custom hardware to extend the instruction set with custom instructions and coprocessors targeted to the application using the processor. This can significantly enhance the performance of the soft processor in targeted applications.
- Designs utilizing soft processors can be migrated to the latest FPGA fabric and onto any device in an FPGA family. This portability increases design reuse and extends the life cycle of soft-processor based designs.
- Soft processors can be deeply embedded into the system architecture for integrated control of the hardware architecture.
- Designs can be implemented by utilizing more than one processor to increase overall system performance. While this is possible with hard processors, in SDR-based systems this can be changed from configuration to configuration of the modem.

Disadvantages

- Soft processors cannot reach the same clock rate as hard processors, generally resulting in lower performance than hard processors (unless custom instructions and coprocessors are used to enhance the soft processor).

- Soft processors do not have the power saving features available in many wireless-focused hard processors such as the ARM processor.
- Soft processors do not have the installed base of software available for mainstream processors like the ARM processor. In general, a soft processor will not have a standardized wireless protocol stack available from a third party vendor.

Soft processors provide a complementary technology to hard processors in SDR applications. In particular, in areas of the SDR system that changes between standards, soft processors can be used as necessary in each configuration with custom instructions, coprocessors, and number of instances, and interconnect optimally configured for each standard.

In many radio applications, the amount of processing performance required to implement the physical-layer processing significantly exceeds the capabilities of DSP processors. In these cases, specialized hardware architectures can be utilized to meet the performance requirements while minimizing cost, power, and the area of the system architecture. There are two key hardware implementation technologies available to implement the hardware portion of the system, application specific integrated circuits (ASICs) and FPGAs.

9.2.3.2. Application Specific Hardware (ASICs)

ASICs have been used for implementation of most digital wireless standards when DSP processing performance itself was insufficient. The characteristics of ASICs are high performance, low power, and low production cost (when produced in volume). These are all valuable characteristics for mobile phones. Unfortunately, there are two major weaknesses of ASICs – cost of development, and inflexibility of solution. The cost of developing ASICs is constantly growing, which further limits the use of ASICs to applications that demand the capabilities of an ASIC or which have huge volumes. However, for SDR, the main challenge with ASICs is the inflexibility of the solution. This limits the use of ASICs to systems that can rely on parameterized hardware or a few very well defined system standards. Parameterized hardware is not adequate for SDR systems, due to the high level of flexibility across multiple existing and undefined future radio standards that must support complete exchange of radio modules and protocols.

9.2.3.3. Field-Programmable Gate Arrays (FPGAs)

The FPGA has all of the computational benefits of an ASIC without its rigidity and high development cost. Figure 9.2 shows a generic composite of modern heterogeneous FPGA architectures in a representative SDR system like that in Figure 9.1.

The FPGA architecture consists of a fabric of logic elements (LEs) with programmable interconnect or configuration control, programmable I/O and high-performance memory. FPGAs are extremely valuable in SDR systems as a means of addressing the high processing demands of wide-band communication systems with the unique benefit of total reconfigurability.

9.2.3.4. Wireless System Architectures

Typical wireless system architectures will implement the system control and configuration and the signal processing datapath across three types of devices, as shown in Figure 9.2. The

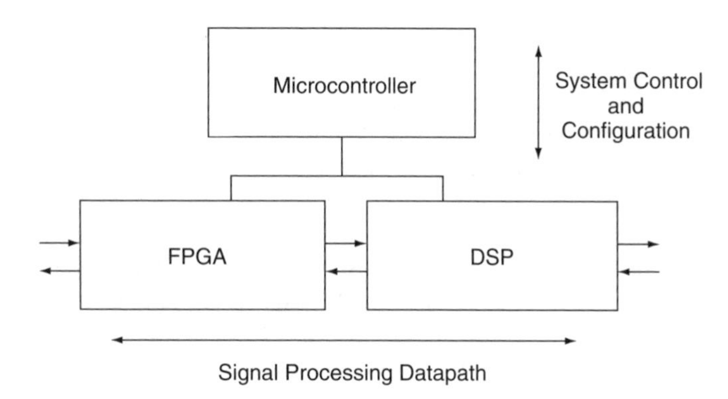

Figure 9.2 Typical wireless system topology

system microcontroller with its memory, peripherals, bus and I/O perform system control. The FPGA and programmable DSP with their memory, coprocessors and cascaded data path functions perform the high-rate data flow processing.

The partitioning between the FPGA and the DSP depends on the bandwidth of the system. For low-bandwidth systems, such as FM and voice TDMA, the DSP can handle all the processing. For wideband systems, the majority of the processing resides on the FPGA, with only some symbol rate processing and estimation processing, as well as high-rate physical layer control, residing on the DSP.

Communications between the three subsystems can be through shared memory or registers, and will be increasingly moving to high-speed switch interconnects supported by all devices. The relevant software operating systems, development tools and intellectual property thus stay with their natural hosts, as do the hardware I/O interfaces and peripherals.

In 3G systems, the signal processing performed by the FPGA had initially, in some cases, been implemented as ASIC implementations. For SDR, however, the demand for multiple standards has made the FPGA the long-term processing solution. It turns out that while commercial radio equipment had moved to ASICs, in many cases the trend is for this equipment to migrate back to FPGAs as the standards continue to evolve and the costs of designing and maintaining ASIC-based systems continues to rise exponentially while FPGA performance, cost, and power continue to improve with Moore's law.

9.3. The Future of FPGAs

9.3.1. What is an FPGA?

FPGAs can be considered in two related contexts. The first context is 'what makes an FPGA an FPGA?' to which the answer is the programmable logic elements and routing. The second context is as the product created by the FPGA semiconductor manufacturers. This second, somewhat wider, context alludes to the fact that a device from an FPGA manufacturer can include any semiconductor structures including memories, control and DSP processors, special purpose hardware and high performance I/O logic, as well as the basic FPGA fabric. Considered within this latter context, FPGAs can be seen to be heterogeneous semiconductor devices that can combine the best processing elements from all semiconductor components into a single heterogeneous processing device. While FPGA technology is 20 years old at the time this book was written, the implementation of heterogeneous components utilizing FPGA technology is in its infancy.

As has been discussed, SDR systems will require a combination of many of the different processing and memory structures available in semiconductor devices, and FPGA manufacturers are uniquely positioned to develop these heterogeneous architectures. The following sections discuss the structures available on existing FPGA architectures, potential system partitionings onto FPGAs plus external components, existing design flows to support such architectural partitionings, and a snapshot of Altera's FPGA product family.

9.3.2. Fabric of Logic Elements with General and DSP Arrays

The foundation technology for FPGA devices is a large array of programmable logic elements interconnected with programmable routing. Programmable logic elements, while relatively simple in nature, represent an area of continued innovation and enhancement. The basic logic element structure consists of a programmable lookup table and a register. In the Stratix device family, for example, the LE can be represented by Figures 9.3 and 9.4; these simple elements can be programmed and integrated from higher-level languages and tools to form nearly any digital system.

As FPGAs have evolved, other structures common in the semiconductor industry have been integrated with the FPGA fabric to leverage the best characteristics of a heterogeneous processing system. This evolution has been enabled by the huge capacity and performance provided by the latest high-speed and high-density fabrication processes. These other structures include memories, advanced I/Os, processors and DSP functions.

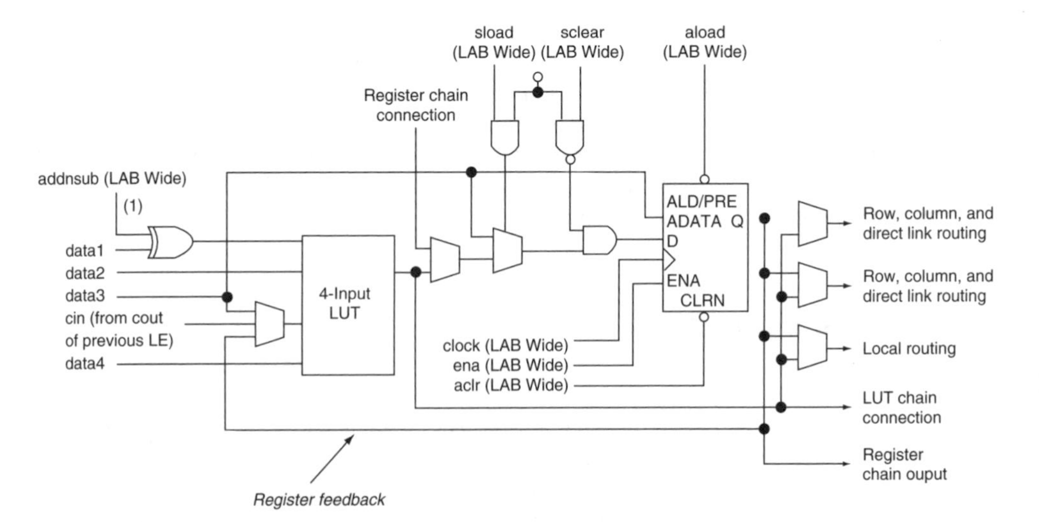

Figure 9.3 Logic element – normal mode

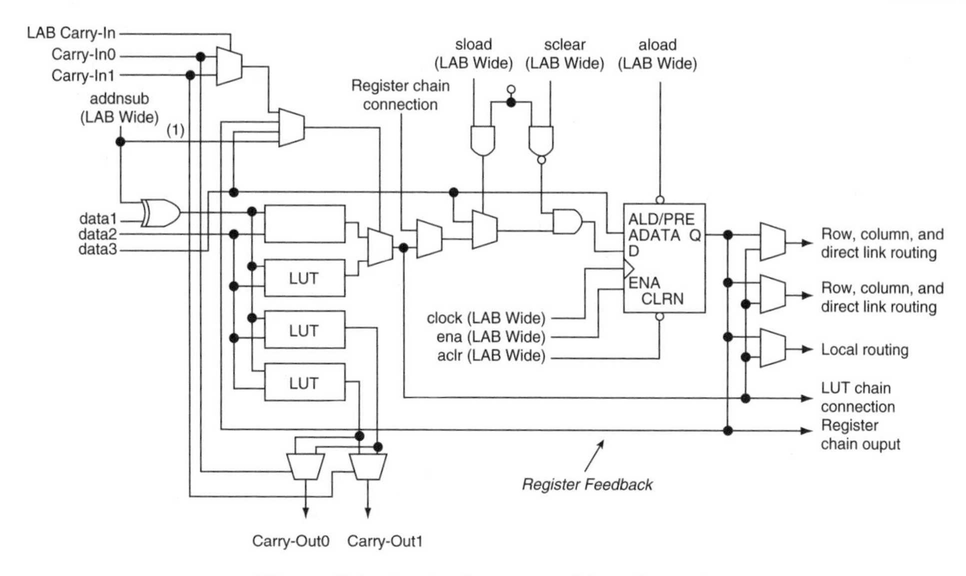

Figure 9.4 Logic element – arithmetic mode

9.3.2.1. Memory

FPGA memories were one of the first structures added amongst the logic element array. Today's FPGA memory architectures include several different sized memories arrayed across the devices. The small memories enable very high bandwidth per bit of storage whereas the large memories are very efficient, area wise, but with low bandwidth per memory bit of storage. Across all memories on an FPGA, aggregate memory bandwidth can exceed 8 terabits per second of throughput. While these memories are very high performance, they are also very flexible to utilize. In most cases, the memories are dual ported and support programmable word size versus depth ratios. FPGAs also support external memories of all sorts.

9.3.2.2. I/O Architectures

Unlike traditional microprocessors that have a variety of proprietary and standard fixed interfaces, FPGAs take software programmability all the way to the physical I/Os [5]. FPGAs have fully programmable I/O that can be programmed to implement a large variety of proprietary or commercial interface standards. These I/Os maintain a continued evolution that enables support for the latest high speed standards (e.g. dedicated serializers and de-serializers) as well as including circuitry that simplifies the board integration process (e.g. dynamic phase alignment) [6]. The programmability of these interfaces enables optimal selection of devices and system architectures across processing components of the SDR system.

9.3.2.3. Processors

A very interesting evolution in the development of FPGAs has been the inclusion of hard processors into the heterogeneous FPGA fabric. There are many implementation issues to consider when integrating hard processors within an FPGA fabric. Two options have been demonstrated in mainstream FPGAs, either a dedicated processor with all its peripherals in dedicated logic or a stand-alone processor core embedded within the FPGA. These permutations can also be built with one or more processors per FPGA die. The key benefit a hard processor can provide is a very tight integration between the processor and the related FPGA logic. Clearly this technology has potential benefits to SDR systems.

There are two key value propositions for having the processor with all its related peripherals implemented in dedicated logic [7, 8]. The first is the silicon efficiency of implementing the baseline peripherals as dedicated logic instead of consuming logic elements unnecessarily; this benefit brings lower power, lower cost and higher performance. The second benefit is that the processor can be booted without loading the FPGA fabric. In an SDR system, this enables the processor to continue operation while the FPGA is being reconfigured. Thus, the dedicated processor with peripherals can take on the role of the processing master in an FPGA based SDR system.

The alternative approach of embedding a processor into the FPGA fabric without its peripherals can be useful where tightly coupled proprietary processing logic needs to be integrated with the processor. It is not able, however, to act as the master processor in an SDR system unless there is no expectation that the FPGA will be reconfigured during operation.

To date, no existing FPGA devices that incorporate a hard processor have been optimized for low-power operation, such that the FPGA can be turned off or such that the processors can enter low-power modes. This presents an evolutionary direction that FPGAs could take that would enable low-power operation. A solution to this today is to utilize external low-power processors for system control (e.g. an ARM processor) and externally control the FPGA operation to minimize power utilization.

9.3.2.4. DSP Processing Blocks

Due to the prevalence of DSP applications and the ever-increasing need for signal processing performance, FPGAs have been augmented with optimized silicon multiply/accumulate (MAC) hardware. This specialized hardware has a low level of granularity, making it easy to infer from a language-based description in the FPGA hardware design flow. The DSP blocks provide a significant boost in MAC capability as compared to LE-based MAC hardware, thus lowering the overall cost of FPGAs for DSP processing. This is a major reason why FPGAs have become so cost effective for SDR systems that require substantial signal processing performance.

There are two major architectures targeting dedicated MAC processing. One technology utilizes a stand-alone hardware multiplier operator while the other, utilized by Altera, more closely resembles a complete ALU pipeline, with registered inputs, outputs and dedicated accumulator logic.

9.3.3. Three Representative FPGA System Partitionings

This section will look at three partitionings that are representative of where FPGAs can be used with particular advantage in baseband SDR systems.

9.3.3.1. Narrow-band Wireless Systems

The SDR system in Figure 9.5 shows a typical programmable narrow-band system where the FPGA is performing the high computational load filtering and digital down conversion that cannot be handled by a DSP. In narrow-band systems, the vast majority of the processing MIPS are consumed by the filtering operations. These types of operation can be computed an order of magnitude more efficiently on dedicated hardware coprocessors as compared to DSPs. The low-MIPS baseband processing can be performed on the DSP.

9.3.3.2. Wideband Wireless System

The SDR system in Figure 9.6 shows a typical programmable wideband system where the FPGA is performing the majority of the physical layer processing, leaving only symbol

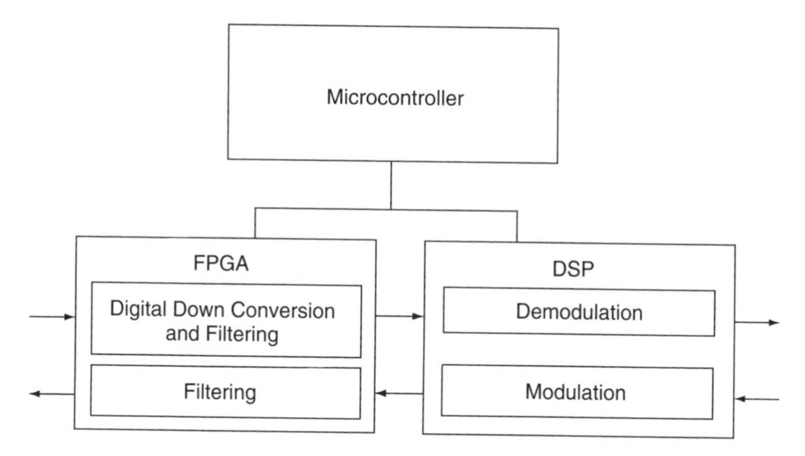

Figure 9.5 Narrow-band wireless system

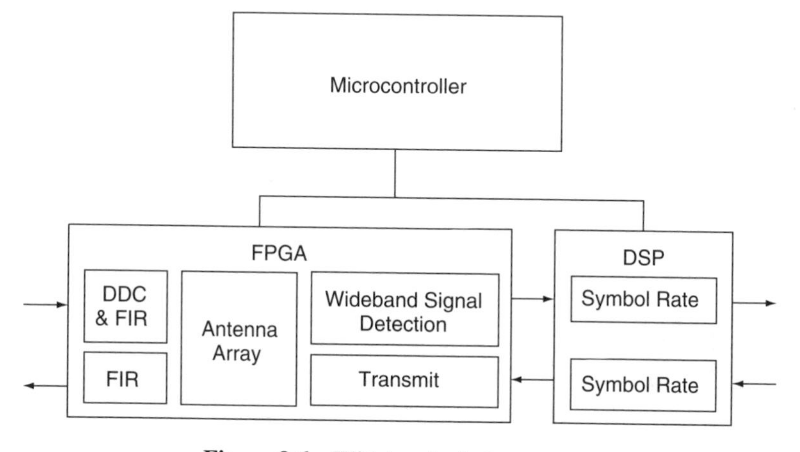

Figure 9.6 Wideband wireless system

processing to the DSP. The signal processing requirements of wideband systems can be delivered by FPGAs over 10-times more efficiently than is possible using a DSP. This means that where a single FPGA is required for processing *N* channels of a given system, a ratio of ten DSP processors or more would be required compared to a single FPGA. Thus utilizing the FPGA results in significant power, cost and area savings.

9.3.3.3. System-on-a-Chip (SoC)

With the rapid evolution of FPGAs, systems can be conceived and implemented that may utilize only FPGAs for the complete system implementation, as shown in Figure 9.7 – a system-on-a-chip solution. The controller function, all input/output, all fixed and programmed signal processing and all memories can be interconnected under program control for the optimum performance in the different operating modes.

This approach results in increased functional partitioning flexibility, because the control and DSP processing can be tightly coupled to the system architecture implemented on the FPGA. The data path processing can be quite advanced, and even under on-chip program control, by using a soft microcontroller core configured into the FPGA.

The challenge for SDR is that the FPGA configuration needs to be changed dynamically and under software control. This is where the on-board ARM processor with its dedicated peripherals becomes valuable, because the ARM can reconfigure the FPGA without being affected itself by the reconfiguration process.

The practical proof of the capability of using FPGAs in these three basic configurations is illustrated in the next section, where we consider the range of architectural solutions developed by Altera Corporation within four types of FPGA family.

9.4. Architectural Solutions

Altera architectures can address all of the baseband SDR system partitioning options shown in Figures 9.5 through 9.7. To illustrate this we will use the composite Altera FPGA architecture in

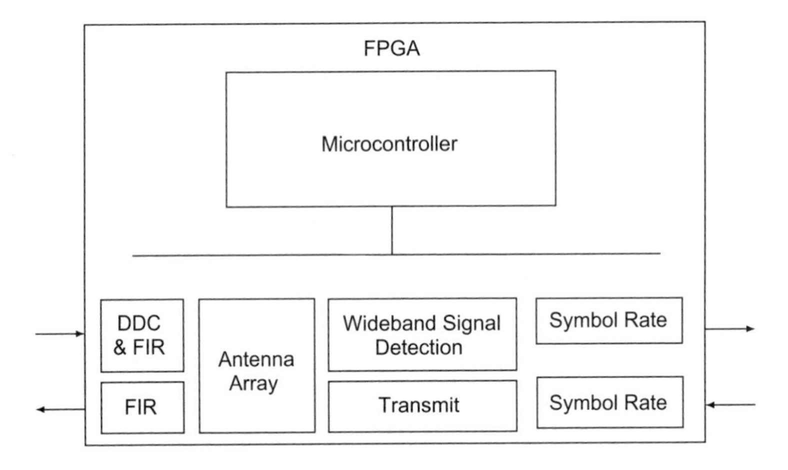

Figure 9.7 A fully FPGA-implemented SDR system

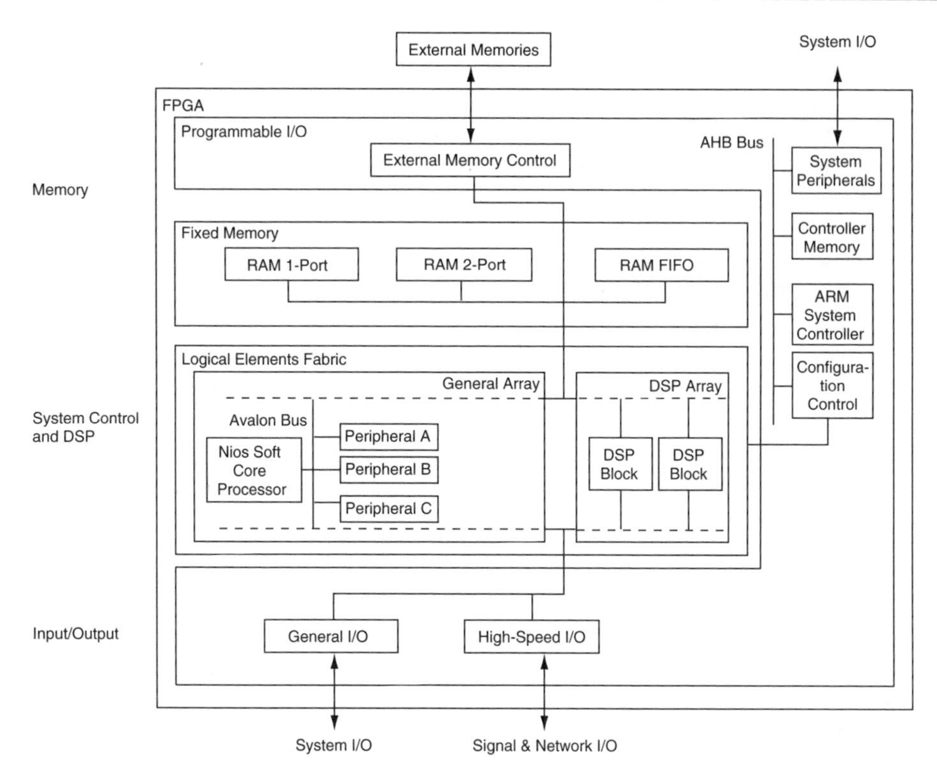

Figure 9.8 Composite system-on-a-programmable-chip (SoPC) architecture

Figure 9.8. This illustrates a complete system-on-a-programmable-chip (SoPC) solution; portions of this may be used separately for the partial FPGA uses, as we will describe.

The fabric of logic elements has both general arrays and DSP arrays with shared interconnect to memories and input/output circuits. The general array is shown with the Nios soft core microcontroller and its Avalon bus, with three user-defined peripherals that can include I/O. The DSP array is shown with two representative DSP algorithm blocks. Both arrays have programmable interconnect, determined by the configuration control, shown connected to the hard core ARM922T microcontroller used when that function is on-chip.

The input/output functions in the fixed I/O portion of the FPGA include programmable high- and low-speed interfaces, an external DRAM controller and the ARM922T with its AHB bus and standard peripherals and I/O.

In addition to possible off-chip DRAM, the TriMatrix on-chip RAM is shown with three representative uses of its modules for single-port, dual-port, and FIFO buffer configurations. Individual devices, design tools, and intellectual property used in the partitioning and design flows, are described in the later sections.

9.4.1. The Baseband SDR Design Process

It is important to look at the SDR system design process with FPGAs because it is there that the real issues of the configurability requirements are uniquely well met. The adequacy and ease-of-use of the design process are crucial to the success of any proposed SDR technology – a power-efficient, optimally flexible architecture is of no value if its design flow is too complex or unwieldy or if there exists no existing pool of trained engineers familiar with the design process concepts and requirements.

We will use the form in Figures 9.9 and 9.10 to illustrate the design process for two representative system partitionings. At the top are the SDR functions as they are expressed in the software domains and hardware units. In the center are the design tools and at the bottom are two SDR system partitionings. The partitioning on the left uses discrete DSPs, microcontrollers, and FPGAs. The partitioning on the right places all processing onto the FPGA.

The design flow in Figure 9.9 is for the discrete component example where the FPGA is used for the majority of the data-flow processing but not the low-rate signal processing or the system control similar to that shown previously in Figure 9.5. The microcontroller and the DSP each have their own software functions in the form of code and pre-defined hardware units for the items shown at the top, along with their own programming environment, including mostly communications IP, for the microcontroller and DSP signal processing functions.

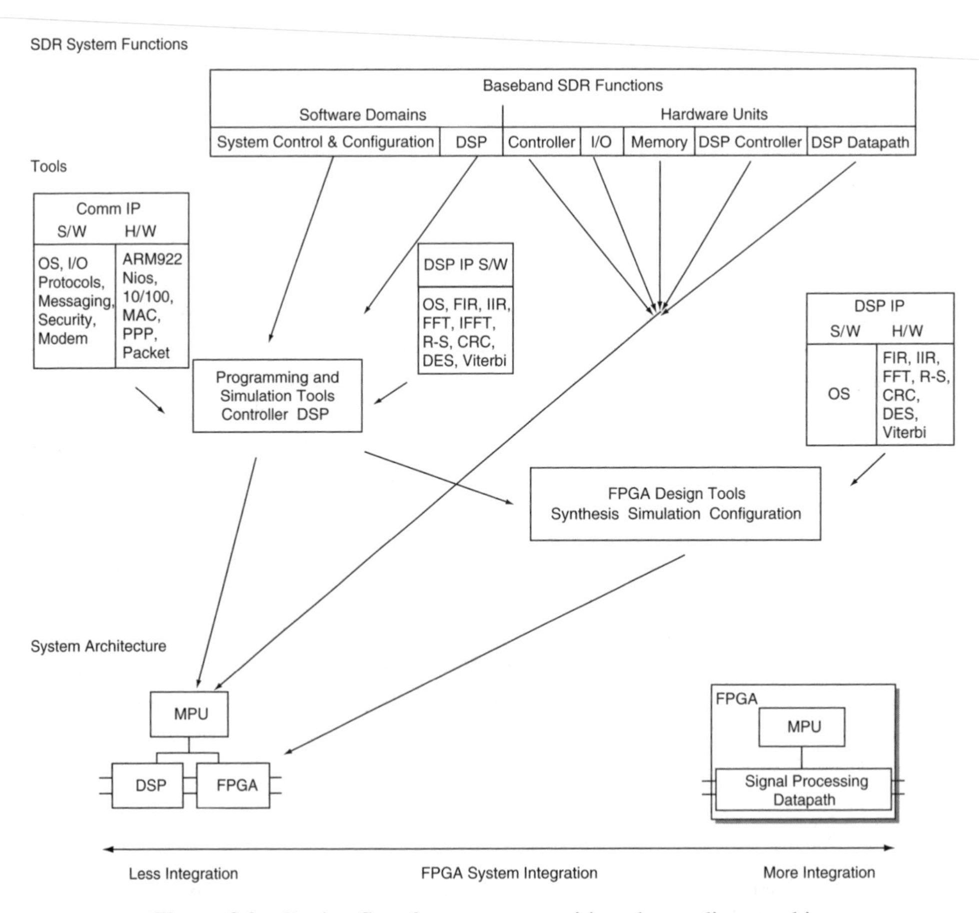

Figure 9.9 Design flow for a system partitioned onto discrete chips

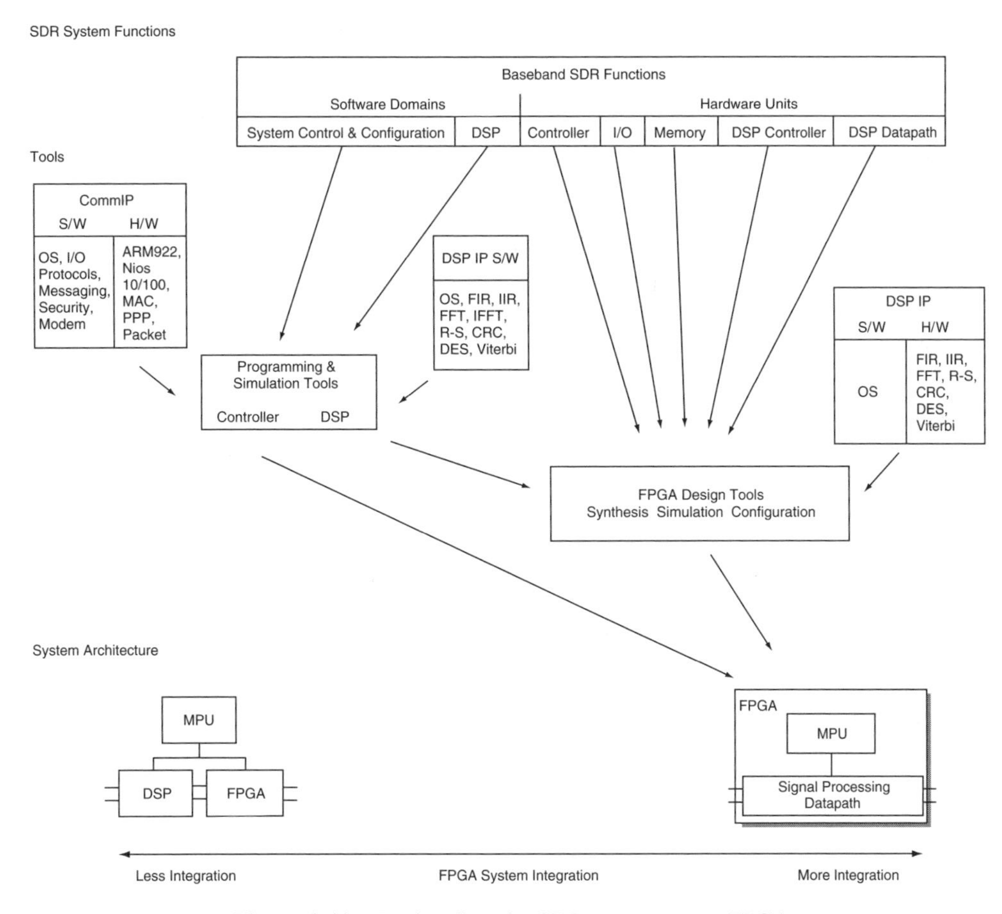

Figure 9.10 Design flow for SDR system on an FPGA

The natural partitioning to an FPGA coprocessor occurs when greater processing bandwidth in the different modes is required for functions in the DSP data path. The hardware design can often come from Altera's MegaCore DSP IP Library for the various functions; some typical functions are shown on the right of the figure. Of even greater value is the possibility of synthesizing the dataflow architecture by the FPGA tools directly from the system DSP simulation. As shown, this path exists in the Altera toolset by providing a direct HDL link for synthesis from the MATLAB DSP system simulation. This makes it easy to identify the most compute-intensive operations from the completed simulation code and to off-load them to the FPGA. The operations, of course, may vary over time with multimode SDR operation and these are easily downloaded into the FPGA in real-time as needed. In the composite FPGA of Figure 9.8 a subsystem would likely use the LE fabric, both general and DSP, the fixed input/output, fixed memories and possibly the hard ARM922T controller or the Nios configurable soft core.

Figure 9.10 illustrates the design flow for a FPGA system-on-a-programmable-chip that includes all system control, DSP, memory and I/O functional units as shown earlier in the third partitioning example of Figure 9.7.

In this case, all of the baseband SDR system resides on a single low-power, economical FPGA without any loss of design or operational flexibility. Now the system controller can move on-chip as a hard core with its standard peripherals fully supported with its tools, operating system and communications software IP as before. All compute-intensive processing can be configured and then reconfigured for each mode with optimum interconnect and operation with memories and I/O. Again, as different modes require different network interfaces, the high-speed I/O can be reconfigured.

In the composite FPGA of Figure 9.8 a complete system-on-a-programmable-chip would use the LE fabric, both general and DSP, all fixed input/output, all internal TriMatrix memories and external DRAMs, the hard core ARM922T controller with peripherals on the AHB, and possibly the Nios configurable soft core with its selected peripherals along its Avalon bus. Not incidentally, there are multiple PLL clock oscillators for the variety of different clock domains that must be maintained in such a system.

9.4.2. Device Families

We have looked at a composite FPGA device in Figure 9.8 for our different architectural mappings of the baseband SDR functions onto FPGA devices. Table 9.1 below provides a succinct summary of the specific Altera device families that are popular at the time of the writing of this book. The table illustrates the breadth of choices that were available at one particular snapshot in time and exemplifying the second of the contextual definitions of 'what is an FPGA?' given at the start of Section 9.3. A more complete description of each family follows later in the chapter.

Table 9.1 Functional features of Altera FPGA families suitable for SDR

Product family	Logic elements	Microcontroller	DSP Array	Memory	Bus	Input/ output
Stratix	10–79 K	Nios – Soft Core	Six 22 Blocks and four 176 Multipliers	TriMatrix: 1–7 Mb RAM single and dual port	Avalon	General and external memory
Stratix GX	10–40 K	Nios – Soft Core	Six 14 Blocks and 48 112 Multipliers	TriMatrix: 0.9–3.4 Mb RAM single and dual port	Avalon	General, high-speed and external memory
Excalibur	4–38 K	ARM922 with I/O and peripherals – Hard Core Nios – Soft Core	–	32–256 KB single port, 16–128 KB dual port	AHB	General and external memory
Cyclone	3–20 K	Nios – Soft Core	–	60–294 Kb RAM	Avalon	General, PCI and external memory

9.4.3. *Intellectual Property Availability*

As mentioned earlier, ease-of-design is critical to the success of a technology. In this respect the IP libraries assembled over time and made available by the FPGA manufacturers to designers have been a major factor in facilitating their take-up for SDR (and other) applications. The extremely wide range of IP available today, some of which was shown being used in the design processes in Figures 9.9 and 9.10, is summarized below. From the perspective of avoiding the need to reinvent the wheel, the availability of such libraries from FPGA manufacturers is of huge practical benefit to the product designer.

9.4.3.1. DSP IP Library

Altera's extensive DSP IP portfolio consists of proven, high-performance, standard algorithms and functions created to help engineers meet today's rapidly evolving technologies. Each MegaCore and AMPP function has been rigorously tested and meets the exacting requirements of various industry standards. The DSP portfolio includes all the blocks of intellectual property required to build system-on-a-programmable-chip (SoPC) solutions. In addition, because these functions are reusable, each function may be instantiated multiple times in different designs.

DSP algorithms can be integrated easily as hardware accelerators for a Nios processor or into the data path as a pre- or post-processor to implement computationally rigorous routines, leaving the digital signal processor at the center of the original design. Adding a high-performance FPGA to address bottlenecks at both ends of a process allows DSP software engineers to leverage existing software code while enjoying the benefits of hardware acceleration.

Example DSP functions included within the comprehensive IP library include:

Color space converter
Constellation mapper/demapper
Correlator
Cyclic redundancy checker
DES encryption processor
FFT/IFFT
FIR compiler
Hadamard transform processor
High-speed Rijndael encryption and decryption algorithms
IIR filter compiler
Low-speed Rijndael encryption/decryption processors
MD5A hash function
Numerically controlled oscillator compiler
Reed–Solomon decoder minicore
Reed–Solomon compiler, decoder
Reed–Solomon compiler, encoder
SHA-1 hash function
Symbol interleaver/deinterleaver
Turbo decoder function
Turbo encoder function
Viterbi compiler, high-speed parallel decoder
Viterbi compiler, low-speed/hybrid serial decoder
CCIR-656 decoder
CCIR-656 encoder
DVB FEC codec
FFT/IFFT high performance 64-point
Multi-standard ADPCM encoder/decoder
Reed–Solomon decoder
Reed–Solomon encoder
Viterbi decoder
2D DCT IDCT
Fast black and white JPEG decoder
Fast color JPEG decoder
AES cryptoprocessor
C32025 digital signal processor
Color space converter
DES (X_DES) digital encryption function
Forward discrete cosine transfer (DCT)

JPEG fast decoder
JPEG fast encoder
Digital PLL synthesizer
Up converter
High-speed AES encryption/decryption cores (RJDCEO/RJDCDO/RJDCED)
RSDECD1 – Reed–Solomon decoder
RSENCD1 – Reed–Solomon encoder
Telephony tone generation
Binary pattern correlator
Complex multiplier/mixer
Digital modulator
Linear feedback shift register
Numerically controlled oscillator
StarFabric link
TC1000 turbo decoder
TC3000 turbo product code decoder
TC3401 low complexity turbo product code decoder
TC3404 very high speed turbo product code decoder
CORDPOL (CORDIC)
Digital down converter reference design
Direct sequence spread spectrum reference design
Logarithm function
QPSK reference design
u-Law and A-Law companders

9.4.3.2. Communications IP Library

A similarly comprehensive range of functionality is included within the communications library:

8B10B encoder/decoder
ATM cell processor compiler
Common switch interface (CSIX-L1)
E3 mapper
POS-PHY level 2 Link
POS-PHY level 2 PHY
POS-PHY level 3 Link
POS-PHY level 3 PHY
POS-PHY level 4
PPP packet processor 622 Mbps
SONET/SDH compiler
T3 framer
T3 mapper
UTOPIA level 2 master
UTOPIA level 2 slave
UTOPIA level 3 master
UTOPIA level 3 slave
UTOPIA L2 master receiver
UTOPIA L2 slave transmitter/receiver
UTOPIA L3 slave transmitter/receiver
ATM deformatter
ATM formatter
T1 deframer
T1 framer
SDLC controller
802.11 MAC
InfiniBand link layer
ATM AAL5 segmentation and reassembly
ATM cell delineator
BERT – bit error rate tester
Data encoder/decoder
General framing procedure (GFP) controller
HDLC, bit-oriented
Packet over SONET controller
HDLC single channel with FIFO buffers
T1 deframer
T1/E1 framer
FlexBUS-3 link layer with FIFOs V1.0
Inverse multiplexing for ATM (IMA) version 1.0/1.1
Multi channel HDLC
SPI-4 phase 1 with FIFOs v1.0 (FlexBUS-4)
Single channel HDLC
10 gigabit Ethernet MAC
10 gigabit Ethernet physical coding sublayer (PCS)
10 gigabit fibre channel FC-1 core
10/100/1000 Mbps full duplex Ethernet MAC
BOOST lite bluetooth baseband core
Cell delineation ATM

T3 framer
UTOPIA L2 master receive
UTOPIA L2 master transmit
UTOPIA L2 slave receive
UTOPIA L2 slave transmit
CSIX interface
UTOPIA level 2 to level 3 MUX
UTOPIA level 3 to level 2 DEMUX

9.5. Design Flow and Tools

9.5.1. Importance of Tools

The flexibility on the FPGA chip must translate into useful flexibility in the design space of the system engineer or else it will remain under utilized. That is the purpose of the design tools.

9.5.2. System-Level Design Products with Quartus II

The Quartus II design software is at present the only programmable logic software to provide a comprehensive framework that advances programmable logic to a true system-level design solution. Designers using these system-level design tools can build and analyze systems at a high level of abstraction. There are two system-level design tools that integrate fully with Quartus II – SoPC Builder and DSP Builder.

9.5.2.1. SoPC Builder

SoPC Builder [9] is an industry leading hardware/software IP architecture design and integration tool. SoPC Builder automates the task of adding, parameterizing and linking intellectual property cores, including multiple embedded processors, for system-on-a-programmable-chip (SoPC) applications. The user defines the switch architecture component interconnect matrix to maximize system performance and then generates the system. SoPC Builder automatically assembles all the hardware components using the high-performance Avalon switch bus architecture. In addition, a .h header file is automatically created that embodies a software view of the entire system with register and memory maps, as well as pre-defined software routines to control all the IP. SoPC Builder enables designers to turn their concepts into working systems in minutes by eliminating the engineering time required for system integration tasks. Figure 9.11 illustrates the SoPC Builder design flow.

9.5.2.2. DSP Builder

DSP Builder is DSP design software that acts as an interface between The MathWorks' industry-leading, system-level digital signal processing (DSP) tool Simulink and the Quartus II design software. DSP Builder facilitates a seamless design flow by which designers can perform algorithmic DSP design in the MATLAB software and system integration in the Simulink software, and then port the design to hardware description language (HDL) files for use in the Quartus II design software. DSP Builder also produces HDL test bench files that can be used in ModelSim – Altera and other third-party HDL simulators. The design flow for the MathWorks/DSP Builder tools is illustrated in Figure 9.12.

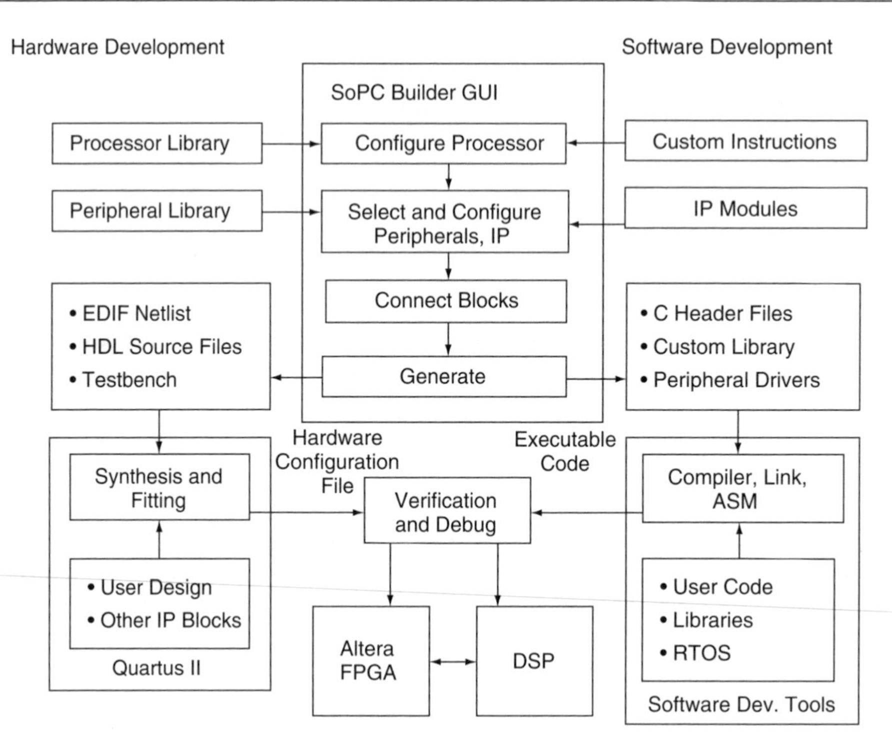

Figure 9.11 SoPC builder design flow

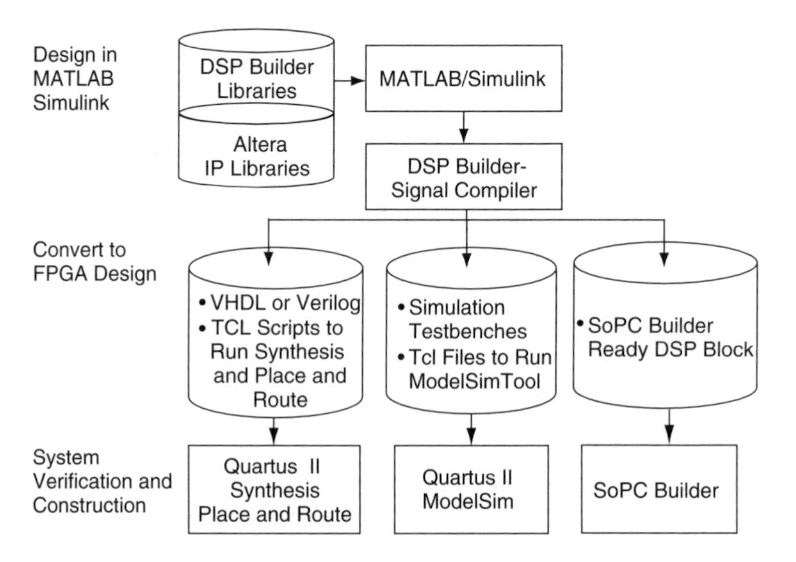

Figure 9.12 The Mathworks/DSP Builder design flow

DSP Builder can be used to build stand-alone systems or it can generate components for further integration by SoPC Builder. As such, DSP Builder is an integral tool for developing components for use as FPGA coprocessors by the Nios CPU or by external DSP or control CPUs.

9.5.3. Quartus II Design Software

The Quartus II software provides a complete flow for creating high-performance system-on-a-programmable-chip (SoPC) designs. The Quartus II software integrates design, synthesis, place-and-route, and verification into a seamless environment, including interfaces to third-party EDA tools.

9.5.3.1. LogicLock Block-Based Design

LogicLock block-based design is a new design methodology, available exclusively through the Quartus II software, which provides the capability to design and implement each functional module independently. Designers can integrate each module into a top-level project while preserving the performance of each module during integration. The software allows block-based design methodologies as a standard feature, helping to increase designer productivity and shorten design and verification cycles. This is possible because each module is optimized only once. For more information, see Ref. [10].

9.5.3.2. Timing Closure Flow

Powerful timing closure flow features allow design optimization. Timing closure flow features include:

- timing closure floorplan;
- cross-probing between the timing analyzer and timing closure floorplan editor;
- path-based assignments;
- netlist optimizations to tune designs for peak performance;
- incremental, block-based placement.

9.5.3.3. Multiple Design Entry Methods

The software supports both VHDL and Verilog hardware description language (HDL) text and graphical-based, design entry methods. It also allows the two methods to be utilized in combination within the same project. Using the block design editor, top-level design information can be edited in graphical format and converted to VHDL or Verilog for use in third-party synthesis and simulation flows.

The MegaWizard plug-in manager allows easy incorporation of pre-existing intellectual property from the available design libraries. It permits parameterization and instantiation of a library of parameterized modules (LPM) functions and of Altera or AMPP partner IP megafunctions in text- or graphical-based designs to reduce design entry time and to improve design performance.

9.5.3.4. Quartus II Synthesis

The Quartus II design software includes integrated VHDL and Verilog hardware description language (HDL) synthesis technology and NativeLink integration to third-party synthesis software from industry-leaders including Mentor Graphics, Synopsys, and Synplicity.

9.6. Representative Devices

In this section we describe four popular FPGA families current at the time of the writing of this book. Whilst specifics may date, given the rapid progress of FPGAs, none the less this section serves an important purpose in exemplifying the specifics of different configurations to readers who are less familiar with modern FPGA devices.

9.6.1. Stratix

9.6.1.1. Summary

The Stratix family of FPGAs (Table 9.2) [11] is based on a 1.5-V, 0.13-μm, all-layer copper SRAM process, with densities up to 79 040 logic elements (LEs) and up to 7.4 Mbits of RAM. Stratix devices may incorporate up to 22 digital signal processing (DSP) blocks with up to 176 (9-bit×9-bit) embedded multipliers, optimized for DSP applications that enable efficient implementation of high-performance filters and multipliers. The devices support various I/O standards and offer a complete clock management solution using a hierarchical clock structure with up to 420-MHz performance and up to 12 phase-locked loops (PLLs).

Table 9.2 Stratix family device summary

Feature	EP1S10	EP1S20	EP1S25	EP1S30	EP1S40	EP1S60	EP1S80
Logic elements	10 570	18 460	25 660	32 470	41 250	57 120	79 040
M512 RAM blocks (32 × 18 bits)	94	194	224	295	384	574	767
M4K RAM blocks (128 × 36 bits)	60	82	138	171	183	292	364
M-RAM blocks (4K × 144 bits)	1	2	2	4	4	6	9
Total RAM bits	920 448	1 669 248	1 944 576	3 317 184	3 423 744	5 215 104	7 427 520
DSP blocks	6	10	10	12	14	18	22
Embedded multipliers	48	80	80	96	112	144	176
PLLs	6	6	6	10	12	12	12
Maximum user I/O pins	426	586	706	726	822	1022	1238

9.6.1.2. Functional Description

Structure. Stratix devices utilize a two-dimensional row- and column-based architecture to implement custom logic. A series of column and row interconnects of varying length and speed provides signal interconnects between logic array blocks (LABs), memory block structures, and DSP blocks. The logic array consists of LABs, with ten Logic Elements (LEs) in each LAB. LABs are grouped into rows and columns across the device.

Memory. M512 RAM blocks are simple dual-port memory blocks with 512 bits plus parity (576 bits). These blocks provide dedicated simple dual-port or single-port memory up to 18-bits wide at up to 318 MHz. M512 blocks are grouped into columns across the device in between certain LABs. M4K RAM blocks are true dual-port memory blocks with 4 kbits plus parity (4608 bits). These blocks provide dedicated true dual-port, simple dual-port or single-port memory up to 36-bits wide at up to 291 MHz. These blocks are grouped into columns across the device in between certain LABs. M-RAM blocks are true dual-port memory blocks with 512 kbits plus parity (589 824 bits). These blocks provide dedicated true dual-port, simple dual-port or single-port memory up to 144-bits wide at up to 269 MHz. Several M-RAM blocks are located individually or in pairs within the device's logic array.

DSP. Digital signal processing (DSP) blocks can implement up to either eight full-precision 9×9-bit multipliers, four full-precision 18×18-bit multipliers, or one full-precision 36×36-bit multiplier with add or subtract features. These blocks also contain 18-bit input shift registers for digital signal processing applications, including FIR and infinite impulse response (IIR) filters. DSP blocks are grouped into two columns in each device.

Input/Output. Each device I/O pin is fed by an I/O element (IOE) located at the end of LAB rows and columns around the periphery of the device. I/O pins support numerous single-ended and differential I/O standards. Each IOE contains a bidirectional I/O buffer and six registers for registering input, output and output-enable signals. When used with dedicated clocks, these registers provide exceptional performance and interface support with external memory devices such as DDR SDRAM, FCRAM, ZBT, and QDR SRAM devices. High-speed serial interface channels support transfers at up to 840 Mbps using LVDS, LVPECL, 3.3V PCML, or HyperTransport technology I/O standards.

9.6.2. Stratix GX

9.6.2.1. Summary

The Stratix GX family of devices (Table 9.3) [12] is Altera's second FPGA family to combine high-speed serial transceivers with a scalable, high-performance logic array. Stratix GX devices include four to 20 high-speed transceiver channels, each incorporating clock data recovery (CDR) technology and embedded SERDES† capability at data rates of up to 3.125 gigabits per second (Gbps). These transceivers are grouped in integrated, four-channel blocks

† SERDES or serializers/de-serializers are components that take high clock rate serial data streams and convert them to lower clock rate parallel data streams and vice versa.

Table 9.3 Stratix GX family device summary

Feature	EP1SGX10C EP1SGX10D	EP1SGX25C EP1SGX25D EP1SGX25F	EP1SGX40D EP1SGX40G
Logic elements	10 570	25 660	41 250
Transceiver channels	4, 8	4, 8, 16	8, 20
Source-synchronous channels	22	39	45
M512 RAM blocks (32 × 18 bits)	94	224	384
M4K RAM blocks (128 × 36 bits)	60	138	183
M-RAM blocks (4K × 144 bits)	1	2	4
Total RAM bits	920 448	1 944 576	3 423 744
Digital signal processing blocks	6	10	14
Embedded multipliers	48	80	112
PLLs	4	4	8
Maximum user I/O pins	330	542	544

and are designed for low power consumption and small die size. The Stratix GX technology is built upon the Stratix architecture and uses a 1.5 V logic array. Its scalable, high-performance architecture is ideally suited for high-speed backplane interface, chip-to-chip, and communications protocol-bridging applications.

9.6.2.2. Functional Description

The Stratix GX device family supports high-speed serial transceiver blocks with CDR circuitry as well as source-synchronous interfaces. The channels on the right side of the device use an embedded circuit dedicated for receiving and transmitting high-speed serial data streams to and from the system board. These channels are clustered in a four-channel serial transceiver building block and deliver high-speed bi-directional point-to-point data transmissions to provide up to 3.125 Gbps of full-duplex data transmission per channel. The channels on the left side of the device support source-synchronous data transfers at up to 1 Gbps using LVDS, LVPECL, 3.3-V PCML or HyperTransport technology I/O standards.

Structure. Stratix GX devices contain a two-dimensional row- and column-based architecture to implement custom logic. A series of column and row interconnects of varying length and speed provides signal interconnects between logic array blocks (LABs), memory block structures and DSP blocks. The logic array consists of LABs, grouped into rows and columns across the device, with ten LEs in each LAB.

Memory. M512 RAM blocks are simple dual-port memory blocks with 512 bits plus parity (576 bits). These blocks provide dedicated simple dual-port or single-port memory up to 18-bits wide at up to 312 MHz. M512 blocks are grouped into columns across the device in between certain LABs. M4K RAM blocks are true dual-port memory blocks with 4 kbits plus parity (4608 bits). These blocks provide dedicated true dual-port, simple dual-port, or single-port memory up to 36-bits wide at up to 312 MHz. These blocks are grouped into columns across the device in between certain LABs. M-RAM blocks are true dual-port memory blocks with 512 kbits plus parity (589 824 bits). These blocks provide dedicated true dual-port, simple dual-port or single-port memory up to 144-bits wide at up to 300 MHz. Several M-RAM blocks are located individually or in pairs within the device's logic array.

DSP. DSP blocks can implement up to either eight full-precision 9×9-bit multipliers, four full-precision 18×18-bit multipliers, or one full-precision 36×36-bit multiplier with add or subtract features. These blocks also contain 18-bit input shift registers for digital signal processing applications, including FIR and infinite impulse response (IIR) filters. DSP blocks are grouped into two columns in each device.

Input/Output. Each device I/O pin is fed by an I/O element (IOE) located at the end of LAB rows and columns around the periphery of the device. I/O pins support numerous single-ended and differential I/O standards. Each IOE contains a bi-directional I/O buffer and six registers for registering input, output, and output-enable signals. When used with dedicated clocks, these registers provide exceptional performance and interface support with external memory devices such as DDR SDRAM, FCRAM, ZBT, and QDR SRAM devices.

9.6.3. Excalibur

9.6.3.1. Summary

The Excalibur family of FPGAs (Table 9.4) [13] has a heterogeneous system architecture (programmable logic arrays, embedded RISC processor with bus structure, on-chip memory, and peripherals) designed to combine the performance advantages of ASIC integration with the flexibility and time-to-market advantages of FPGAs.

9.6.3.2. Functional Description

Embedded RISC Processor. The ARM922T is a member of the ARM9 family of RISC processor cores. Its Harvard architecture, implemented using a five-stage pipeline, allows single clock-cycle instruction operation through simultaneous fetch, decode, execute, memory, and write stages. Independent of the logic array configuration, the embedded processor can undertake the following activities: boot from external memory, execute embedded software, communicate with the external world, run a real-time operating system, run interactive embedded software debugging sessions, configure/reconfigure the logic array and detect errors, and restart/reboot/reconfigure the entire system as necessary.

Table 9.4 Excalibur family device summary

Feature	EPXA1	EPXA4	EPXA10
RISC processor	ARM922T	ARM922T	ARM922T
Maximum operating frequency	200 MHz	200 MHz	200 MHz
Single-port SRAM	32 kbytes	128 kbytes	256 kbytes
Dual-port SRAM	16 kbytes	64 kbytes	128 kbytes
Typical gate count	100 000	400 000	1 000 000
Logic elements	4160	16 640	38 400
Embedded system blocks	26	104	160
Maximum system gates	263 000	1 052 000	1 772 000
Maximum user I/Os	246	488	711
UART, timer, watchdog timer	Yes	Yes	Yes
JTAG debug module	Yes	Yes	Yes
Embedded trace module		Yes	Yes
General purpose I/O port	4 bits	8 bits	–
Low-power PLL	Yes	–	–

The Logic Array and Extension Support. The logic array can be configured to implement various extensions:

- additional soft-core peripherals such as a UART, Ethernet, MAC, CAN controllers, PCI, or any other IP core;
- peripherals that are AHB bus masters and ones that are slaves, controlled by the embedded processor;
- on-chip and off-chip memories, including peripherals that exchange data using the on-chip dual-port RAM;
- high speed data paths under embedded processor control, and
- multi-processor systems, using multiple Nios embedded processors.

Additional embedded processor interrupt sources and controls allow FPGA designers to use the full range of available intellectual property functions to implement complex system-on-a-programmable-chip (SoPC) designs in minimal time but with maximum customization. The bi-directional bridges and dual-port memory interfaces between the embedded stripe and the logic array are synchronous to the clock domain that drives them; however, the embedded processor domain and the logic array domains are asynchronous. The clock domain for each side of the interfaces can be optimized for performance. The bi-directional bridges handle the resynchronization across the domains and are capable of supporting 32-bit data accesses to the entire 4-Gbyte address range (32-bit address bus). The SDRAM memory controller PLL allows users to tune the frequency of the system clock to the speed of the external memory implemented in their systems.

Embedded Memory. The embedded stripe contains both single-port and dual-port SRAM. There are two blocks of single-port SRAM; both are accessible to the AHB masters via an arbitrated interface within memory. Each block is independently arbitrated, allowing one block to be accessed by one bus master while the other block is accessed by the other bus master. Up to 256 kbytes of single-port SRAM are available, as two blocks of 2×128 kbytes. Each single-port SRAM block is byte addressable. Byte, half-word and word accesses are allowed and are enabled by the slave interface. In addition, there are either one or two blocks of dual-port SRAM in the embedded stripe, depending on the device type. The outputs of the dual-port memories can be registered. One of the ports gives dedicated access to the logic array; the other port can be configured for access by AHB masters or by the logic array. The width of the data port to the logic array is configurable as ×8, ×16 or ×32 bits. For the larger devices, the dual-port SRAM blocks can be combined to form a ×64-bit data-width interface. This allows the designer to build deeper and wider memories and multiplex the data outputs within the stripe.

Memory Control. The Excalibur family provides two embedded memory controllers that can be accessed by any of the bus masters: one for external SDRAM and a second for external flash memory or SRAM. The SDRAM memory controller supports the following commonly available memory standards, without the addition of any logic: single-data rate (SDR) 133-MHz data rates, or double-data rate (DDR) 266-MHz data rates. An embedded stripe PLL supplies the appropriate timing to the SDRAM memory controller subsystem. Users can program the frequency to match the chosen memory components. The expansion bus interface (EBI) supports system ROM, allowing external flash memory access and reprogramming. In addition, static RAM and simple peripherals can be connected to this interface externally.

A single 16-kbyte memory region in the embedded stripe contains configuration and control registers, plus status and control registers for the embedded peripherals. The region contains the following modules: configuration registers, PLLs, a UART, a timer, a watchdog timer, a general purpose I/O Port and an interrupt controller.

9.6.4. Cyclone

9.6.4.1. Summary

The Cyclone field programmable gate array family [14] is based on a 1.5-V, 0.13-μm, all-layer copper SRAM process, with densities up to 20 060 logic elements and up to 288 kbits of RAM. With features like phase locked loops (PLLs) for clocking and a dedicated double data rate (DDR) interface to meet DDR SDRAM and fast cycle RAM (FCRAM) memory requirements, Cyclone devices are a cost-effective solution for data-path intensive applications. Cyclone devices support various I/O standards, including LVDS at data rates up to 311 megabits per second (Mbps) and 66-MHz, 32-bit peripheral component interconnect (PCI), for interfacing with and supporting ASSP and ASIC devices. Separate low-cost serial configuration devices are available to configure Cyclone devices.

Table 9.5 Cyclone family device summary

Feature	EP1C3	EP1C6	EP1C12	EP1C20
Logic elements	2910	5980	12 060	20 060
M4K RAM blocks (128 × 36 bits)	13	20	52	64
Total RAM bits	59 904	92 160	239 616	294 912
PLLs	1	2	2	2
Maximum user I/O pins	104	185	249	301

9.6.4.2. Functional Description

Structure. Cyclone devices contain a two-dimensional row- and column-based architecture to implement custom logic. Column and row interconnects of varying speeds provide signal interconnects between LABs and embedded memory blocks. The logic array consists of LABs grouped into rows and columns across the device, with ten LEs in each LAB. Cyclone devices range between 2910 to 20 060 LEs.

Memory. M4K RAM blocks are true dual-port memory blocks with 4 kbits of memory plus parity (4608 bits). These blocks provide dedicated true dual-port, simple dual-port or single-port memory up to 36-bits wide at up to 200 MHz. These blocks are grouped into columns across the device in between certain LABs. Cyclone devices offer between 60 and 288 kbits of embedded RAM.

Input/Output. Each Cyclone device I/O pin is fed by an I/O element (IOE) located at the ends of LAB rows and columns around the periphery of the device. I/O pins support various single-ended and differential I/O standards, such as the 66-MHz, 32-bit PCI standard and the LVDS I/O standard at up to 311 Mbps. Each IOE contains a bi-directional I/O buffer and three registers for registering input, output, and output-enable signals. Dual-purpose DQS, DQ, and DM pins along with delay chains (used to phase-align DDR signals) provide interface support with external memory devices such as DDR SDRAM, and FCRAM devices at up to 133 MHz (266 Mbps).

Clocking. Cyclone devices provide a global clock network and up to two PLLs. The global clock network consists of eight global clock lines that drive throughout the entire device. The global clock network can provide clocks for all resources within the device, such as IOEs, LEs, and memory blocks. The global clock lines can also be used for control signals. Cyclone PLLs provide general-purpose clocking with clock multiplication and phase shifting as well as external outputs for high-speed differential I/O support.

9.7. Conclusions

In this chapter we have shown how modern FPGAs can play a variety of roles in implementing cellular baseband software defined radios. FPGAs today have come a very long way from their humble origins, to the point that their name could perhaps be considered a misnomer – their composition and their capabilities far exceed the FPGAs of former days. Further, the pace of progress continues to be rapid. For these reasons FPGA devices can be critical in meeting the central requirement of SDR of configurability to changing needs without compromising speed, power or cost. In fact they often excel in these very characteristics.

9.7.1. Current Confirmation

9.7.1.1. Architectures, IP, Tools and Devices

By way of example, we have illustrated how the various existing products of Altera Corporation provide an integrated set of devices and design support for SDR solutions today. The devices provide a wide choice of architectural features and sizes while the tools aid every aspect of the design, including sources of communications- and DSP-related intellectual property.

Hence, there is at the time of the writing of this book, existing proof in device, tool or intellectual property that confirms the benefits of FPGAs to be part of a current SDR design.

9.7.2. Inevitable Evolution

9.7.2.1. Architectures, IP, Tools and Devices

Even within the relatively short life of practical FPGA designs, the trends for the future are apparent. Architectures will continue to become more heterogeneous by adding particular processors, blocks, I/O, memory and peripherals as clusters of applications demand them. Intellectual property will continue to be expanded, through the involvement of a large body of independent developers. Design tools can easily expand within the framework of existing tools to support the devices and more complex system design. Because of the economic importance of FPGAs they will continue to demand the leading-edge device technologies with ever increasing speed, power and cost benefits. As densities increase, more and more of the system will reside on a single chip, with resulting lower testing and packaging costs.

9.7.2.2. SDR Application Specific FPGAs

With the continuing success of SDRs application to baseband cellular communications, it is easy to foresee that FPGAs will become available that are increasingly configured to meet the specific needs of SDRs. These will only enhance the advantages of the FPGA solutions for SDR.

References

[1] W.H.W. Tuttlebee (Ed.), *Software Defined Radio: Enabling Technologies*, John Wiley & Sons, Ltd, 2002.
[2] *Joint Tactical Radio System (JTRS) SCA Developer's Guide*, Rev 1.1, Raytheon Company, http://jtrs.army.mil/pages/sections/technicalinformation/fset_technical.html?technical_SCACurrent
[3] *Nios 3.0 CPU – Datasheet*, Altera Corporation, http://www.altera.com/literature

[4] *Delivering RISC Processors in an FPGA for $2.00*, Altera Corporation, http://www.altera.com/literature
[5] *The Evolution of High-Speed Transceiver Technology*, Altera Corporation, http://www.altera.com/literature
[6] *The Need for Dynamic Phase Alignment in High-Speed FPGAs*, Altera Corporation, http://www.altera.com/literature
[7] *The Advantages of Hard Subsystems in Embedded Processor PLDs*, Altera Corporation, http://www.altera.com/literature
[8] *Excalibur Embedded Processor PLD Stripe Power Consumption*, Altera Corporation, http://www.altera.com/literature
[9] *SOPC Builder – Datasheet*, Altera Corporation, http://www.altera.com/literature
[10] Altera Application Note AN 161: *Using the LogicLock Methodology in the Quartus II Design Software*, http://www.altera.com/literature
[11] *Stratix FPGA Family – Datasheet*, Altera Corporation, http://www.altera.com/literature
[12] *StratixGX FPGA Family – Datasheet*, Altera Corporation, http://www.altera.com/literature
[13] *Excalibur Device Overview – Datasheet*, Altera Corporation, http://www.altera.com/literature
[14] *Cyclone FPGA Family – Datasheet*, Altera Corporation, http://www.altera.com/literature

10

FPGAs: A Platform-Based Approach to Software Radios

Chris Dick and Jim Hwang

Xilinx, Inc., San Jose, CA and fred harris
San Diego State University, CA

This chapter provides an overview of the field programmable gate array (FPGA) configured as a real-time signal processor for use in software defined radio (SDR) systems. We focus on signal processing requirements in the physical layer (PHY), which are particularly instructive for gaining insight into the computational power and suitability of FPGA technologies. While there are many complex aspects to a SDR, for example waveform management using the software communication architecture (SCA) and system interoperability based on CORBA, the PHY of an advanced radio system presents significant implementation challenges – and this is an area where FPGAs are frequently employed in configurable radios with great success.

We provide an overview of FPGA device architecture, highlighting features of interest to the signal processing practitioner. We then examine state-of-the-art design methodologies for realizing FPGA signal processing systems and present three representative examples of functions implemented in a SDR PHY: digital down conversion, radio spectrum channelization, and adaptive equalization for a quadrature amplitude modulation (QAM) receiver.

10.1. The FPGA as Signal Processor

To date, the traditional technology choices for real-time signal processing have been DSP microprocessors, ASSPs (application specific standard parts) and custom ASIC (application specific integrated circuit) solutions. However, the silicon landscape for real-time signal processing is changing.

10.1.1. Semiconductor Trends

Semiconductor process technology continues to advance according to Moore's Law [1], device geometries continue to shrink, and this is likely to be the situation for the next 15+ years.

Software Defined Radio: Baseband Technologies for 3G Handsets and Basestations. Edited by W. Tuttlebee
 ISBN: 0-470-86770-1

As highlighted by Bass and Christensen [2], this has produced the interesting situation whereby semiconductor fabrication facilities now offer the vast majority of circuit design teams more transistors than they need. This so-called design gap has been widening for some time. In fact, the National Technology Roadmap for Semiconductors noted it 5 years ago, observing that while the number of transistors that could be put on a die was increasing at a rate of about 60% a year, the number of transistors that circuit designers could design into new independent circuits was going up by only 20% a year [2]. This trend is observed in the DSP processor space where we note that even high-end DSP microprocessors do not push the transistor densities described by Moore's Law. Why is this the case?

10.1.2. Computing Architectures

One answer to this question has its roots in the computing paradigm and computing architectures on which these devices are based – the von Neumann machine. When von Neumann and his colleagues were developing their early computers, not only was the design optimization criterion different but the base technology used to construct these machines bore little resemblance to that available today. The original machines were constructed using heterogeneous materials for each major subsystem. The storage devices were made of wire and magnets (relays), or mercury and glass tubes (acoustic-wave-based mercury delay lines). The computing components were constructed out of glass and electric fields guiding electron beams (vacuum tubes). The diversity of the materials required the architectural separation of the various subsystems, for example the memory and computation sections.

Today we construct silicon based computing machines. From a materials perspective they are homogeneous systems in which the memory and computing components are constructed from the same material – silicon. Yet much of the time the architecture developed by von Neumann still persists, even when it is far from an optimal use of silicon for a desired set of processing tasks. Given a large transistor budget it is difficult to use these resources effectively to evolve a DSP microprocessor along its traditional trajectory. How many multiply-accumulate (MAC) functional units can be incorporated into an instruction set architecture (ISA) computing machine without introducing significant issues with functional unit scheduling and compiler writing?

10.1.3. Architectural Customization

To bring value to state-of-the-art semiconductor products, the transistor budget must be used in a different way. This is precisely what the FPGA does. As highlighted in Ref. [2], while price and performance are still key metrics valued in the market, there are signs that a seismic shift is occurring, giving way to a new era in which in which customization matters as much, if not more. FPGAs are the ultimate commercial device for architectural customization; signal processing systems can be constructed that are limited only by the imagination of the designer. FPGAs spend the transistor budget in a fundamentally different way to ISA machines. To see this consider the architecture of the Xilinx Virtex-II Pro [4] FPGA shown in Figure 10.1.

10.1.3.1. FPGA Structures

The device is organized as an array of configurable logic blocks (CLBs) and programmable routing resources used to provide the connectivity between the logic tiles, FPGA I/O pins and

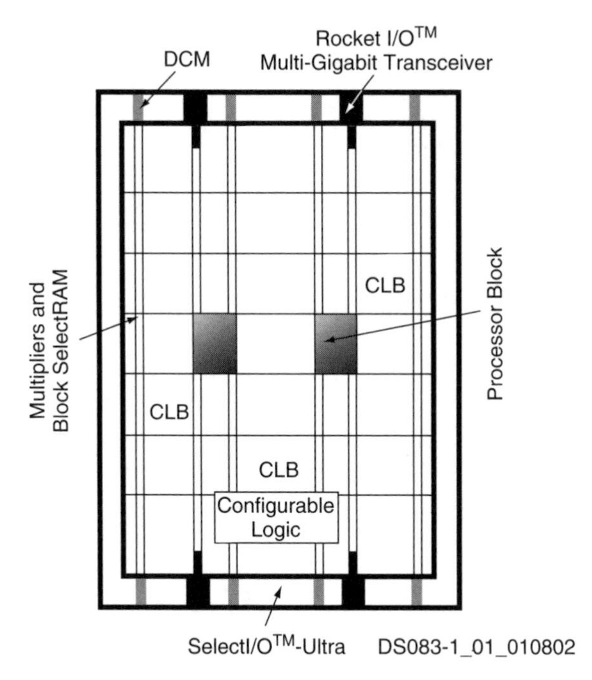

Figure 10.1 The Virtex-II Pro FPGA architecture. This FPGA family provides an array of 18×18-bit multipliers, each accompanied by an 18-kbit memory block (556 in total in the XC2VP125) overlaid on the CLB array

other resources such as on-chip memory, digital clock managers (DCMs), embedded hardware multipliers, processors and multi-gigabit transceivers (MGTs). Part of the transistor budget is used to support the programmable routing resources. Having invested part of our transistor budget in this manner, we have a highly configurable device that can be used to construct customized, and yet configurable, data paths to address computational problems. The FPGA approach to computing is like having a desktop silicon foundry with a turn-around time measured in minutes or hours, instead of months or years as it is for many complex ASICs. The device personality is held as a configuration bitstream in static RAM (SRAM) so that modifications, functional extensions and bug fixes can be easily applied – even after the system has been deployed in the field. For example, a network like the Internet could be employed to supply new FPGA configuration data to remote equipment.

10.1.3.2. The Configurable Logic Block

Each CLB consists of four logic slices. Figure 10.2 provides high-level view of a logic slice, while a detailed diagram is shown in Figure 10.3. Each slice essentially comprises two lookup tables (LUTs), two registers and additional circuitry to enable high-speed arithmetic.

The lookup table component of a logic slice is particularly interesting because it can be configured in multiple modes. It can be utilized as a 16×1-bit RAM or ROM, and can also be used in a mode referred to as *shift register logic 16* (SRL16) mode. Functionally, the SRL16

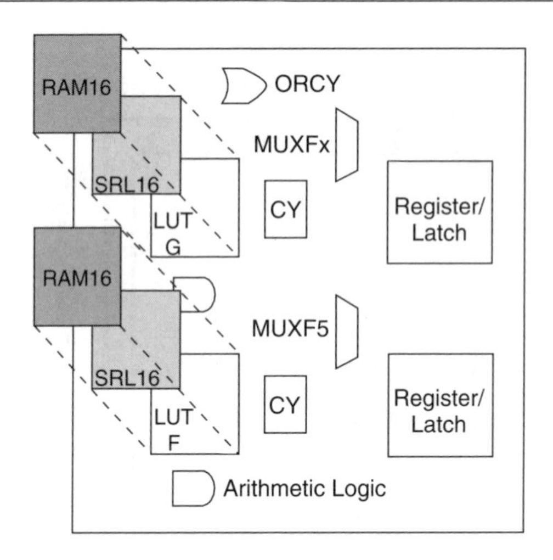

Figure 10.2 Virtex-II (Pro) logic slice – high-level view

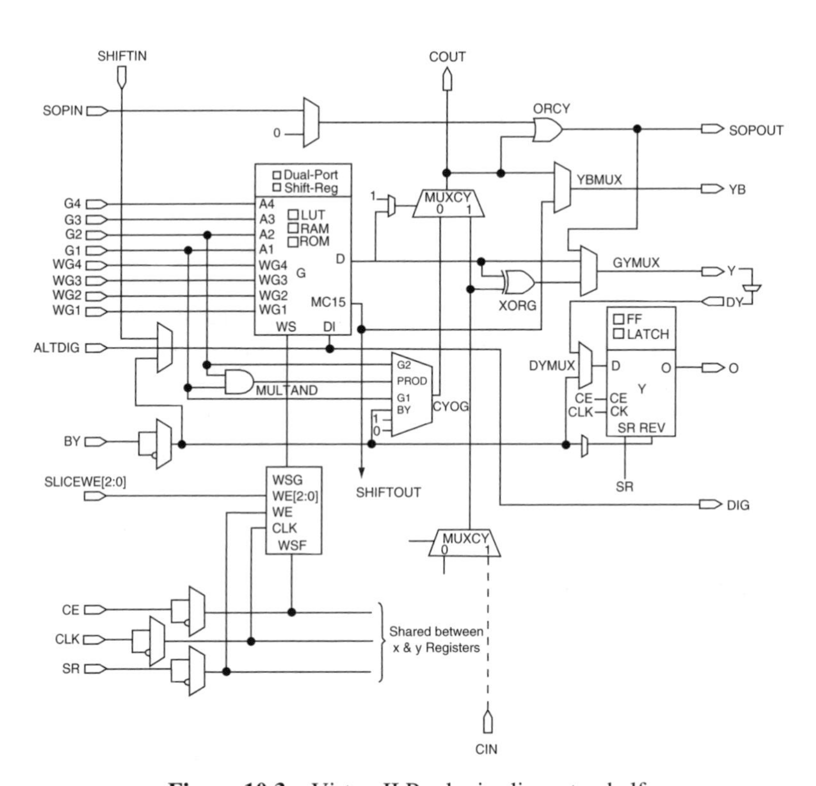

Figure 10.3 Virtex-II Pro logic slice – top half

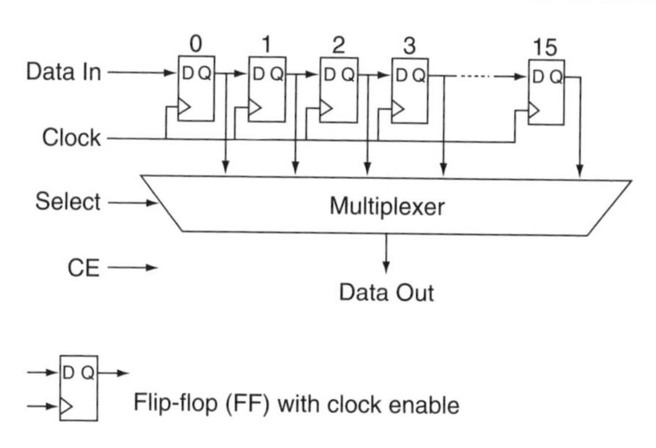

Figure 10.4 Functional view of the SRL16 configuration of a slice LUT. The multiplexer provides dynamic (run-time) access to any of the flip-flop data outputs

can be viewed as a series arrangement of 16 flip-flops with a dynamically programmable tap point, as shown in Figure 10.4.

The RAM and ROM configurations can be used for storing filter coefficients or data vectors in a signal processing system. The utility and versatility of the SRL16 configuration may not immediately be obvious with respect to signal processing applications, but this unique aspect of Xilinx FPGAs is extremely powerful for building very efficient time-division multiplexed hardware that, for example, can be used to process multiple channels of data. An example is the pulse shaping and up-sampling of the in-phase (I) and quadrature (Q) components of the baseband signal in a quadrature amplitude modulator (QAM) modulator or 3G base transceiver station transmit filter.

10.1.3.3. Resource Options

Table 10.1 provides a summary of the resources available in the various members of the Virtex-II Pro FPGA family [3].

The differing resources supplied by the various devices permits system designers to match the computational requirements of the processing task. Later in this chapter we describe example implementations, illustrating the ways in which the resource options listed in the table may be used to service radio requirements of varying complexity.

10.2. Designing for the Architecture

The FPGA device architecture is rich with features that, when engaged appropriately, are the keys to building high-performance and cost-effective signal processing systems. Of most obvious interest to signal processing is the high-density array of multipliers. Other features include the highly parallel memory system (an FPGA running at 200 MHz can have aggregate memory bandwidth of hundreds of terabytes/s), double data-rate registers (DDR) for interfacing to high-speed I/O devices such as multi-gigahertz sample rate analog-to-digital converters (ADC). The primary logic fabric itself is used to implement arithmetic functions ranging from something as simple as an adder, to arctangents, square-roots and logarithms, all potentially realized with a CORDIC [4] processing engine.

Table 10.1 Virtex-II Pro devices and resources [3]

Device	CLB Array			Slices	Distributed RAM bits	Block RAM bits	No. of Blocks	No. of Multipliers	No. of DCMs/ DLLs	No. of BUFGs	No. of Max I/O	No. of Max diff. pairs	No. of Rocket I/Os	No. of PPCs
Virtex-II Pro 1.5V	4 slices per CLB				4 slices per CLB	18 kbits per block		18 × 18	DCMs					
XC2VP2	16	×	22	1408	45 056	221 184	12	12	4	16	204	100	4	0
XC2VP4	40	×	22	3008	96 256	516 096	28	28	4	16	348	172	4	1
XC2VP7	40	×	34	4928	157 696	811 008	44	44	4	16	396	196	8	1
XC2VP20	56	×	46	9280	296 960	1 622 016	88	88	8	16	564	276	8	2
XC2VP30	80	×	46	13 696	438 272	2 506 752	136	136	8	16	692	340	8	2
XC2VP40	88	×	58	19 392	620 544	3 538 944	192	192	8	16	804	402	0 or 12	2
XC2VP50	88	×	70	23 616	755 712	4 276 224	232	232	8	16	852	426	0 or 16	2
XC2VP70	104	×	82	33 088	1 058 816	6 045 696	328	328	8	16	996	498	20	2
XC2VP100	120	×	94	44 096	1 411 072	8 183 808	444	444	12	16	1164	582	0 or 20	2
XC2VP125	136	×	106	55 616	1 779 712	10 248 192	556	556	12	16	1200	600	0 or 20	4

10.2.1. Appropriate Architecture allows Silicon Efficiency

The shift register logic 16 (SRL16) component, mentioned earlier, is unique to the Xilinx range of FPGAs and serves to illustrate how the flexibility to implement alternative and novel architectural approaches allows significant gains in silicon efficiency. As indicated in Figure 10.2, the SRL16 is actually a configuration of the slice LUT. In the same FPGA footprint that is used to realize a 16×1 RAM or ROM, an addressable series interconnection of flip-flops (FF) can be implemented. In addition to its obvious applicability for implementing tapped delay lines, this resource turns out to be extremely powerful for constructing highly efficient time-division multiplexed (TDM) hardware for processing multiple data streams. It finds use in building multichannel direct digital synthesizers (DDS [5]), FIR and IIR filters, and Viterbi decoders to name a few.

We highlight this functionality with the Reed–Solomon (RS) encoder [6] shown in Figure 10.5. A systematic code is generated by forwarding the input message sequence to the output node followed by parity-check digits that are computed using a transposed filter operating over a Galois field (GF) [6]. The GF multipliers and adders are constructed using exclusive-OR gates and realized in the logic fabric, primarily using the logic slice LUTs. The delay components can be any memory resource capable of storing a field element from GF(2^m) [6]. The field element might typically be an 8-bit quantity realized using the registers in four logic slices. The add-delay path is pipelined to minimize the critical path in the circuit. Clock rates in excess of 200 MHz can be achieved in a Virtex-II Pro FPGA.

To service multiple channels, one could step-and-repeat the RS encoding circuit, in which case, 16 copies of the single channel implementation would be required to service 16 channels. For a specific example, consider the RS(128,122) code used in the J.83 Annex B cable modem standard, which uses six parity symbols per codeword. A single channel configuration can be implemented using 56 logic slices. Sixteen copies of the design would cost 896 slices. However, in this example there is a large disparity between the input symbol rate (roughly 5 to 6 mega-symbols/s) and the maximum clock rate of the design (roughly 200 MHz). There is sufficient compute capacity to timeshare the RS encoder functional units across a large number of input streams. A traditional approach to this task employs arrangements of

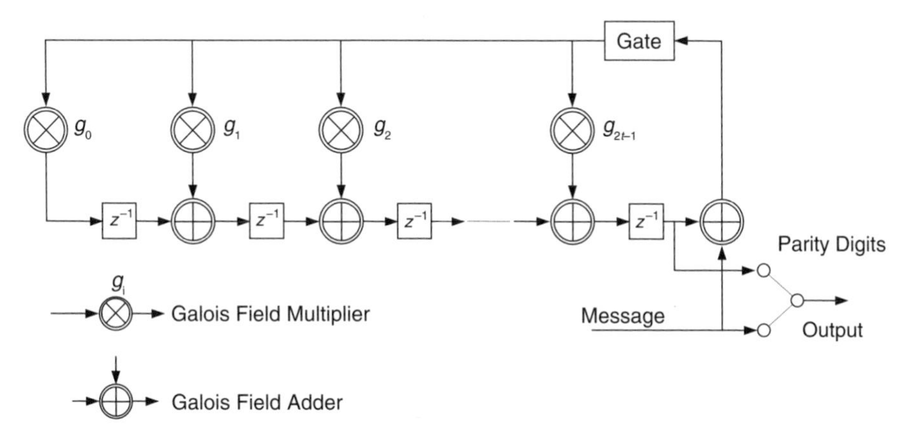

Figure 10.5 Reed–Solomon encoder generating $2t$ parity-check digits

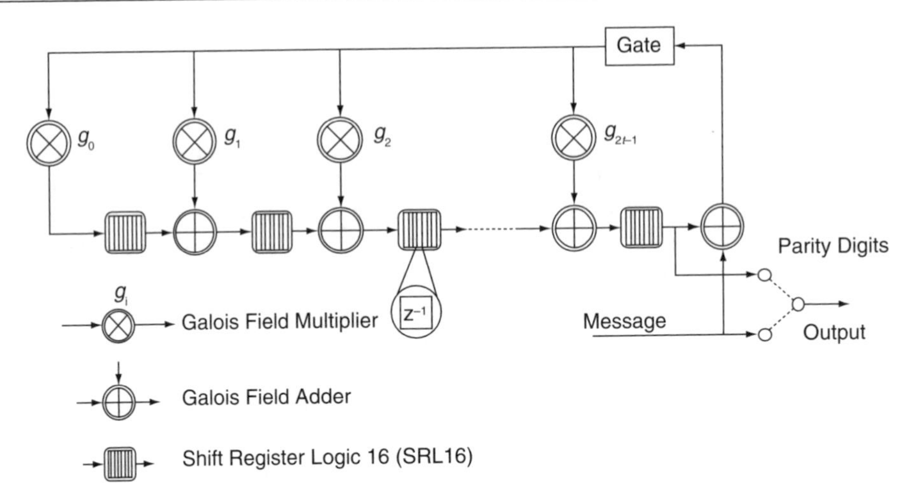

Figure 10.6 Multi-channel Reed–Solomon encoder. The SRL16 FPGA device primitive provides an extremely efficient method for time division multiplexing hardware resources

multiplexers and de-multiplexers to accommodate the multiple data streams. However, an SRL16 achieves this objective in a far more efficient and elegant manner.

The SRL16 multichannel RS encoder is shown in Figure 10.6. Each of the state components in Figure 10.5 has been replaced with SRL16s. In the same area that previously supported the state information for a single channel, it is possible to resource 17 channels – 16 storage elements by virtue of the SRL16 itself with the 17th accommodated by the FF sitting behind the SRL16. There is a small additional hardware overhead due to a slight increase in control complexity. A 16-channel design implemented this way occupies only 89 slices, which represents $89/(16\times56)\times100=10\%$ of the area of the step-and-repeat solution.

Similar efficiencies can be achieved for many other multichannel signal processing functions, such as multichannel FIR filters suitable for pulse shaping in a UMTS base-transceiver (BTS) or in a microwave modem.

The intellectual property (IP) libraries [7] that support Xilinx FPGAs in fact utilize this unique capability to timeshare hardware across multiple data streams. For example, a 32-channel Viterbi decoder [8] and multi-channel multi-rate filters [9, 10] and DDS [11] to name a few.

10.3. A New Approach to FPGA Programming

FPGAs provide an extremely flexible platform for architecting signal processing systems. However, the many degrees of freedom allowed in specifying system architecture create challenges for the application programmer. In this section, we describe a new approach to programming FPGAs that is particularly well suited to the signal processing engineer who may not be familiar with custom chip design or conventional approaches to programming FPGAs.

10.3.1. The Complexity of Traditional FPGA Design

Mainstream design practice for FPGAs mirrors that for application-specific integrated circuits (ASICs). In essence, ‘programming’ means specifying algorithms in a hardware description language (HDL) like VHDL or Verilog, that is used both for simulation and synthesis into vendor-specific FPGA and intellectual property (IP) libraries. Although HDLs provide significant control over the implementation, the level of design abstraction is quite low. Furthermore, HDLs are unfamiliar, indeed imposing, to many DSP engineers conversant with programming in C or assembly language. An HDL design flow provides little resemblance to the familiar 'edit–compile–debug' programming model that underlies the success of application programmers for modern microprocessor-based systems. As a result, the complexity of implementing a sophisticated, high-performance FPGA has historically been a significant barrier to wide adoption of FPGAs in signal processing [12].

10.3.2. New FPGA Design Flows

Emerging new FPGA design flows address this major obstacle to the wider adoption of FPGAs, by providing a programming model that is natural to a signal processing engineer. The design flows are based on extensions to well-established, system-level modelling environments that provide high level abstractions and automatic code generation into efficient FPGA realizations. Much of the success of the microprocessor can be traced to the attractiveness of its programming model. Programming languages provide a rich tapestry for specifying computation that can be mapped onto a von Neumann architecture. New FPGA design flows aim to provide a similar model for programming FPGAs. However, since the FPGA fabric is inherently different from the von Neumann machine – its strength lies in its ability to support highly parallel computation – it is natural that the programming abstractions will differ as well.

10.3.3. Programming Custom Hardware

10.3.3.1. Compiling to Hardware – Limitations

Recent commercial and research activities aim to provide familiar programming abstractions for FPGAs through code generation from languages like C and Matlab [12–14]. Such languages provide significant expressiveness and flexibility in expressing algorithms to be implemented in a DSP processor. Unfortunately, they are poorly suited to targeting custom hardware, which has no requirements for fetching, decoding and executing instructions for a fixed data path and complex memory architecture. The programmer is required to work around semantic shortcomings in the language to achieve efficient, high performance implementations. Ultimately, such programs either resemble synthesizeable HDL code (hence lose the intended benefit of an imperative programming model), or require copious helpings of in-line pragmas to instruct the compiler in how language constructs should be mapped into hardware (thereby losing the intended ease-of-programming and code portability). So while it is tempting to think that compiling C or Matlab programs into gates will allow software programmers to create high performance systems in FPGAs (or ASICs), there remains a significant gap between what can be done (in theory as well as in practice) and what needs to be done to make this a reality.

10.3.3.2. Compiling to Hardware to Exploit Parallelism

Problems with compiling imperative programming languages into parallel architectures are akin to problems that arose in past decades of research in distributed systems [15], which shares with hardware design fundamental issues of concurrency, parallelism, and synchronization. Imperative sequential languages do not provide adequate mechanisms for representing concurrency and timed (let alone real-time) behavior. The opportunity for extracting parallelism is limited to loop level and instruction level, which does not take full advantage of the capabilities of the FPGA to implement parallel architectures. Moreover, many signal processing algorithms are best implemented in hardware using multirate techniques to exploit parallelism while reducing the need for extremely high clock rates. This has the benefit not only of increased performance, but reduced power consumption. However, there are currently no means whatsoever for implementing sophisticated multirate systems in hardware using today's compiler technology.

10.3.4. Programming for FPGAs

With an FPGA, we abandon many of the limitations of the von Neumann processor architecture, but also the comfort of a mature programming model. There appears to be no single language suitable for specifying both the embedded hardware and embedded software of a complex signal-processing system (we assume automatic code generation as axiomatic, and hence dismiss the use of distinct languages for modelling and implementing a single component). The abstractions to describe a data path are not ideal for describing a control path, and neither is well suited to specifying firmware (e.g. network layer protocols) that can reasonably reside in an FPGA.

10.3.4.1. Simulation to Code Generation – Simulink

Consequently, we are naturally led to component-based design environments that support heterogeneous specification and simulation [16–18]. These systems are particularly well suited for system modelling at many levels of abstraction. Emerging design flows now provide code generation for FPGAs that make such abstractions imminently practical. For example, Simulink is a modelling environment that provides high-level simulation of dynamical systems, with companion software for automatic code generation (for microprocessors) from the system model [18]. Simulink is library based and supports a visual data flow model of computation† that makes it well suited for exposing parallelism that can be exploited in an FPGA.

10.3.4.2. Control and Branching – Stateflow

Although good for specifying data paths, this approach is less convenient for control and branching constructs. Fortunately, a complementary software tool called Stateflow provides an implementation of Statecharts [19], that are well suited for control structures. Simulink and Stateflow interact through signals at sample rates defined in Simulink, by Stateflow events, and through variables (i.e. parameters) defined in the Matlab workspace. However, while Simulink provides a rich modelling environment, other key interfaces are required to use it to implement high-performance FPGA signal processing systems.

---

† This model is distinct from synchronous data flow [20].

10.3.4.3. Extensions to Simulink

For the remainder of this section, we describe extensions to Simulink that form the basis of an emerging software platform for signal processing using FPGAs. We describe customization mechanisms that are part of the framework (i.e. are intrinsic to Simulink), but have surprisingly expressive capabilities for designing systems. The richness of customization interfaces reinforces our belief that Simulink (with appropriate extensions in the form of plug-in software and new interfaces) can provide a uniform environment for platform-based design in FPGAs.

10.3.5. System Generator for DSP

10.3.5.1. Concepts

'System Generator for DSP' is a software tool that extends Simulink with interfaces and abstractions that make Simulink a powerful hardware design environment [14]. System Generator provides hardware-centric Simulink libraries for system-level simulation and code-generation software that translates Simulink subsystems constructed from the libraries into efficient FPGA realizations (in the form of RTL VHDL and IP cores).

10.3.5.2. Capabilities

In System Generator, the user has the ability to 'right-size' data paths throughout a design to meet system criteria, since the tool supports arbitrary-precision, fixed-point arithmetic. This allows an algorithm to be very efficiently mapped into hardware. The software provides polymorphic operators, automatic type and sample-rate propagation, and high-level functions (e.g. FFTs), all of which have system-level simulation models that are faithful (bit and cycle accurate) to efficient hardware realizations. By providing both high-level functions as well as arithmetic primitives, the tools encourage and support an 'edit–simulate' stepwise refinement of a system, similar to programming practice for microprocessor systems. System Generator also extends the simulation capabilities of Simulink through co-simulation interfaces to FPGA hardware platforms and HDL simulators. Both of these extensions contribute significantly to making Simulink a hardware design framework.

10.3.5.3. Data Type and Sample Time Propagation

In System Generator, components communicate via signals (logical connections) through unidirectional ports. Supported signal types include arbitrary precision fixed-point (signed or unsigned) and Boolean arithmetic. Fixed-point types support rounding and truncation on quantization and saturation or truncation on overflow. Each block defines its semantic rules for data-type propagation according to the arithmetic or logic function it computes. Arithmetic blocks either produce full precision outputs (i.e. lossless, based on the precision of their inputs) or can perform quantization under the control of block parameters.

System Generator blocks implement Simulink interfaces for data-type propagation, which means that data types do not have to be set explicitly on each block. A change of precision at a primary input or data source suffices to customize a model in relative scale. Similarly, every System Generator is sampled, and modifying a rate-changing block (e.g. up/down samplers, serial/parallel converters, and multirate filters) will adjust system sample rates according to Simulink propagation rules.

Propagation rules for data types and sample rates are intrinsic to Simulink, which makes these system parameters much easier to adjust during modeling than in a programming language or HDL environment, where clocks and data types are part of the functional interface (and hence, less transparent).

10.3.6. Interaction with Matlab

Each component in a Simulink model has access to the underlying Matlab interpreter, which can be used for defining its behavior. A common use of Matlab is the M-code based simulation model, or *S-function* [18]. In addition, Matlab can be used as a scripting language that supports sophisticated customization interfaces. Some of these concepts may be familiar to users of other visual graphical environments, but some are, to all intents, unique to Simulink.

10.3.6.1. Customization via Parameters

An S-function *mask* defines a scope in which the function's variables can be declared, defined, and used. The initialization of these variables can be specified using Matlab expressions that are evaluated by Simulink during initialization of a simulation (actually, Simulink supports run-time parameterization, but in most instances this does not make sense in hardware and is generally not allowed with System Generator libraries). This interaction with Matlab allows sophisticated customization, bringing largely the full power of a Turing-complete programming language to bear in specifying system attributes.

10.3.6.2. Automated Adjustments

Recent work has applied techniques for automatic precision adjustment to modify the mask variables in a System Generator model using Matlab functions [21, 22]. In addition, latency parameters are specified as mask variables, introducing opportunities for automatic retiming [23, 24].

10.3.6.3. Example – FIR Filter Engine

As an example of using Matlab to customize System Generator parameters, consider a multiply-accumulate engine for an FIR filter. Unlike a DSP processor, the data-path width in an FPGA can be chosen to minimize resources without loss of precision. This can reduce power and increase the achievable clock rate of the overall system. It is not difficult to see that an accumulator with $m + \lceil \log_2 \Sigma |h_i| \rceil$ bits will never overflow, for impulse response vector h and m-bit input. This expression can be written in Matlab code as follows:

```
m+ceil (log2 (sum (abs (h))))
```

which can be bound to the bit precision variable in the accumulator mask dialog. Since this expression involves only the input bit precision and the impulse response, changes to either independent variable will result in the appropriate rescaling of the accumulator precision to avoid loss of precision, while minimizing resource usage.

It is typical to define key system parameters as variables in the Matlab workspace, and customize a System Generator model in terms of these variables. Although similar customization is possible with language-based system environments, in System Generator this is particularly natural given the interactions between Simulink and the Matlab interpreter.

10.3.7. Programmatic Customization using M-Code

The ability to invoke Matlab functions at various stages of a simulation is extremely powerful. The previous sections described customization mechanisms that automatically modify model parameters. Matlab also provides mechanisms to modify the *structure* of a model in response to system parameters.

10.3.7.1. Model Customization

While the Simulink graphical block editor is commonly used for schematic capture of a system, it appears less widely appreciated that Simulink models can be constructed *programmatically* through a Matlab API that supports block and signal instantiation, customization, deletion, and other construction methods [18]. It is particularly intriguing to use the Matlab API to customize Simulink models *in situ*. Simulink supports block-specific callback functions during model initialization, the start of simulation, at every simulation step, and when parameters are changed (there are in fact many others). By judicious invocation of the Matlab API, the topology of a Simulink model can be customized during the initialization of a subsystem. This allows the user to customize a model in ways normally considered impossible in a graphical environment.

10.3.7.2. Example – CORDIC Rectangular-to-Polar Conversion

Consider for example a rectangular-to-polar conversion block implemented using a CORDIC [3] processor (Figure 10.7). The CORDIC algorithm is well suited to an FPGA, with the ability to trade-off hardware resources for precision.

The computation is defined by the following set of iteration equations:

$$
\begin{aligned}
&x_{i+1} = x_i + y_i \delta_i 2^{-i} \\
&y_{i+1} = y_i - x_i \delta_i 2^{-i} \\
&z_{i+1} = z_i + \delta_i \tan(2^{-i}) \\
&y \to 0 \\
&\delta \in \{-1, 1\}
\end{aligned}
$$

Each iteration can be implemented by a processing element (PE) as shown in Figure 10.8; the variable δ_i is chosen to be either ± 1, so that y_i tends monotonically to zero, by examining the sign bit of the input y_{i-1}.

The CORDIC processor consists of a cascaded chain of such PEs (Figure 10.9). At the output of the final PE in the chain, z_i goes to $\tan^{-1}\left(\frac{y_0}{x_0}\right)$, and x_i converges to $K\sqrt{x_0^2 + y_0^2}$, where constant $K = 1.646760\ldots$. Excluding the scale factor K, we observe that the equations can be implemented solely with shift-and-add operations.

It can be shown that the bit precision of the output is a function of the number of iterations [25], so we can trade-off area for precision by adjusting the number of processing elements under the control of a system parameter.

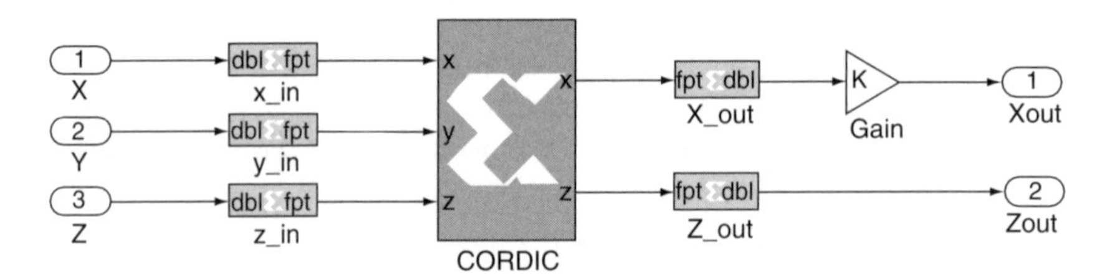

Figure 10.7 CORDIC rectangular-to-polar processor

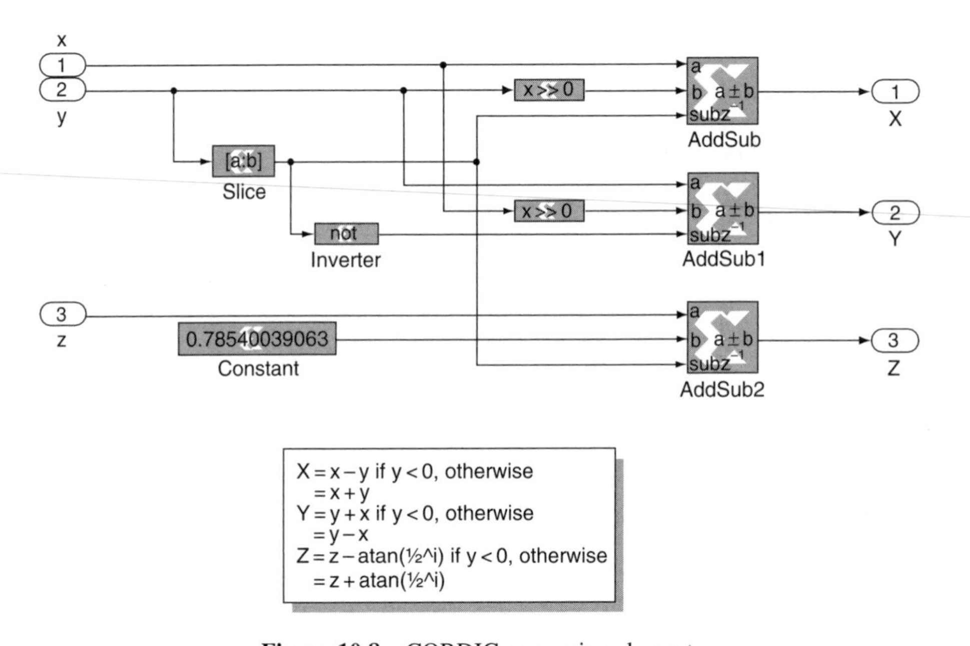

Figure 10.8 CORDIC processing element

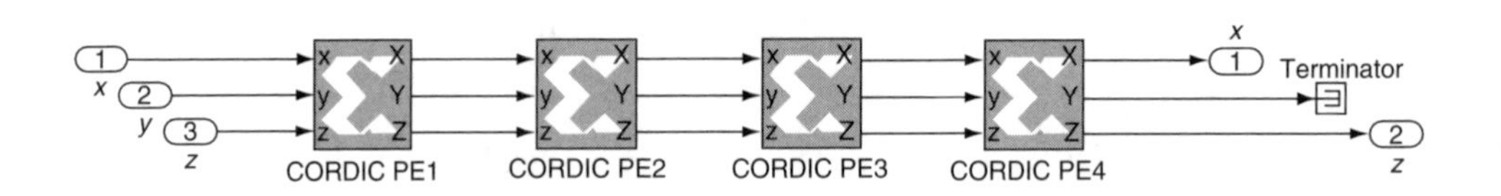

Figure 10.9 CORDIC processing element chain

Iterated block instantiation would typically be accomplished in VHDL with a FOR GENERATE statement. In System Generator, this can be accomplished by invoking Matlab API functions from the mask initialization block for the subsystem shown in Figure 10.3.† The function call:

```
a=find_system(gcb, 'lookUnderMasks', 'all', 'FollowLinks',
'on', 'masktype', 'CORDIC iteration PE');
```

returns a vector (cell array) of all contained CORDIC PE subsystems, each of which can be deleted as follows:

```
delete_block(a{i});
```

The function call

```
delete_line(gcb, ['CORDIC PE' int2str(i-1) '/' int2str(j)],
['CORDIC PE' int2str(i) '/' int2str(j)]);
```

deletes the inter-PE signals; similar calls delete the signals to the subsystem input/output ports. The desired number of PE blocks can be reinserted into the subsystem as follows:

```
add_block('sysgenCORDIC_lib/CORDIC PE', [gcb '/CORDIC PE'
int2str(i)], 'ii', int2str(i-1), 'pe_nbits', 'pe_nbits',
'pe_binpt', 'pe_binpt','position',[150+(i-1)*125, 70,
205+(i-1)*125, 130]);
```

where variables `pe_nbits` and `pe_binpt` are used in the manner described in a previous section. The PEs are connected using the following function:

```
add_line(gcb, ['CORDIC PE' int2str(i-1) '/' int2str(j)],
['CORDIC PE' int2str(i) '/' int2str(j)], 'autorouting', 'on');
```

For more information on the syntax of the API, we refer the reader to the Simulink documentation [18]. The Matlab API syntax is idiosyncratic, for instance, requiring the user to specify the location in graphics coordinates for the instanced block. Recent work provides abstractions that greatly simplify several aspects of model construction using the API [26].

The resulting implementation of the CORDIC processor is extremely efficient. With 12-bit inputs and outputs, the model with eight PEs providing 8 bits of fractional precision runs in excess of 200 MHz using only 304 slices (59%) of a Xilinx xc2v80-5 device††.

Because the callback functions are transparent to the user of the model, this structural customization is similarly transparent. To one familiar with hardware design using module generators (e.g. Lola [27]), it is useful and a bit surprising that a similar richness of programming constructs is available in Simulink, which is normally thought of strictly as a 'schematic' environment. A significant advantage to the System Generator approach is the hardware-faithful, high-level simulation environment – contrast the ease of constructing realistic test harnesses in Simulink against other environments – while maintaining automatic code generation from the model. Although self-modifying code is considered to be an archaic

† The CORDIC processor source code is included in the System Generator v2.3 software release.
†† Device speed data version: PRODUCTION 1.113 2002-09-20, ISE 5.1.02i software.

(and generally dangerous) practice in traditional programming, programmatic model construction and customization takes Simulink beyond its traditional role as a graphical capture environment. Visual capture environments usually require the user to alter explicitly the topology of a block diagram. Conditional and iterated instantiation of system components is usually considered to be the exclusive purview of programming languages and HDLs. Work in progress is using extensions of the technique described here to build high-level IP functions.

10.3.8. Co-simulation Interfaces

Simulink provides rich simulation semantics for dynamical systems defined by differential and difference equations. System Generator extends Simulink to provide hardware faithful simulation through the Xilinx Blockset libraries. We have already commented on the need for heterogeneous simulation semantics in system-level design framework, and described how Simulink and Stateflow provide distinct models of computation.

10.3.8.1. HDL Co-simulation

There are instances where it may be preferable to specify a system component in an HDL (with event-based simulation semantics). For example, it may be desirable to incorporate legacy IP blocks into a system model. It can also be necessary to implement a function that requires fine-grained access to the underlying FPGA fabric available only through an HDL. Although it is possible to create bit and cycle accurate Simulink models (e.g. C++-compiled, dynamic linked libraries) for HDL modules, this can be a time-consuming task requiring significant effort. Algorithmic approaches that do not require a separate model and verification of model equivalence are highly preferred.

System Generator provides an HDL co-simulation interface that allows the user to import HDL entities into Simulink as a black box, with simulation provided by the ModelSim simulator [28]. The System Generator black box interface customizes a block icon according to the HDL entity interface, and will automatically invoke the HDL simulator during simulation. If desired, the HDL simulator provides detailed visibility into the internals of the HDL black box during co-simulation. If such visibility is not required (e.g. when the IP module is well tested and known to function correctly), the user can incorporate the module into Simulink using System Generator hardware co-simulation interfaces [29].

10.3.8.2. Hardware Co-simulation

System Generator provides hardware co-simulation interfaces that allow a user to represent an FPGA platform as a Simulink S-function that can run in lockstep with Simulink, or asynchronously with inputs and outputs sampled at the times defined by Simulink. Several commercial hardware platforms available today have the ability to import a System Generator created hardware implementation into Simulink, where it can be co-simulated with the software representation of other hardware and software components [31, 32]. This form of 'hardware-in-the-loop' co-simulation has many additional benefits, including a dramatic acceleration of Simulink simulation. Using System Generator's black box interface, it provides the ability to import an HDL implementation directly into Simulink without writing an equivalent simulation model. Hardware co-simulation also enables access to Matlab and Simulink for data analysis and visualization of the detailed behavior of working hardware. Other benefits of hardware co-simulation include incremental design verification and an ability to perform real-time signal processing within Simulink.

10.3.9. Embedded Systems

A platform-based approach to system design is increasingly important when expanding scope beyond the embedded hardware components of a signal processing system to encompass firmware and software components. Discrete-event and Simulink-based simulation are effective for hardware components, but not particularly well suited for modeling software processes.

In signal processing systems, there is seldom ambiguity about partitioning between hardware and software; high-performance data paths must be implemented in hardware. The important issue to resolve in embedded hardware and software design is how to provide mechanisms that allow system components to be specified in a language appropriate for the task (VHDL, Matlab, Simulink, C, Python, etc.). And of course, specification is not sufficient: we require automatic code generation for each type of component, including inference of the interface logic between system components.

Many finite state machines are naturally described with discrete-event semantics in an HDL, and can be incorporated into System Generator through co-simulation. System Generator provides a simple 8-bit controller that is useful for real-time control of hardware and implementing state machines in the FPGA. More sophisticated controllers (e.g. the PowerPC405 included in the Virtex-II Pro FPGAs) using complex embedded processors are less conveniently viewed in the same bit and cycle accurate manner as other System Generator blocks. For these kinds of embedded processors, what is germane from the perspective of the DSP hardware coprocessor, are the bus transactions, which inherently involve latency (possibly indeterminate due to interrupts). These transactions can be modelled elegantly in Stateflow. Recent work extends System Generator to design embedded software as well as hardware, employing Stateflow to model concurrent hierarchical finite state machines within the Simulink environment [33]. In this approach, bus interfaces (e.g. IBM's CoreConnect) are described in Stateflow, allowing the user to maintain bit and cycle accurate abstract modelling in System Generator.

10.4. FPGA DSP Usage in the Radio PHY

For several years now, FPGAs have held a prominent position in the radio PHY. Indeed, as FPGA device technology has evolved, so too has the role and importance of these devices in radio-system physical-layer processing significantly expanded. FPGAs are used in the PHY for many functions, including those listed in Table 10.2. This is not an exhaustive list.

To provide the reader with a quantitative perspective on the capabilities of modern FPGA DSP in demanding state-of-the-art wireless systems, three examples are considered in this section that highlight not only the signal processing, but also the new generation design methodologies that enable engineers to produce high quality FPGA-based signal processing systems very quickly. These examples also illustrate the ways in which the architectural flexibility inherent in FPGAs allows alternative algorithmic thinking and transformations to

Table 10.2 PHY signal processing functions implemented using FPGAs

PHY signal processing functions
Multirate filters, including • Pulse shaping and interpolation in wireless transmitters, e.g. 3G basestation equipment and military SDRs (software defined radios) • Matched filtering and decimation in digital receivers (modems), e.g. 3G wireless basestations, satellite modems, cable modems, microwave communication links and military SDRs
Digital down- and up-conversion (DDC, DUC) Direct digital frequency synthesis Radio spectrum channelization Rake receiver Beam forming Adaptive interference cancellers in 3G UMTS radios Digital pre-distortion for power amplifier linearization in 3G basestation equipment Sinc correction for the digital-to-analog (DAC) in a wireless transmitter
Entire M-ary QAM modulators and demodulators including • High data rate fractionally spaced feed-forward equalizers (FFE) • Decision feedback equalizers (DFE) • Blind equalizers • Carrier Recovery • Timing Recovery
Multiple–input–multiple–output (MIMO) signal processing functions Fast Fourier transforms for OFDM and spectrum channelization Source coding (MPEG4/7) for image transport streams Space–time coding Multi-user detection in 3G UMTS radios Encryption/decryption Peak-to-average control in OFDM transmitters, e.g. 4G wireless transmitters
Forward error correction • Convolutional codes • Viterbi decoding • Reed–Solomon encoding and decoding • Turbo convolutional code encoding/decoding • Turbo product code encoding/decoding

be employed, permiting processing simplifications, that are not possible in conventional DSP architectures.

10.4.1. Case Study – Digital Down Conversion

The first example of signal processing in the radio PHY is a case study for a very high-performance digital down converter (DDC). The required signal processing task is to access a 50-MHz wide channel that is located at a center frequency of 1 GHz (Figure 10.10).

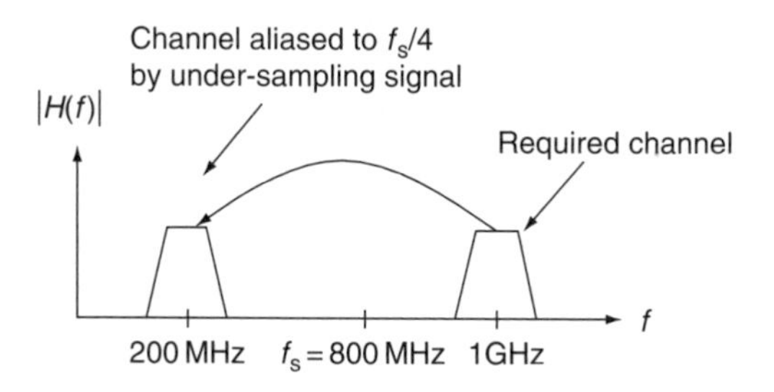

Figure 10.10 Down converting a 1-GHz-centered channel using bandpass sampling

10.4.1.1. Receiver Design Options

A simple approach might be to sample at $f_s = 2.2 \times 1.025$ GHz (20% guard interval) and then use a digital mixer and direct digital synthesizer to translate the desired channel to baseband. Even though the converter technology is available to support this sample rate, it is not a simple matter to bring the data stream into the FPGA processing fabric through the input–output blocks (IOBs) of, say, a Virtex-II Pro device at this high rate.

A more elegant solution is to employ a combination of under-sampling and $f_s/4$-centered (f_s designating the sample rate) polyphase transform to shape, re-sample and translate the channel so that is centered on 0 Hz. Further, if a judicious choice of sample rate is made, the heterodyne can be combined with the polyphase transform and no explicit mixers (multipliers) or direct digital synthesizer is required. It is convenient to select a sample rate of 800 MHz and intentionally alias the channel to a center frequency of 200 MHz, which is the quarter sample rate. The heterodyne is now simple since the samples of the complex heterodyning sinusoid are [+1 0 −1 0 +1 0 −1 . . .] and [0 −j 0 +j 0 −j 0 +j 0 . . .] for the real and imaginary components respectively. The complete IF sampled radio is shown in Figure 10.11. The input data is supplied at a 4 GHz first IF (intermediate frequency) mixed to a second IF with a 3.0 GHz local oscillator, filtered and then sampled at $f_s = 800$ MHz.

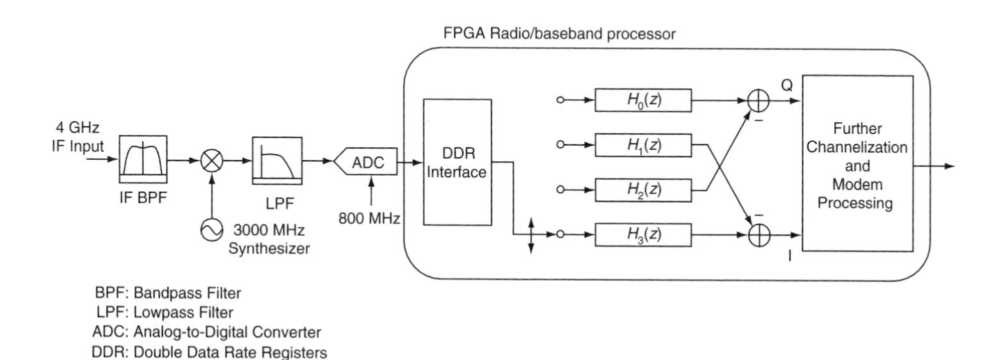

Figure 10.11 IF-sampled radio and FPGA signal processor for channelizing and downconverting a 1-GHz-centered signal. The double data-rate (DDR) capability of the Virtex-II Pro FPGA family is used to accommodate the high ADC sample rate

10.4.1.2. The High Rate ADC-FPGA Interface

One of the design challenges is to deliver the 800 MHz sample stream from the ADC into the FPGA signal processing datapath. The Max-108 ADC [34] used in the design provides an on-chip de-multiplexer that partitions the 800 MHz digitized data stream into odd and even indexed streams, each running at one half the sample rate, which in this case is 400 MHz. State-of-the-art FPGAs like Virtex-II and Virtex-II Pro not only provide hardware resourcing for supporting demanding arithmetic requirements, but also for servicing challenging I/O specifications. In this case the double-data-rate (DDR) [3] capability of the FPGA IOBs and slice-based registers will be employed to implement the ADC-FPGA interface and to deliver the sample stream to the polyphase filter. A DDR register captures data on both the rising and falling edges of the clock. The even-indexed sample stream from the ADC is brought into the FPGA using one set of 8 DDR IOBs clocked at 400 MHz, while the odd sample stream is supported in the same manner using a separate suite of eight DDR IOBs. The data captured in the even IOB register bank is further de-multiplexed into two 200-MHz data streams. Each one of these streams provides data to the phases $H_1(z)$ and $H_3(z)$ of the polyphase filter in Figure 10.11. The odd data stream is processed in a similar manner, with the de-multiplexed samples servicing phases $H_0(z)$ and $H_2(z)$ of the polyphase filter. Using this approach, the data path need only operate at a frequency of 200 MHz, which is straightforward for the Virtex-II Pro family of devices.

10.4.1.3. Matched Filtering

In this example, the matched filter can be performed by the same structure that performs the channelization. This is permissible if the input signal has had any significant carrier frequency offset removed; it will be assumed that this is in fact the case here. The matched filter requirements specify a root-raised-cosine (RRC) filter with excess bandwidth $\alpha = 0.25$ and an integration interval of eight symbols. A 128-tap filter will support these requirements. Figure 10.12 (a)–(d) show the filter infinite precision impulse response, magnitude frequency response, filter coefficients quantized to 14 bit precision, and the magnitude frequency response of the quantized filter respectively.

The coefficients from the prototype 128-tap RRC filter are distributed across the four-path polyphase filter in a column-wise manner, while the data is processed in a row-wise fashion. Each of the segments will implement a 32-tap convolution. After delivering the input samples to the four filter phases, the sample rate that each filter segment must support is $800/4 = 200$ MHz. The computation rate for each filter segment is $32 \times 200e^6 = 6.4e^9$, or 6.4 billion MACs. The aggregate performance is 25.6 gigaMACs. To place this figure in context, consider the performance of a state-of-the-art DSP processor. A 4-MAC 600 MHz DSP processor can theoretically achieve a computation rate of 2.4 GMACs (16-bit precision operations). If the problem could be partitioned across multiple processors so as not to impact the peak performance of the devices, $\lceil 25.6/2.4 \rceil = 11$ DSP processors would be required to implement the channelizer. The FPGA solution is a single chip implementation realized in a Virtex-II Pro device. The overall system implementation is significantly simplified by not requiring a large number of chips, and because the channelizer FPGA resource requirements does not challenge FPGA device density, the option is available to use a large device like a Virtex-II Pro XC2VP70 and realize most, if not all, of the remainder of the physical layer processing in the one chip.

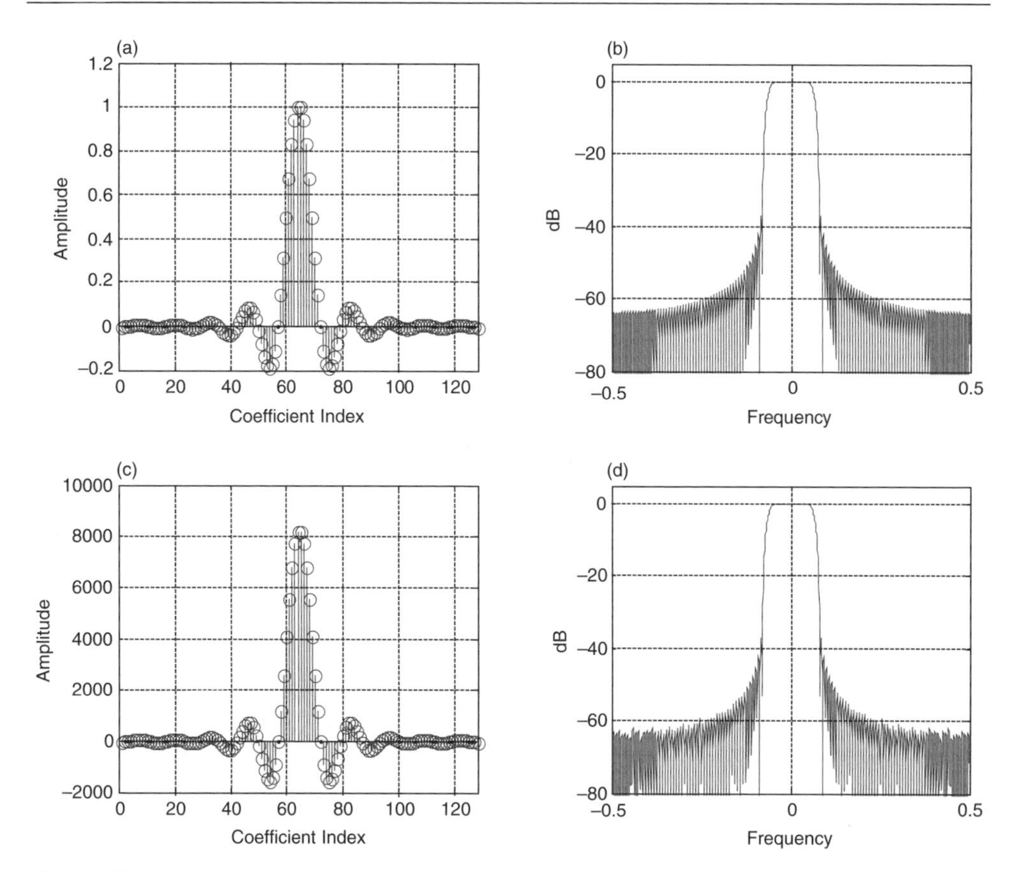

Figure 10.12 (a) RRC impulse response, infinite precision; (b) magnitude frequency response for infinite precision coefficients; (c) RRC quantized coefficients, and (d) magnitude frequency response for 14-b precision coefficients

10.4.1.4. Implementation Architecture

The FPGA implementation will target the Virtex-II Pro [3] family. These devices are rich in arithmetic resources, with the largest device (XC2VP125) supplying 556 embedded 18×18-bit precision multipliers. A fully parallel architecture will be employed with a unique embedded multiplier servicing each of the 128 filter taps. Even though the input data is streaming into the FPGA at a rate of 800 MHz, the sample commutation operation at the input of the polyphase filter means that each filter segment need only support the low output rate of 200 MHz. The transposed FIR architecture [35] shown in Figure 10.13 will be used in this design.

The inherent pipelining in this filter architecture is particularly useful for minimizing the critical path (with respect to timing) in the hardware realization. Recall the structure of the Virtex-II Pro logic slice shown in Figure 10.3. Arithmetic operations like addition and subtraction are constructed using the LUTs in a slice. The output of each LUT is followed by a register, which means that a 1-bit registered addition can be realized using half of a logic slice, with an N-bit wide registered addition requiring $N/2$ slices. Each of the add-delay elements

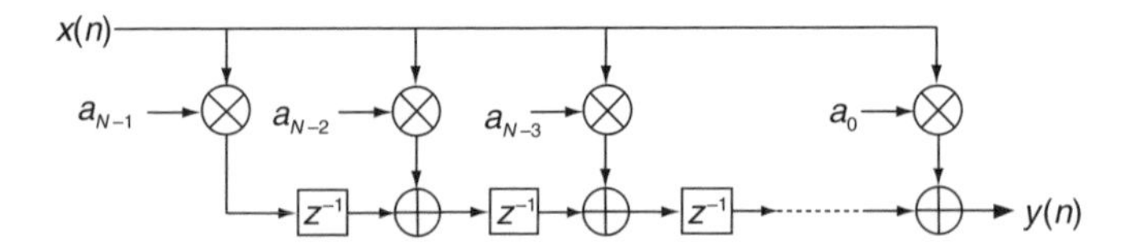

Figure 10.13 Transposed FIR filter

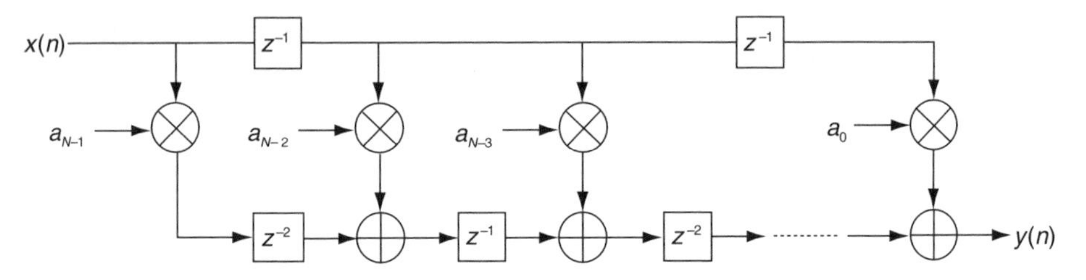

Figure 10.14 Transposed FIR structure pipelined to minimize the load on the input data broadcast net $x(n)$

in the partial product summation path in the transposed FIR filter are supported in this manner. One of the potential limitations with respect to achieving timing closure of the transposed FIR is the input data broadcast net. This arc can be shortened by introducing additional pipeline registers in the design. In this case, the broadcast net in each polyphase segment was effectively reduced by a factor of two by the insertion of a pipeline register after the sixteenth filter tap of each polyphase filter segment as shown in Figure 10.14.

10.4.1.5. Design Using System Generator

The channelizer was implemented using System Generator (version 3.1) [36]. Figure 10.15 is a screen shot of the System Generator design and also shows the details of one of the 32-tap filter segments. Each polyphase filter segment is parameterized with respect to data and coefficient precision in addition to the number of filter cells. The graph for each filter arm is conveniently constructed using the programmatic approach described in Section 10.3.7.

To meet the 200-MHz target clock frequency, each multiplier in the design has registered input operands and an output register. The internal register in the embedded mutliplier itself was also engaged. A fixed 22-bit wide accumulation path was used for each filter. After processing the design the FPGA technology mapping and place-and-route tool suite (version ISE 5.2.03i [37], speedfile: ADVANCED 1.78 2003-05-08) and targeting a Virtex-II Pro XC2VP501152-6 device, the design easily met the 200-MHz clock frequency requirement, occupying 3680 logic slices and consuming 128 multipliers – one for each filter tap. The '-6' suffix on the device part name designates the speed grade of the FPGA. A '-6 part' is the middle speed grade, sitting between the slowest (-5) and the fastest (-7). Faster devices tend to be more expensive and, even though the processing performed in the channelizer is significant, it is comfortably supported in the less expensive mid-speed-grade FPGA.

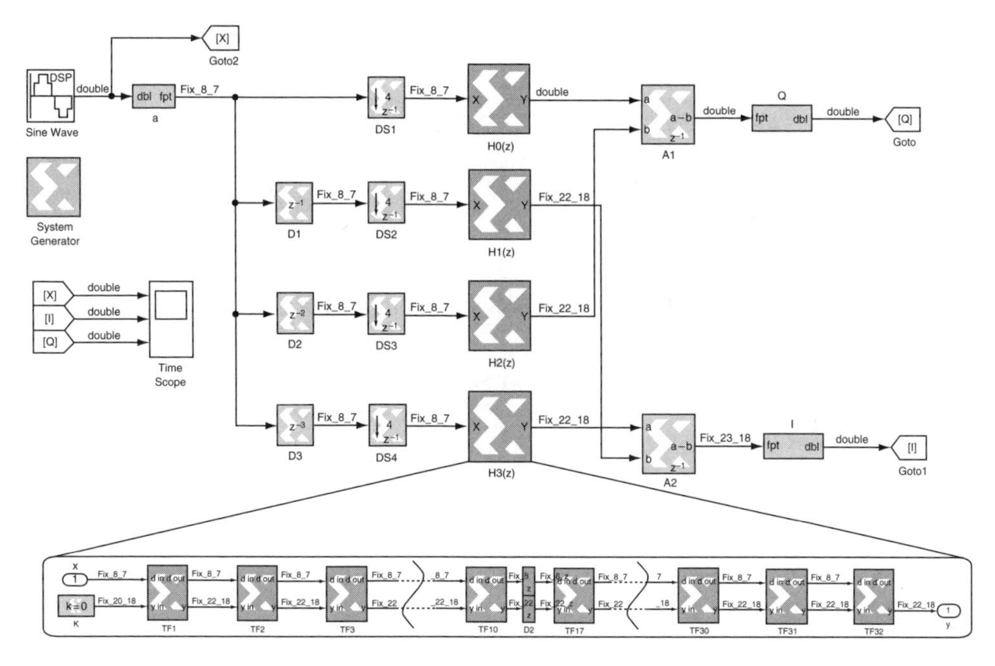

Figure 10.15 $f_s/4$ channelizer implemented using system generator. Each filter segment contains a 32-tap transposed form FIR filter running at 200 MHz

10.4.1.6. A Practical Hardware Example

There are several examples of systems that use the DDR-based ADC-FPGA interface scheme. Figure 10.16 shows a FPGA-based signal processing module developed by Sandia National Laboratory. The MAX-108 ADC on the board is interfaced to the FPGA signal processing elements using the double data-rate techniques previously described.

10.4.1.7. Design Complexity and Resource Usage

To calibrate the reader with respect to the amount of FPGA resources used in this example, consider the XC2VP30, a mid-density device in the Virtex-II Pro family. As shown in Table 10.1, it comprises 13,696 logic slices and 136 embedded multipliers. Realized in this device, the channelizer consumes 27% of the logic slices and 95% of the multiplier resources. A higher density member of the Virtex-II Pro family is the XC2VP70, with 33,088 slices and 328 multipliers. Targeting this device, the channelizer consumes 11% of the available slices and 39% of the multipliers, which would leave ample resources available to implement other sophisticated signal processing functions in the same device as the high-performance channelizer.

If the transposed filter is more deeply pipelined, with a register after every fourth tap, the load on the data broadcast net is further reduced and the design achieves a clock frequency of 287 MHz in the fastest speed grade (-7) Virtex-II Pro device. This design occupies 4158 logic slices. The computation performance is $128 \times 287e^6 = 37$ billion MACs, which is equivalent to approximately 16,600-MHz, 16-bit precision, 4-MAC high-end DSP processors. Using the

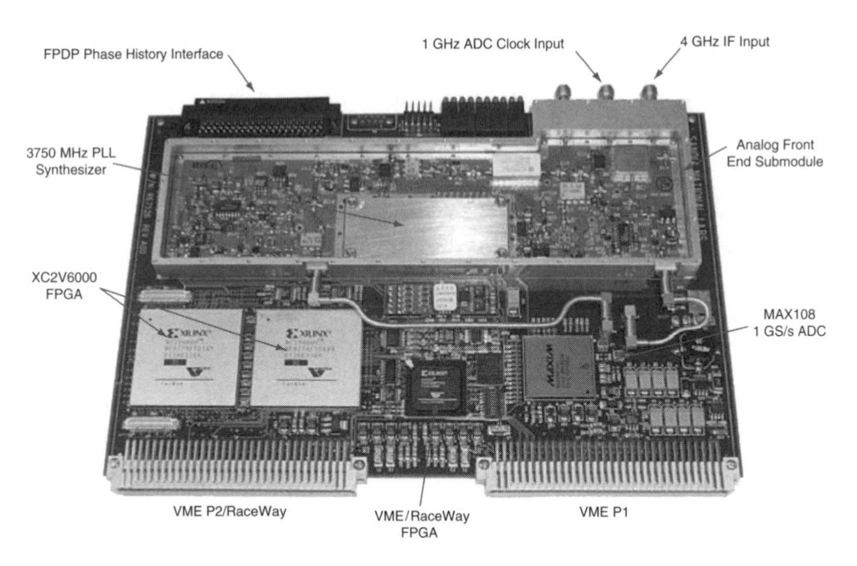

Figure 10.16 FPGA signal processing module produced by Sandia National Laboratory. The system samples data at 1 giga-sample/s rates, using FPGA double data-rate register capability (image courtesy of Sandia National Laboratory, New Mexico, USA)

DDR-based ADC-FPGA interface previously described, the DDC can support samples rates in excess of 1 giga-sample/s.

10.4.2. Case Study – MultiChannel Receiver

A key component of all digital radio systems is channelization [38] of the radio spectrum, whether it be to support 11-carrier cdma2000 or three-carrier UMTS in the commercial sector, or surveillance drop-receivers, satellite or cross disciplinary (army navy, airforce) software radios in a military context. Unlike the previous example, the problem addressed in this section is to develop efficient implementations for down converting multiple communication channels.

10.4.2.1. The Requirement

We consider the input signal to be composed of many equal-bandwidth, equally spaced, frequency division multiplexed (FDM) channels as shown in Figure 10.17. These many channels are to be digitally down converted to baseband, bandwidth constrained by digital filters, and subjected to a sample-rate reduction commensurate with the bandwidth reduction.

10.4.2.2. Channelizer Design Approach

The block diagram of a single channel of a conventional channelizer is shown in Figure 10.18. This structure performs the standard operations of down conversion of the selected channel, with a complex heterodyne, low-pass filtering to reduce bandwidth to the channel bandwidth, and down sampling to a reduced rate commensurate with the reduced bandwidth.

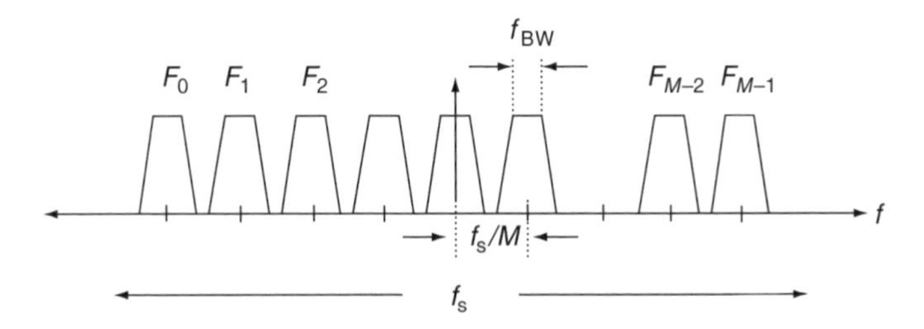

Figure 10.17 Input spectrum of frequency division multiplexed signal to be channelized

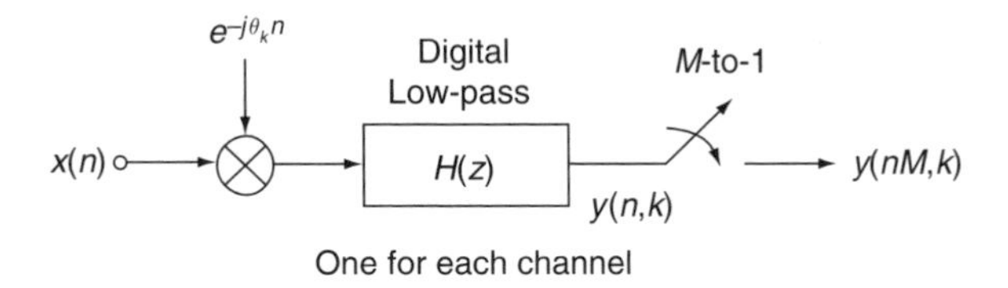

Figure 10.18 *k*th channel of conventional channelizer

The expression for $y(n,k)$, the time series output from the kth channel, prior to resampling, is a simple convolution as shown below.

$$\begin{aligned} y(n,k) &= [x(n)e^{-j\theta_k n}] * h(n) \\ &= \sum_{r=0}^{N-1} x(n-r)e^{-j\theta_k(n-r)}h(r) \end{aligned}$$

The output data from the complex mixer is complex and, hence, is represented by two time series, I(n) and Q(n). The filter with real impulse response $h(n)$ is implemented as two identical filters, each processing one of the quadrature time series. We can rearrange the summation to obtain a related summation reflecting the *equivalency theorem* [39]. The equivalency theorem states that the operations of down conversion followed by a low-pass filter are totally equivalent to the operations of a band-pass filter followed by a down conversion. The block diagram demonstrating this relationship is shown in Figure 10.19. Note here, that the up-converted filter, $h(n)\ \exp(j\theta_k n)$, is complex and as such, its spectrum resides only on the positive frequency axis without a negative frequency image.

Recognizing that there is no need to down convert the samples we discard in the down sample operation, we choose to down sample only the retained samples. This is shown in

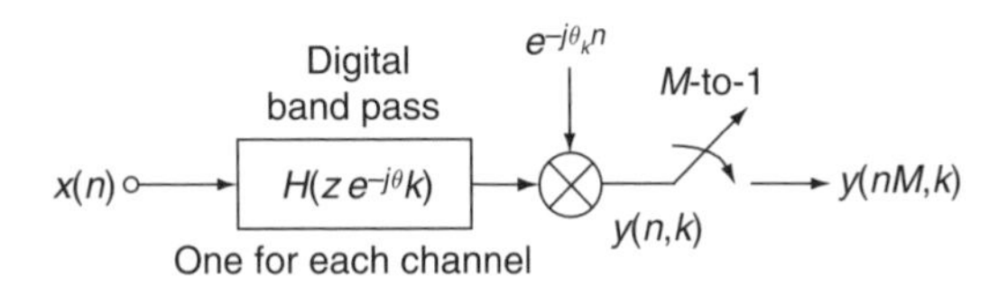

Figure 10.19 Band-pass filter, *k*th channel of channelizer

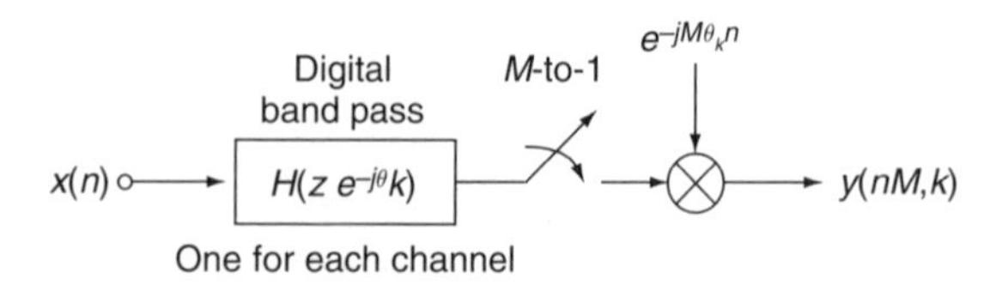

Figure 10.20 Down sampled down converter, bandpass kth channel

Figure 10.20. We note that when we bring the down converter to the low-data-rate side of the resampler, we are in fact also down sampling the time series of the complex sinusoid. The rotation rate of the sampled complex sinusoid is Θ_k and $M\Theta_k$ radians per sample at the input and output respectively of the M-to-1 resampler.

This change in rotation rate is an aliasing affect, a sinusoid at one frequency or phase slope appears at another phase slope when resampled. We now invoke a constraint on the sampled data center frequency of the down converted channel. We choose center frequencies θ_k which will alias to DC (zero frequency) as a result of the down sampling to $M\theta_k$. This condition is assured if $M\theta_k$ is congruent to 2π, which occurs when $M\theta_k = k2\pi$, or more specifically, when $\theta_k = k2\pi/M$. The modification to Figure 10.20 to reflect this provision is seen in Figure 10.21. The constraint, that the center frequencies be integer multiples of the output sample rate, assures aliasing to baseband by the sample rate change. When a channel aliases to baseband by the resampling operation, the resampled related heterodyne defaults to a unity-valued scalar, which consequently is removed from the signal processing path. Frequency offsets of the channel center frequencies, due to oscillator drift or Doppler effects, are removed after the down conversion by a baseband phase locked loop (PLL)-controlled mixer. This baseband mixer operates at the output sample rate rather than at the input sample rate for a conventional down converter. We consider this required final mixing operation to be a post conversion task and allocate it to the next processing block.

Examining Figure 10.21, we note that the current configuration of the single channel down converter involves a band-pass filtering operation, followed by a down sampling of the filtered data to alias the output spectrum to baseband. Following the idea developed in the previous section that led us to down convert only those samples retained by the down sampler, we similarly conclude that there is no need to compute the output samples from the passband filter that will be discarded by the down sampler. We now interchange the operations of filter and down sampler with the operations of down sampler and filter. The process that accomplishes this interchange is known as the *Noble Identity* [40].

We now complete the final steps of the transform that changes a standard mixer down converter to a resampling M-path down converter by applying the frequency translation property of the z-transform to produce the signal flow-graph of Figure 10.22.

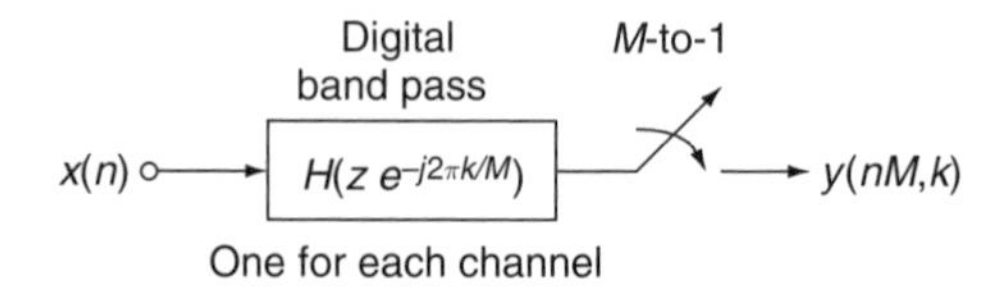

Figure 10.21 Alias to baseband, down sampled down converter, bandpass kth channel

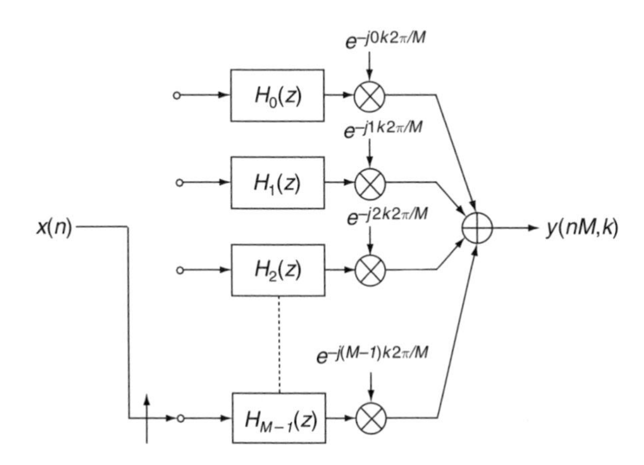

Figure 10.22 Resampling M-path down converter

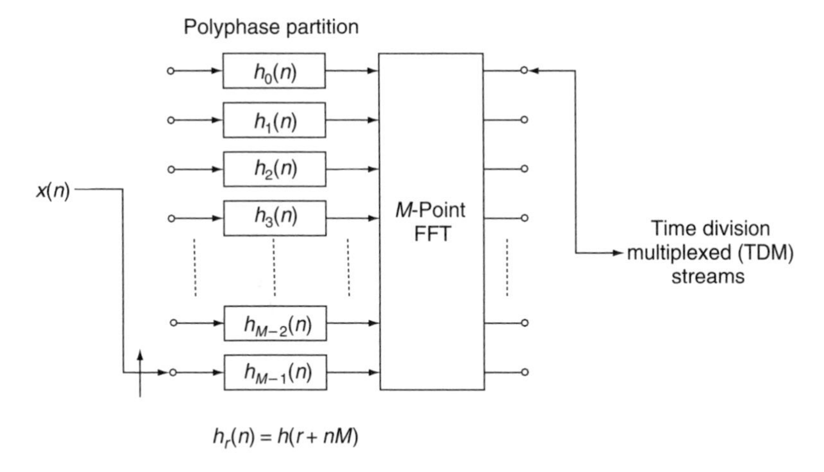

Figure 10.23 Polyphase channelizer: resampler, all-pass partition and FFT phase shifters

When the frequency index k is iterated over all M possible values, the phase rotators generate a Butler matrix, and the implied signal processing function is a DFT, implemented, of course, as a fast Fourier transform [35]. The completed channelizer is now a single merged process called a polyphase M-path filter bank, as shown in Figure 10.23. The polyphase filter-bank partition offers a number of significant advantages relative to the set of individual down-conversion receivers. The primary advantage is reduced cost due to the major reduction in system resources required to perform the multichannel processing.

10.4.2.3. Design Implementation Options

We will now consider the FPGA implementation of a 16-channel version of the channelizer in Figure 10.23. The requirements are to channelize a 400-MHz wide section of spectrum. The filter requirements for each of the 16 channels is a passband edge frequency of 1 MHz, a 0.875 MHz transition region with 60 dB sidelobes and 0.2 dB passband ripple. The 160-tap equiripple filter satisfies the filter specifications. The channelizer will use a different approach for realizing the filters to that employed in the previous digital down-converter example. While multiply accumulate-based architectures (using explicit multipliers like those found in Virtex-II Pro) are an excellent choice for implementing filter operations in many cases, there is a rich body of algorithms for implementing convolutions that do not directly rely on the support of an explicit multiplier. For the past 20 or so years, these algorithms have, for the most part, been ignored in the context of configurable, real-time signal processing solutions. The DSP microprocessor has been the dominant technology in this space and these devices include a data path that utilizes one, or a small number, of ASIC style multipliers, and so any reduced complexity or multiplier-free algorithms are typically penalized in these implementations – the MAC(s) are available so they must be used to achieve effective processing. This approach to data-path realization is of course not true with the FPGA. With FPGA DSP, the designer specifies a customized, and optimized, data path to solve the problem at hand. A suitable number, possibly hundreds, of functional units of appropriate type, precision and connectivity, are employed. The rich set of algorithmic options for constructing computing machines once again becomes available in a very practical setting, and it makes sense to use them for FPGA signal processing. For example, canonic sign-digit (CSD) arithmetic [35], sigma-delta modulation based filters [41, 42], and distributed arithmetic (DA) [10, 43, 44] are a few techniques that have effectively been employed in FPGA DSP systems.

10.4.2.4. The Filter Bank – A Distributed Arithmetic Approach

A distributed arithmetic approach will be employed to implement the filter bank for the channelizer. The details of DA will not be covered here – the reader is referred to the excellent tutorial by White [44] for an overview of the method. A DA filter does not require an explicit multiplier, instead the convolution is re-cast as a table-lookup based procedure. One interesting property of DA filters is the inbuilt scalability of the algorithm. It offers a very natural approach to making a trade-off between filter performance (input sample rate) and hardware resource requirements. Further, DA algorithms are extremely well matched to the Xilinx FPGA architecture with the SRL16 [3] device primitive implementing access to the filter regressor vector in the manner required with DA and logic slice based distributed memory supplying the parallel array of pre-computed partial product tables. DA filters offer the option of implementing word-slice processing, where a word slice can be at either of the extremes of 1-bit, resulting in a bit serial processor, or a complete sample, producing a fully parallel implementation, or at any slice width between these two end-points. Increasing the processing word width implies an increase in hardware resource requirements, but results in a proportional increase in filter performance (sample rate).

The output rate for each polyphase segment of the channelizer is 800/16=50 MHz. A completely bit serial design of the segment filters would require a master processing clock frequency of $8 \times 50e^6 = 400$ MHz – each bit of a data sample needs to be indexed in turn, which requires eight clock cycles. While certain tasks can actually be performed at this frequency in recent generation FPGAs, it is a little too high for running a DA FIR filter. The 8-bit input samples will be partitioned into 2-bit slices, each 2-bit slice being allocated its own processing engine. With this hardware folding approach, a 200-MHz processing clock is required, and this can easily be achieved. During each period of the 200 MHz clock a 2-bit slice of each regressor vector sample is processed in parallel. Four clock cycles are required to process the 8-bit precision data.

10.4.2.5. FPGA Implementation

The design was implemented using System Generator and realized in a Virtex-II Pro device using the ISE 5.2i implementation tools. The complex filter occupies 5392 logic slices. A technology view of the filter is shown in Figure 10.24. The more densely populated regions of the floorplan is the filter logic while the vertical and horizontal structures are the interconnect between the various components. The computation performance of the complex filter is equivalent to $2 \times 160 \times 50e^6 = 16$ billion MACs/s. An interesting figure of merit is the amount of equivalent work performed by each logic slice in the design, which is $16e^9/5392 \approx 3$ billion MACs/s/slice. This is a particularly interesting, i.e. large, number given the modest amount of logic in the slice architecture shown in Figure 10.2.

In addition to the complex filter bank that shapes the signal and performs a bandwidth limiting function, an FFT is required to complete the channelization process and generate the required 16 complex output time series. The performance requirements placed on the FFT are quite high. If the transform is viewed as a streaming data process, with the complex input samples supplied on a time-division multiplexed bus, the FFT would need to support an I/O rate equal to that of the ADC sample rate, or 800 MHz. This is too high a clock rate to operate the FPGA signal processing engine. Instead, the inherent parallelism of the FFT and the natural parallelism of the FPGA itself will be exploited to construct a vector-based transform processor.

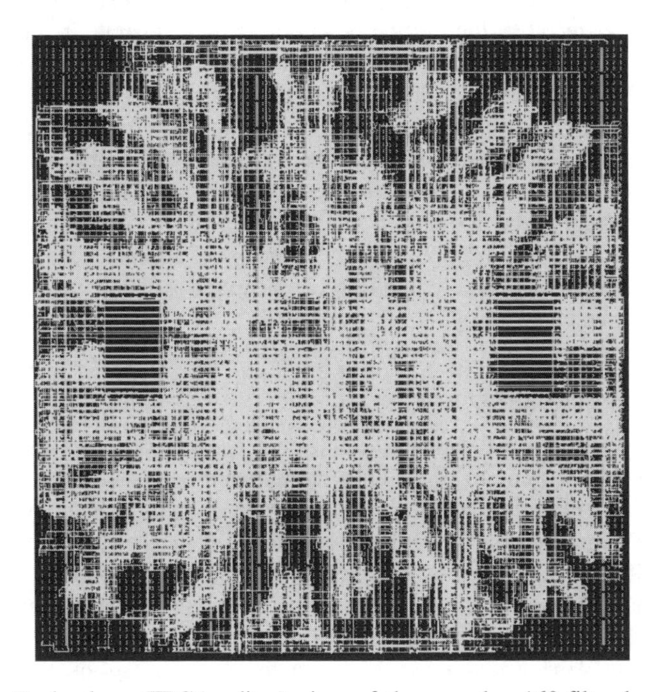

Figure 10.24 Technology (FPGA editor) view of the complex 160-filter bank implemented using distributed arithmetic techniques

The FFT will accept a 16-element vector of input samples, and generate a new 16-element output vector during each processing clock period. A Cooley–Tukey radix-4 algorithm [45] was implemented using System Generator for DSP. The FFT occupied 1812 logic slices and 36 multipliers. In this case each complex multiplication was implemented using four real multiplies and two additions. If the 3-multiply algorithm [45] were employed the multiplier count could be reduced to 27 at a cost of a modest increase in the number of additions. The FFT supports a clock rate of 210 MHz, which means that a new complex input/output vector is delivered/produced at the rate of 150 million vectors per second. In fact, to support the polyphase filter bank, the transform rate is 50 million transform/s. This means that the FFT could in fact support four polyphase filter banks.

10.4.3. Case Study – Adaptive Systems

Adaptive signal processing is an important aspect of a modern communication system whether it is for supporting space-time coding in adaptive antenna arrays for 3G wireless or military radio environments, or for mitigating the effects of inter-symbol interference (ISI) in a high data-rate microwave modem for example.

10.4.3.1. Adaptive Channel Equalizer

This final example of FPGA signal processing demonstrates the implementation of an adaptive channel equalizer. Many practical narrowband communication receivers require not just one equalizer, but typically three. A blind equalizer is typically used initially to acquire the channel and assist the carrier-recovery loop to the point where it can operate in a decision directed mode. A practical approach that is used for blind channel acquisition is the constant modulus algorithm (CMA) [46]. A decision directed fractionally spaced (FS) feed-forward equalizer (FFE) and decision feedback equalizer (DFE), for mitigating the effects of pre- and post-cursor ISI respectively, are also employed. In each of the equalizer configurations, the most common algorithm used for coefficient adaption is the least mean squares algorithm (LMS) [47] or a derivative thereof, such as the normalized LMS or leaky LMS. Although we are describing this adaptive signal processing example within the context of an equalized QAM demodulator, the LMS algorithm is in fact equally applicable to adaptive antenna arrays and systems that employ space–time coding.

10.4.3.2. Equalizer Design Options

DFE-based equalizers are challenging to implement because they are nonlinear recursive systems that tend to be difficult to pipeline. In many situations this is motivation to use a fractionally spaced CMA/FFE equalizer alone. In an FPGA context there are many architectural options available depending on the required bit rate of the system. The FS (two samples/ symbol) FFE will be implemented using a polyphase filter decomposition. To achieve very high data rates, a fully parallel implementation can be considered, where each filter tap and LMS engine in the FFE is allocated individual processing elements (PEs).

The design was implemented with System Generator. An eight-tap equalizer, with four-taps allocated to each of two polyphase filter segments, using 18-bit precision filter coefficients and 24-bit precision accumulators in the filter and LMS processors occupies 1037 logic slices, uses 66 embedded multipliers, and supports a clock frequency of 240 MHz in a Virtex-II Pro

XC2VP501152-7 FPGA. Using a 16-QAM alphabet this is equivalent to a 480 Mbit/s data rate, and for a 1024-QAM configuration the data rate is an impressive 1.2 Gbit/s. Another way to spend the high clock rates afforded by FPGAs is to time share a set of resources to effect a computation. In this case we might allocate one filter and one LMS PE to each of the polyphase sub-filters. Each of the PEs will be responsible for servicing four inner product computations (for the actual equalizer filter) and four LMS updates for the coefficients. This equalizer configuration has an FPGA footprint of 833 logic slices and 18 embedded multipliers. The time-shared nature of the implementation, means of course, that the bit rate is lower than that of the fully parallel implementation, but it does occupy fewer FPGA resources. The register files used in both the FIR and LMS PEs are based on SRL16s, and the filter length can be increased to a total of 32-taps (16 in each of the filter segments) with no increase in area.

10.4.3.3. The Fractionally Spaced FFE/DFE/CMA Equalizer

The System Generator implementation of the more challenging fractionally spaced FFE/DFE/CMA equalizer is shown in Figure 10.25. The long combinatorial path in the feedback loop limits the clock frequency to 120 MHz. The FFE consists of 16 taps while the DFE has six taps. The design occupies 2596 logic slices and 68 embedded multipliers. The recursive nature of the DFEs also require the data-path precision to be increased above what is adequate for a FFE. In this case all of the accumulators in the FFE and DFE are kept to 28-bit precision.

10.4.3.4. Design Approach

The extremely powerful Simulink programmatic interface described in Section 10.3.7 was used extensively in this design for exploring and evaluating the design space. For example, the DFE interface is shown in Figure 10.26. The bit-field precision of the key nodes in the DFE graph are propagated to the functional units via variables that are defined in the Matlab

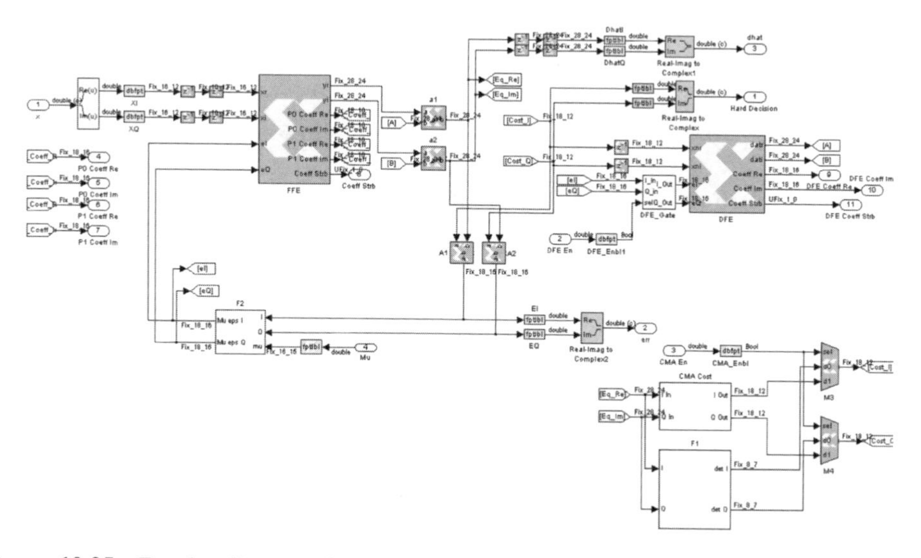

Figure 10.25 Fractionally spaced equalizer implemented in System Generator for DSP. The design includes a FFE, DFE and CMA-based blind equalizer

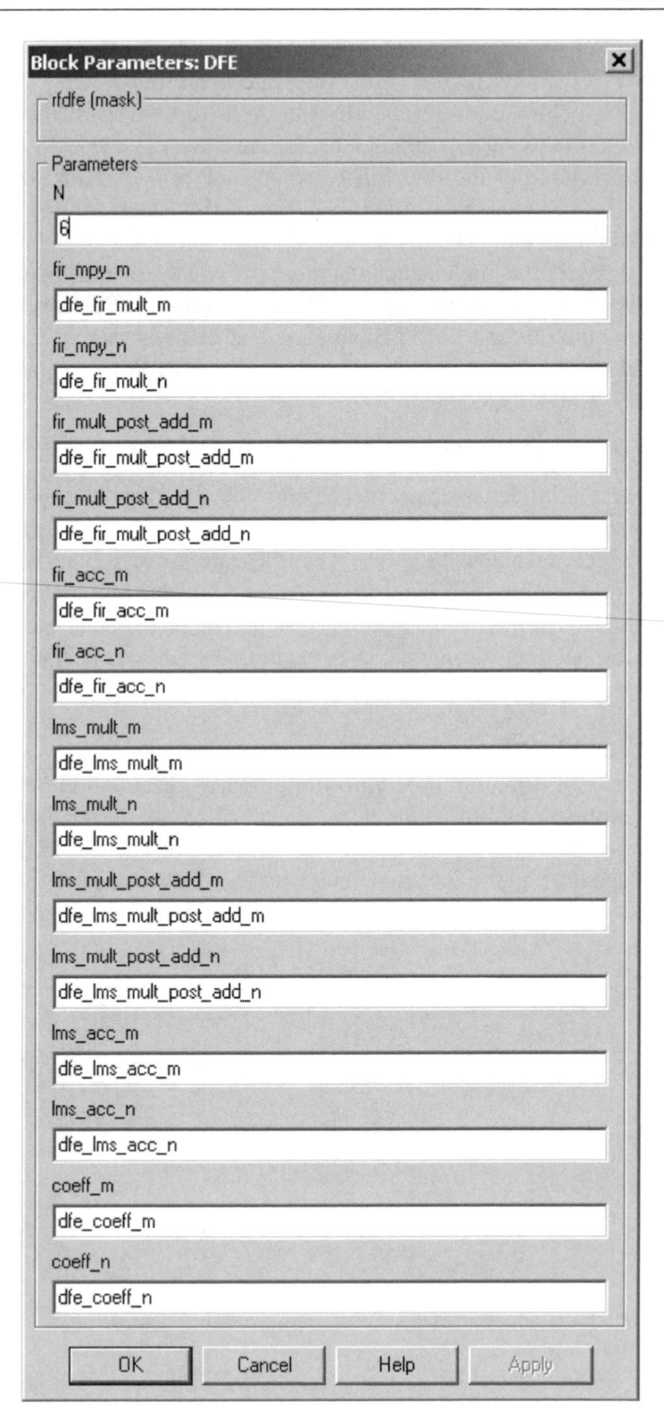

Figure 10.26 Decision feedback equalizer interface. In addition to defining the bit-field precision of the key nodes in the graph, the number of processing elements is supplied through this interface and the physical structure of the subsystem is accordingly adapted

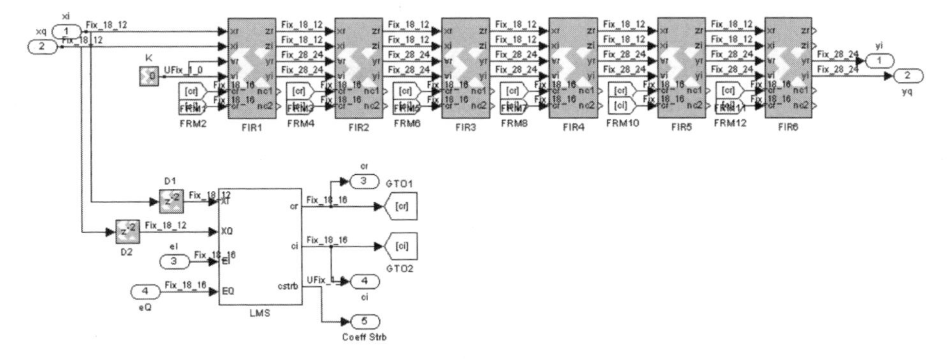

Figure 10.27 6-tap DFE

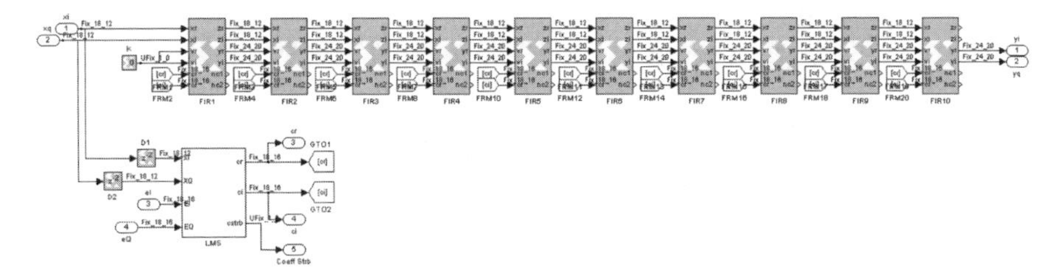

Figure 10.28 10-tap DFE. Generation of different order DFEs is accomplished programmatically, and not through the manual process of the designer adding (deleting) and modifying the graph topology

workspace. In addition, the actual physical structure of the graph, in terms of the number of processing elements, is also defined and the DFE subsystem is reconstructed in response to changes in the parameter set. In migrating from the six-tap DFE shown in Figure 10.27 to the 10-tap DFE of Figure 10.28, no manual work in terms of adding (deleting) blocks or redrawing the graph connectivity is required, it is all accomplished programmatically. This is an extremely powerful feature for generating different design customizations and verifying their performance in an automated manner, for example through the use of Matlab scripts.

10.4.3.5. Simulation and Verification

One final comment is in order with respect to simulation and verification of the adaptive equalizer. The recursive nature of the DFE requires careful consideration of finite arithmetic effects and, in particular, subtle effects that are attributed to truncation versus rounding arithmetic in the FIR filter taps and the LMS PEs. It can require many tens of millions of input vectors for numerical stability issues to present themselves. Simulations of this order require many tens of hours in HDL simulators. Even though the simulation of this equalizer is significantly faster in the Simulink framework when compared with an HDL simulation, thorough simulations still require many hours. The hardware-in-the-loop capability of the design flow (outlined in Section 10.3.8) was engaged to debug and verify this design, and demonstrated a reduction in the design/verification cycle to minutes, rather than hours.

10.5. Conclusion

Reconfigurable logic, and in particular FPGAs, have already staked their claim in signal processing history. At any particular instant during the past decade, FPGAs have provided performance levels from one to three orders of magnitude superior to other competing programmable technologies, depending on the exact functionality considered. All systems that involve a human variable grow in complexity and there is no better example of this than the intricate communication networks that mankind has constructed from the dawn of time.

10.5.1. New Century, New Requirements

At the start of the Twenty-first century, the electronic communication technologies that have served us well for the past decade no longer satisfy the requirements of the commercial, social or defense components of our community. The solution we look to is the so-called software radio that provides multimode high-speed (bits per second) communication without geographical constraints. While the architecture of the overall system is itself complex, in both a business and technical sense, the demands placed on the physical-layer processing of these communication systems continues to grow almost exponentially. The arithmetic requirements cannot be satisfied with micro-evolutions of the von Neumann machine that has served us well for so long. This is the sweet spot for FPGAs, which can provide the high-MOPs resourcing for everything from advanced radio spectrum channelization functions, with sample rates in the GHz region, through to advanced forward error correction functions providing throughput from UMTS 2 Mbit/s data rates to 500 Mbit/s throughput in the most advanced systems.

It is worthwhile commenting on emerging trends that may indicate the role of reconfigurable logic in future SDRs. Leading up to the year 2000, many communication infrastructure providers were on an accelerated development schedule to deploy UMTS compliant basestations, with significant pressure to get equipment into the market place quickly. Many providers employed FPGAs in early deployment because of the time-to-market advantages provided by the technology. Their strategy was to migrate some of the FPGA functionality into ASICs for full production, while preserving configurability to accommodate fluidity of the UMTS specification, and to permit future in-field upgrades. As the major economies of the world cooled and the communication market entered a depression, deployment plans had to be significantly adjusted.

10.5.2. Increasing Demands for Reprogrammability

At the time of this writing, time-to-market pressures have abated somewhat and there is a greater consideration for reprogrammability. There has emerged a much greater focus on cost over the entire product lifetime rather than simply on initial component cost – a field basestation upgrade or an ASIC re-spin can be extremely costly.

10.5.2.1. Technology Complexity and Ongoing Evolution

Most people would agree that the complexity of the technology required to deliver on the promise of 3G communication systems was under estimated. Some well-known technologists would go so far as to claim that the technology was itself too little too early. The 3G specification continues to evolve, with a particular emphasis being placed on increasing the maximum data rate for a subscriber from 2 Mbps to at least 10 Mbps (HSDPA [48]). So again requirement dynamics, coupled with performance, give FPGAs an opportunity in an area once considered the purview of ASICs.

However, the considerations are considerably more complicated today. In the year 2000, an ASIC to service the digital, high-speed, front-end of a system and the baseband processing requirements might have been targeted at a 150-nm silicon process technology node. Today leading edge for production is 90 nm, with a move to 65 nm occurring very soon. Both of these geometries present significant fabrication challenges, far beyond those experienced with 150-nm devices.

10.5.2.2. The Challenge for ASIC Solutions

An analogy for the consequences of Moore's law is the story of the king, the peasant and the chessboard. The peasant has done a favor for the king and is asked to name his reward. He replies simply a single grain of rice on the first square of a chessboard, and twice as many on each succeeding square. Since this sounds simple the king agrees. How much rice does this require? In the beginning very little, but by the time the 32nd square is reached, the rice occupies a room. To get to the 64th, and final square, requires 2×10^{19} grains – variously estimated as requiring the entire area of the earth to produce, and weighing 100 billion tons. That is the way exponentials work. At first they are easy, but later they become overwhelming. Since the invention of the transistor there have been about 32 doublings of the technology – the first half of a chessboard. The overwhelming implications await us now. How does this relate to relationship between FPGAs and software radios?

Radio requirements are estimated to be increasing at a rate that, today, is outpacing even Moore's law. In other words, the sophistication of the radio system, and in particular the requirements of physical-layer processing, are exceptional. The challenge of building an ASIC of the implied complexity, using new generation fabrication technology, in a timely manner, is significant. This is where we refer back to the FPGA, and in particular the primary value propositions of the device, namely accelerated time-to-market, product future proofing and flexibility, and high-performance.

10.5.3. Future Requirements and Solutions

10.5.3.1. Requirements Beyond 3G

Now more than ever before, as we look ahead to beyond 3G, where MIMO, MIMO/OFDM and high bit-rate Turbo codes to mention a few dimensions of the radio PHY, will be deployed, FPGA signal processing is almost the only contending technology on which to base a system.

10.5.3.2. The Need for Complete SoC Solutions

As semiconductor fabrication technology and communication system design both increase in complexity, it will become less useful to produce devices that solve only one part of the problem, for example a DSP processor. Increasingly, sales of stand-alone digital signal processing and embedded chips will give way to SoCs that incorporate DSP or other functions. For example, in a soft radio system, the PHY is dominated by sophisticated and arithmetically demanding DSP, but what about the other parts of the technology that are required to make soft radio a reality?

As we progress up the hierarchy of soft radio we encounter problems that are best addressed using a RISC (reduced instruction set architecture) processor – for example, the TCP/IP stack or the software communication architecture (SCA) for a military SDR. We will also need to communicate with a broader network such as the Internet. Integrated components like the Power-PC 405 and the multi-gigabit transceivers in the Virtex-II Pro FPGA address these considerations by providing a platform-based approach to system implementation.

10.5.3.3. Changing Market Dynamics

Continually decreasing market windows and shorter product life cycles mean that it is more difficult to achieve the high volume sales seen in the past, making the ASIC proposition less attractive. Each year the transistor budget dramatically increases, while design methodologies and software systems for integrated circuit development continue to lag.

10.5.3.4. The Platform FPGA Solution

The Platform FPGA approach to system development and product deployment is a solution to this complex set of market and technology dynamics. Platform FPGAs utilize the transistor budget in ways that are being increasingly rewarded by the market, including timeliness, flexibility, customization, and performance.

References

[1] G.E. Moore, 'Cramming more components onto integrated circuits', *Electronics*, 1965, **38**(8).

[2] M.J. Bass and C.M. Christensen, 'The future of the microprocessor business', *IEEE Spectrum*, 2002 (April), 34–39.

[3] Xilinx Inc., Virtex-II Pro Platform FPGAs, http://www.xilinx.com/xlnx/xil_prodcat_landingpage.jsp?title=Virtex-II+Pro+FPGAs

[4] Yu Hen Hu, 'CORDIC-based VLSI architectures for digital signal processing', *IEEE Signal Processing Magazine*, 1992 (July), 16–35. See also the chapter by T. Henschel and G. Fetweiss in *Software Defined Radio: Enabling Technologies*, W.H.W. Tuttlebee (Ed.), John Wiley & Sons Ltd, Chichester, 2002.

[5] C.H. Dick and f.j. harris, 'Direct digital synthesis – some options for FPGA implementation', *SPIE International Symposium On Voice Video and Data Communication: Reconfigurable Technology: FPGAs for Computing and Applications Stream*, Boston, MA, USA, September 20–21, 1999, pp. 2–10.

[6] S. Lin and D.J. Costello, Jr., *Error Control Coding: Fundamentals and Applications*, Prentice-Hall, Englewood Cliffs, New Jersey, 1983.

[7] Xilinx Inc., Xilinx Intellectual Property libraries, http://www.xilinx.com/xlnx/xil_prodcat_landingpage.jsp?title=Intellectual+Property

[8] Xilinx Inc., http://www.xilinx.com/ipcenter/catalog/search/logicore/viterbi_decoder.htm

[9] Xilinx Inc., http://www.xilinx.com/ipcenter/catalog/search/logicore/xilinx_mac_fir.htm

[10] Xilinx Inc., http://www.xilinx.com/ipcenter/catalog/search/logicore/distributed_arithmetic_fir_filter.htm

[11] Xilinx Inc., http://www.xilinx.com/ipcenter/catalog/search/logicore/direct_digital_synthesizer.htm

[12] P. Banerjee *et al.* 'A MATLAB compiler for distributed, heterogeneous, reconfigurable computing systems', *Proceedings of FCCM '00*, ACM Press, 2001.

[13] Celoxica, Handel-C, http://www.celoxica.com/methodology/default.asp

[14] J. Hwang, B. Milne, N. Shirazi and J. Stroomer, 'System Level Tools for FPGAs,' *Proceedings FPL 2001*, Springer-Verlag, 2001.
[15] H.E. Bal, J.G. Steiner and A.S. Tanenbaum, 'Programming languages for distributed computing systems', *ACM Computing Surveys*, 1989, **21**, 261–322.
[16] J. Buck, S. Ha, E.A. Lee and D.G. Messerschmitt, 'Ptolemy: a framework for simulating and prototyping heterogeneous systems,' *International Journal of Computer Simulation*, 1994 (April).
[17] W.-T. Chang, S. Ha and E.A. Lee, 'Heterogeneous Simulation: Mixing Discrete-Even Models with Dataflow,' *The Journal of VLSI Signal Processing*, 1998 (April), **18**(3), 297–315.
[18] The Mathworks, Inc., *Using Simulink*, 2002.
[19] D. Harel and M. Politi, *Modelling Reactive Systems Using Statecharts*, McGraw Hill, 1998.
[20] E.A. Lee and D. Messerschmitt, 'Synchronous Data Flow,' *Proceedings of the IEEE*, 1987, pp. 55–64.
[21] C. Shi and R. Brodersen, 'An automated floating-point to fixed-point conversion methodology,' *Proceedings ICASSP*, 2003.
[22] A.A. Gaffar, O. Mencer, W. Luk, P.Y. Cheung, N. Shirazi and J. Hwang, 'Floating-point bitwidth analysis via automatic differentiation,' *Proceedings IEEE ICFPT*, 2002.
[23] N. Weaver, 'C-Slow Retiming for FPGAs,' *Proceedings FPGA '03*, IEEE Press, 2003.
[24] Y. Yi and R. Woods, 'FPGA-based System-level design framework based on the IRIS synthesis tool and System Generator,' *Proceedings IEEE ICFPT*, 2002.
[25] J. Walther, 'A unified algorithm for elementary functions,' *AFIPS Joint Computer Conference Proceedings*, 1971, **38**, pp. 379–85.
[26] J. Stroomer, *et al.* 'Simplifying programmatic construction of system generator models using jg,' *Proceedings of FCCM 2003*, ACM Press, 2003.
[27] N. Wirth, 'Lola: A formal notation for synchronous circuits,' in *Digital Circuit Design*, Springer, 1995, pp. 81–92.
[28] Model Technologies, Inc., *ModelSim User's Manual*.
[29] V. Singh, A. Root, E. Hemphill, N. Shirazi and J. Hwang, 'Accelerating a bit error rate tester with a system level tool,' *Proceedings of FCCM 2003*, ACM Press, 2003.
[30] Xilinx XtremeDSP Development Kit.
[31] Alpha Data Systems, 'Simulink board support blockset', http://www.alpha-data.com/xrc-sysgen.html
[32] Lyr Signal Processing, 'DSP link and FPGA link product brief', http://www.signal-lsp.com/documents/PDF/DSPLink_FPGALink_Simulink_System_Generator.pdf
[33] J. Ballagh, J. Hwang, P. James-Roxby and E. Keller, 'Building an OPB peripheral using system generator for DSP,' *Application Note XAPP264*, http://support.xilinx.com/AppNote/XAPP264.pdf
[34] Maxim Semiconductor, http://www.maxim-ic.com/quick_view2.cfm/qv_pk/2092
[35] K.K. Parhi, *VLSI Signal Processing Systems: Design and Implementation*, John Wiley & Sons, Inc., NY, 1999.
[36] Xilinx Inc., System Generator for DSP, http://www.xilinx.com/xlnx/xil_prodcat_product.jsp?title=system_generato
[37] Xilinx Inc., *ISE Logic Design Tools*, http://www.xilinx.com/ise/design_tools/index.htm
[38] f.j. harris, C. Dick and M. Rice, 'Digital receivers and transmitters using polyphase filter banks for wireless communications', *IEEE Transactions. Microwave Theory and Techniques*, 2003, **51**, 1395–412.
[39] J. Wozencraft and I.M. Jacobs, *Principles of Communication Engineering*, John Wiley & Sons, Inc., NY, 1967.
[40] P.P. Vaidyanathan, *Multirate Systems and Filter Banks*, Prentice-Hall, 1993.
[41] C. Dick and f. harris, 'FPGA signal processing using sigma-delta modulation', *IEEE Signal Processing Magazine*, 2000 (January), **17**, 20–35.
[42] C.H. Dick and f.j. harris, 'Configurable logic for digital communications: some signal processing perspectives', *IEEE Communications Magazine: Special Issue on Topics in Software and DSP in Radio*, J. Mitola (Ed.), 1999 (August), **37**, 107–11.
[43] A. Peled and B. Liu, 'A new hardware realization of digital filters', *IEEE Transactions on Acoustic, Speech, Signal Processing*, 1974, **22**, 456–62.
[44] S.A. White, 'Applications of distributed arithmetic to digital signal processing', *IEEE ASSP Magazine*, 1989 (July), **6**, 4–19.
[45] R.E. Blahut, *Fast Algorithms for Digital Signal Processing*, Addison-Wesley Publishing Company, Reading, Mass., 1987.
[46] J.R. Treichler, M.G. Larimore and J.C. Harp, 'Practical blind demodulators for high-order QAM signals', *Proceedings of IEEE*, 1998, **86**, 1907–26.

[47] S. Haykin, *Adaptive Filter Theory*, 3rd Ed., Prentice Hall, New Jersey, 1996.
[48] H. Holma and A. Toskala (Eds), *Radio Access for Third Generation Mobile Communications*, 2nd edition, John Wiley & Sons, Inc., NY, 2002.
[49] S. Haykin, *Adaptive Filter Theory*, Prentice Hall, New Jersey, 1996.
[50] Berkeley Design Technology, Inc. 'FPGAs for DSP,' *Focus Report*, July 2002, http://www.bdti.com/products/reports_focus.html
[51] C. Dick, 'Reinventing the signal processor,' *Proceedings of the SDR Forum* (Winter 2002), 453–69.
[52] The Mathworks, Inc., *Stateflow User's Guide, Version 5*, 2002.
[53] Xilinx, Inc., *Embedded Development Kit*, http://www.xilinx.com/ise/embedded/edk.html

11

Reconfigurable Parallel DSP – *r*DSP

Behzad Mohebbi and Fadi J. Kurdahi

Morpho Technologies, Irvine, CA

11.1. Introduction

The majority of today's commercial digital signal processors (DSPs) can be placed into one of two main categories: programmable DSP or dedicated DSP. While programmable DSPs have the advantage of using the same computational kernels for all algorithms, dedicated DSPs are hardwired for specific algorithms, or classes of algorithms. Even though both have special instruction sets, dedicated DSPs are usually faster and consume less power than general-purpose programmable DSPs. The level and the extent of the architectural specialization of a dedicated DSP is defined by the application algorithms for which the processor is targeted, which consequently reduces the general applicability of the DSP.

By far the largest application area for DSPs, in recent years, has been in the support and implementation of digital communication baseband algorithms. There are overwhelming commercial reasons for the selection and the increased usage of DSPs in such systems. The third generation (3G) wireless communication systems are designed for increased spectral efficiency and high data rate transmission. These emerging systems are based on advanced forms of wideband multiple access techniques, such as CDMA, and broadband modulation techniques, such as spread spectrum and OFDM, which demand a much higher signal processing load than the previous second generation (2G) systems. This increased processing load cannot be supported by traditional DSP platforms at a reasonable cost. As a result, new DSP architectures, based on the parallel processing paradigm, are emerging. Compared to conventional single processor DSPs, these architectures increase the processing power of the DSP with only a marginal increase in power consumption and complexity. The increased performance of parallel processing, at a reasonable cost, stems from the simple fact that the 'inherent' parallelism in a given algorithm allows for lower processing clock rates and data movement, much the same as dedicated DSPs. The lower clock rate effectively means lower power consumption, and a less complex design and manufacturing process of the hardware platform.

Software Defined Radio: Baseband Technologies for 3G Handsets and Basestations. Edited by W. Tuttlebee
 ISBN: 0-470-86770-1

11.1.1. Types of Parallel Architecture

The original parallel architectures were usually set at 'bit' and 'instruction' levels, where the data dependency inhibits the extraction of the algorithm parallelism. With only a limited number of algorithms exhibiting good instruction level parallelism, ILP has become expensive to exploit. Other DSP architectures exploit the 'loop', 'thread' and 'process' level parallelism, where a number of coarse-grain processors interconnected by various techniques, such as dedicated bus, crossbar switch, shared memory or shared network, provide the required processing platform. Although all multiprocessor DSPs aim to exploit some type of inherent process parallelism, there exists a large variety of parallel processing architectures that differ, based on the generality level, interconnect topology, nature of coupling, communication and control mechanism of the processing fabric. In general, a 'tight' coupling of the processors is required for an optimized performance, which, depending on the target application, can be provided by dedicated interconnection between the processing elements, which again leads to the dedicated DSP architecture.

11.1.2. Reconfigurable Architectures

The wide variety of communications algorithms in a typical wireless system means that no single dedicated DSP platform is capable of providing the required spatial mapping of the algorithm to the processing fabric. The need for the restructuring of the computational elements necessitates a 'reconfigurable' approach to the coupling between processors. Such a reconfigurability feature allows the same processing elements to be optimally mapped to different algorithms, thus retaining the high performance of dedicated DSPs, whilst alleviating the need for a multiplicity of dedicated fabrics for algorithm coverage. While the 'compile time' reconfigurable platforms allow the execution of different algorithms on the same platform, the 'run-time' dynamic reconfiguration offers the highest flexibility for software defined radio (SDR) applications, as more functionalities of the transceiver chain can be supported by the same processing resource.

In this chapter, we examine the suitability of a reconfigurable multiprocessor DSP architecture for supporting baseband processing of current and future wireless systems. We also look at the design and available programming tools of one such processor, and describe the real-time performance capabilities of an example WCDMA application for SDR.

11.2. Baseband Algorithms and Parallelism

Fundamentally, there are two major models of parallel processing: heterogeneous processing or multiple instruction–multiple data (MIMD), and homogeneous processing or single instruction–multiple data (SIMD) [1, 2]. The first model assumes relatively independent processors running their own programs and communicating through a variety of synchronization mechanisms. The second model assumes a single sequencer (or controller that broadcasts an instruction, or set of instructions) to all the processors to execute. Figure 11.1 depicts the two architecture, high-level block diagrams, for a linear processing array.

While MIMD architecture, with a similar number of processors, offers the same level of parallelism, it has several disadvantages in processing highly regular operations at high throughputs. These disadvantages include code size, performance degradation (due to synchronization

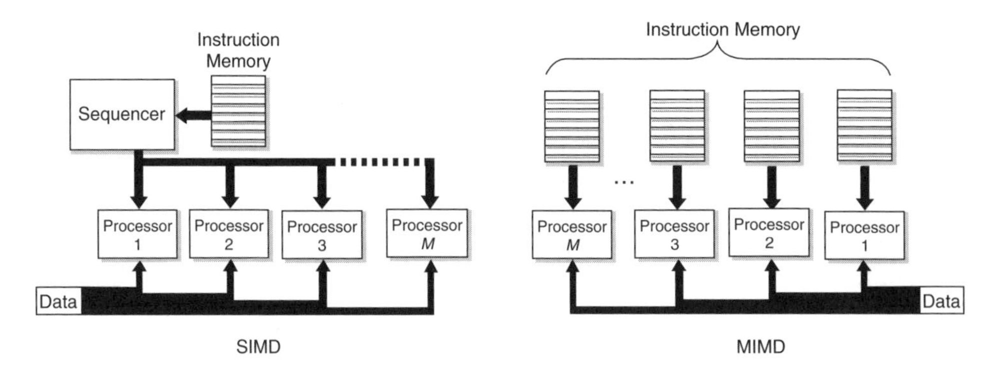

Figure 11.1 SIMD and MIMD architectures

overhead), and relative difficulty of programmability. On the other hand, a pure SIMD architecture may not be best suited for all phases of processing. An example of this arises when computing the sum of N numbers using M processors. An SIMD mode allows each processor to sum N/M numbers with full utilization of the hardware. However, that would not be the case when the partial products in each processor are to be added together. In this case, different processors would need to perform different operations, thereby operating almost in an MIMD fashion. From this discussion, it is clear that a flexible processing paradigm that allows grouping of processors performing similar operations, and a run-time capability of dynamically varying the size and composition of these groups, would result in the most efficient way of performing an application since the mix of processing modes may be custom tailored for each application.

With the rapid advances in chip fabrication technologies, parallel processing architectures have now found a place in computation-intensive, high-throughput tasks such as image, video and multimedia applications that have been emerging in recent years. The image processing applications such as discrete cosine transformation (DCT) and 'motion estimation' all have inherent data parallelism, require high throughput and are highly regular. Similar properties are also exhibited by the baseband algorithms of most wireless systems, as will be explained in this section. Most algorithms in a modern digital modem exhibit some degree of inherent parallelism, which can be exploited for fast implementation by a parallel processing fabric. In this section we examine a range of transceiver baseband algorithms for computation parallelism, and their suitability for implementation on a reconfigurable array of processors that can operate in different modes that can be dynamically varied at run-time. Note that the limit on the dimensions of the processing array is set by data bandwidth of the array and the level of inherent parallelism of the target algorithms.

Figure 11.2 shows a typical transceiver block diagram for a single-input/single-output (SISO) wireless system. The block diagram is an example and as such can be modified for a specific application. Nevertheless, most wireless communication links, regardless of their deployed modulation schemes, coding, etc., require most of the functional blocks shown in Figure 11.2. The specified baseband region in the receiver chain includes conventional, as well as new functional blocks such as digital IF and baseband AGC, which are increasingly seen in software defined radio designs. Overall, DSP implementation of the illustrated system

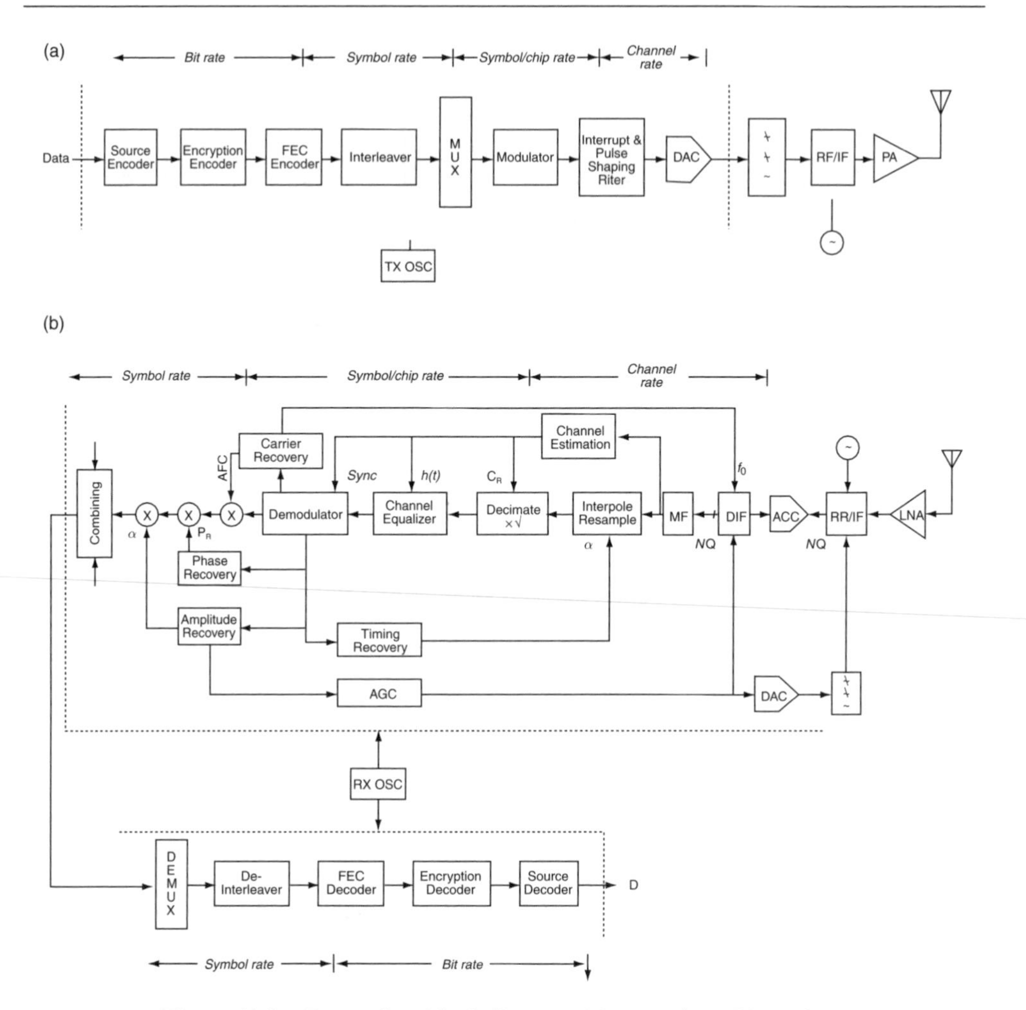

Figure 11.2 Transceiver block diagram: (a) transmitter (b) receiver

raises two basic design issues: firstly, the required processing power and, secondly, possible data bottlenecks in the overall system flow. The processing power requirement is usually a function of the system bandwidth, which itself is set by the supported maximum data rate and, in cellular systems, by the deployed access scheme. Since wireless systems require a specific modulation and access scheme for operation in a cellular environment, the signaling bandwidth of these systems is usually considerably higher than the maximum supported data rate of the system. For example, the TDMA access scheme of GSM/EDGE requires channel transmission rates 8-times higher than the coded symbol rates, and in the case of WCDMA, the speech channel transmission rates can be as much as 128 times (64 on the uplink) the coded symbol rate of an adaptive multirate (AMR) channel (30 ks/s). Regarding the second point, the data flow through the processor requires careful design to avoid data bottlenecks. For example, while the chipping rate of a WCDMA system sets the input at a much higher rate than the output of, say, a despreader, the opposite happens with systems with high-order modulations. The recent adoption of high-order modulation in wireless data systems such as 802.11a and cellular systems such as GSM/EDGE and even WCDMA, where each symbol represents a number of information bits, requires the data bandwidth at the output of the baseband processor to be larger than the input bandwidth. Therefore, it is essential to ensure the availability of adequate data transfer bandwidth at the input and output port of each individual functional block shown in Figure 11.2.

11.2.1. Source Encoder/Decoder

Although, traditionally, the transmitter requires considerably less baseband processing than its associated receiver, recent source encoding algorithms such as vocoders and image compression, have somewhat balanced the processing loads between the two ends of the link, by placing most of the computational requirements of the algorithm at the encoder end, i.e. the transmitter. Although image and voice source encoders use a number of small irregular computation kernels, nevertheless considerable parallelism exists in the overall algorithm.

11.2.1.1. Voice Coding

As an example, consider the GSM enhance full rate (GSM EFR) vocoder which has the same input and the encoded parameters as the AMR-NB vocoder in mode MR122 [3]. The EFR is the third defined vocoder for GSM system, and belongs to the code excited linear prediction (CELP) speech compression family. The EFR with a bit rate of 12.2 kbits/s is an algebraic CELP (ACELP) vocoder, where an excitation signal is built for each pulse, rather than the selection from a code book as a code vector, as was the case with classical CELP vocoders. In the EFR vocoder, the code vector is defined as ten pulses of 1 or −1, out of 40 optional subframe locations, where the remaining 30 are set to zero. Thus, the search operation is based on finding the best ten locations for the pulses, and establishing whether each pulse is a 1 or a −1. This task is performed four times per 20 ms. Prior to the search and determination of the code vector, two other speech analyses are performed. These are linear prediction coding (LPC) analysis and pitch estimation, respectively. These are the fundamental components of the encoder that give the decoder the ability to reproduce the synthesized speech at the receiver [4]. There are more than 24 functions for the encoder and more than 12 functions for the decoder part provided by the standard C code [5]. Amongst the encoder functions, the search algorithm (search_10i40) by far consumes most of the encoder processing, which is more than 22% of the total required cycle count for the encoder. This function alone has considerable inherent parallelism, and speed-ups of more than 4-times have been achieved in executing this function on a 16-element reconfigurable fabric. Other EFR sub-algorithms such as correlation and filtering also exhibit considerable parallelism; these are considered in the following discussions.

11.2.1.2. MPEG Image Coding

The next generation mobile devices must support video I/O in order to enable multimedia content delivery. Video compression is an integral part of these multimedia applications. MPEG standards for video compression are important for realization of mobile digital video

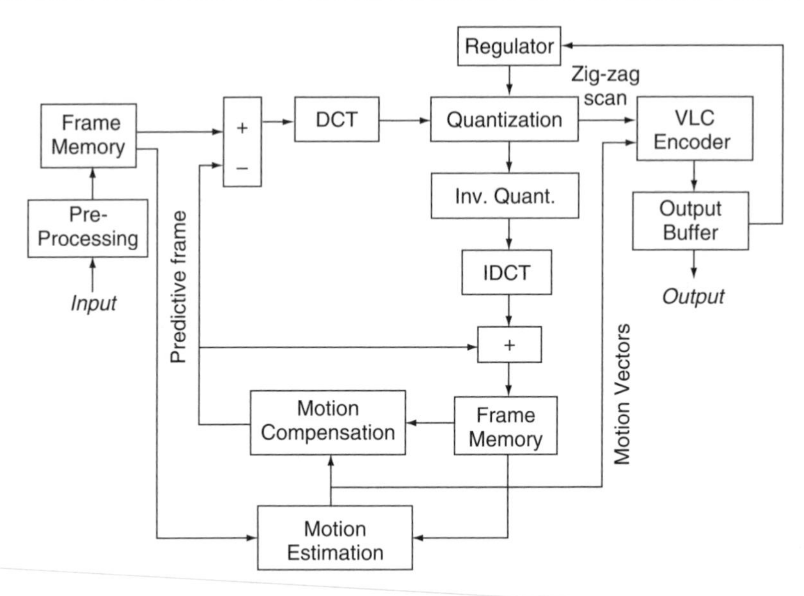

Figure 11.3 Block diagram of an MPEG encoder

services, such as video conferencing, streaming video, and digital video broadcasting (DVB) over a cellular network. Figure 11.3 depicts the block diagram of a typical MPEG encoder.

The required functions for the encoder algorithm are:

Preprocessing: Operations such as color space (CS) conversion from RGB to YcbCr, pre-filtering and subsampling.

Temporal Redundancy Removal: After preprocessing, motion estimation of image pixels is performed to remove temporal redundancies in successive frames (predictive coding) of P type and B type picture frames. Algorithms such as full search block matching (FSBM) may be used for motion estimation. SIMD architecture is particularly well suited for FSBM algorithm implementation. The basic operations involve summing of absolute difference (SAD) between pixels of a reference block and the source frame. A reconfigurable array of processors is capable of performing SAD of several reference blocks in parallel while parallelizing the operation of each SAD.

Spatial Redundancy Removal: Each macro-block (typically consisting of six blocks of size 8×8 pixels) is then transformed using the discrete cosine transform (DCT). The resulting DCT coefficients are quantized to enable compression. The quantized coefficients are rearranged in a zigzag manner (in order of low to high spatial frequency) and compressed using variable length encoding. The quantized blocks of I and P type frames are inverse quantized and transformed back by the inverse discrete cosine transform (IDCT). This operation yields a copy of the picture, which is used for future predictive coding. A 2-D DCT on a reconfigurable fabric can be separated into two 1-D DCT operations, one across rows, the other across columns. While a traditional DSP may be capable of parallelizing the 1-D DCT steps, it would require significant data movement to transpose the 8×8 matrix between the two 1-D DCT steps. A reconfigurable fabric can reduce or eliminate the transposition operation, reducing the cycle count considerably.

11.2.2. Encryption Encoder/Decoder

With the increased commercial use of wireless devices, link security and privacy have become an important issue and thus an integrated part of a wireless link design. Stream ciphering, which is often used for circuit switch services such as voice, operates on a bit-by-bit basis, producing a single encrypted bit for a single plain text input bit. Stream ciphers are commonly implemented as the exclusive-or (XOR) of the data stream with the keystream. Although GSM A5/1 and A5/2 ciphering algorithms, placed just before the GMSK modulator, were designed for a highly secure link, third generation systems enjoy superior algorithms, which are publicly known, but difficult to decrypt without the appropriate ciphering key. The initial step of any ciphering operation involves a key management process, known as authentication and key agreement (AKA), where a cipher key is generated for the block or stream ciphering. The current suggestion for 3GPP AKA algorithm is called MILENAGE. The MILENAGE algorithms set uses as its kernel the well-known 'Rijndael' algorithm [6] which was selected by the National Institute for Standards and Technology (NIST) in the United States as the Advanced Encryption Standard (AES). This process is performed once at the beginning of each session, using layer-3 messages. On the other hand, the encryption algorithm is applied to every transmitted information bit and, as such, application specific hardware has been used for the encryption/decryption processing so far. The 3GPP air interface encryption and integrity algorithm are both built around a block cipher called KASUMI [7]. KASUMI produces a 64-bit output from a 64-bit input, under the control of a 128-bit cipher key. The confidentiality algorithm *f8* is a stream cipher that is used to encrypt/decrypt blocks of data between 1 to 5114 bits, under a confidentiality key. The block diagram of *f8* algorithm [7] is shown in Figure 11.4.

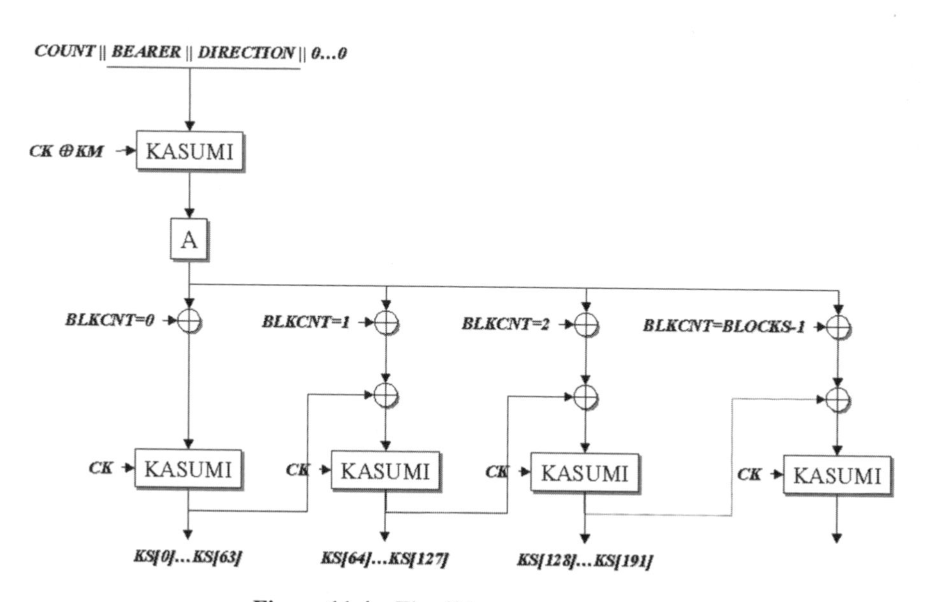

Figure 11.4 The *f8* keystream generator

There are two levels of parallelism exhibited by the *f8* algorithm. The less obvious one is the inherent parallelism in the KASUMI algorithm that can be exploited at word level, and the more obvious one is the algorithm level parallelism apparent in Figure 11.4. Since each parallel block of KASUMI performs similar operations on a different data block, it is evident that the *f8* algorithm exhibits high level of parallelism and is best implemented on a reconfigurable array fabric platform.

11.2.3. FEC Encoder/Decoder

With the exception of the IEEE 802.11b BPSK, QPSK and CCK modulations, almost all modern digital wireless modems use forward error correction (FEC) coding for data protection. FEC (or channel) coding adds structural redundancy to the transmitted information, and, provided the information symbol energy E_b to noise spectral density N_0 is sufficiently high, there is considerable coding gain that justifies the impairment caused by the inclusion of the redundancy. Coding gain is defined as the reduction in required E_b/N_0, to achieve a specific error rate in a coded system, over one without coding [8]. By far the most widely used coding scheme for voice and low data rates is convolutional encoding. Since the introduction of the concept of parallel concatenated coding with recursive convolutional encoders in 1993 [9], turbo codes have found widespread acceptance, among others, for high data rate applications, offering higher coding gain.

A convolutional code maps successive binary information k-tuples to a series of n-tuples and can be implemented with k stage shift register (where k is equal to the coder constraint length) and n modulo-2 adders [8]. The encoder shifts in a new bit of data, while sampling the output of the adders for the encoded output. The decoding of convolutionally encoded sequences of data is based on a maximum likelihood decoding algorithm that takes advantage of the trellis structure of the encoder, to reduce the evaluation complexity. This algorithm is known as the Viterbi algorithm (VA) [10]. With each received n-tuple, the decoder computes a metric or measure of likelihood for all paths that could have been taken during that interval and discards all but the most likely to terminate on each node. An arbitrary decision is made in the unlikely event that the path metrics are equal. The metrics can be formed using either hard or soft decision information with little difference in implementation complexity; modern systems typically use a soft decision VA, many use an MAP-based algorithm. As most of the computational complexity resides in the decoder, it is this algorithm that we examine here for parallel implementation.

One of the major steps in the VA is to update the path metrics (*pm*), by the appropriate branch metrics (*bm*) according to equations (1) and (2), which are executed in the nodes of a 'butterfly' structure shown in Figure 11.5. The output at each butterfly node is the lesser of

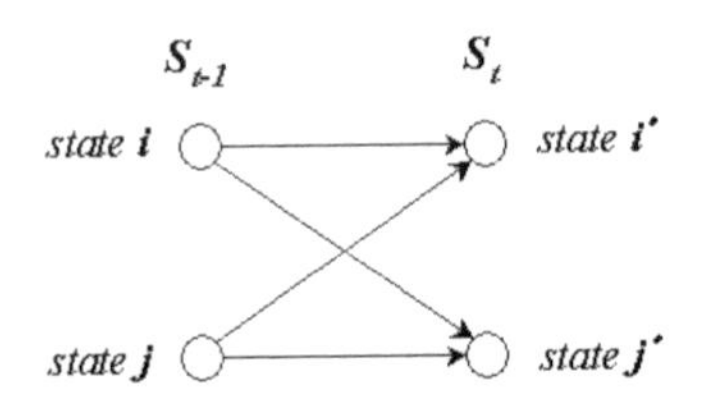

Figure 11.5 Viterbi algorithm (VA) butterfly operation

the two path metrics, summed with the corresponding branch metric. This process is known as add–compare–select (ACS).

$$pm_{i,t+1} = \min(pm_{i,t} + bm_{ii'}, pm_{j,t} + bm_{ji'}) \tag{1}$$

$$pm_{j',t+1} = \min(pm_{i,t} + bm_{ij'}, pm_{j,t} + bm_{jj'}) \tag{2}$$

Figure 11.6 shows the implementation block diagram of the Viterbi decoder which can be divided into three major units. The received input data (soft or hard noisy observation of the transmitted data) is fed to the branch metric unit (BMU) which calculates the set of branch metrics ($bm_{ij'}$) recursively. These metrics are then fed to the add–compare–select unit (ACSU) that updates the path metrics (pm_i) for each new time instant ($t+1$). Finally, the survivor memory unit (SMU) processes the decisions being made in the ACSU and outputs the estimated path (with some latency). The first and the last units of a Viterbi decoder are purely feedforward. Therefore, high-speed parallel implementation can be easily found for the BMU and SMU. The bottleneck in decoder implementation is the ACSU, which includes both addition and comparison in a recursive manner. More specifically, the maximum selection is both nonlinear and data dependent. Also, calculation of each path metric at instant $t+1$ requires knowledge of path metrics at instant t. As detailed in Ref. [11], this bottleneck can be eliminated by algebraic transformations resulting in a linear scale solution for the Viterbi algorithm at the word level.

Fundamentally, Viterbi decoder speed can be increased either by pipelining (or super-pipelining [12]) or by parallel computation. Pipelining methods are mostly suitable for application specific hardware, while both computing devices and ASICs can utilize the inherent parallelism. In general, additional parallelism can be introduced into the VA at the bit level, the word level and the algorithm level [13]:

- At the bit level, transformations allow introduction of bit level pipelining, such that the critical path for each implementation runs only through a few bit levels resulting in the possibility of significant increase in the clock rate [14].
- At the word level, using algebraic transformations, the two operations of the ACS recursion may be identified to form an algebraic structure called *semi-ring* which can represent the VA as linear vector recursion [14]. Using this algebraic transformation, the bottleneck of the VA (the nonlinear feedback in the ACSU) is eliminated. At the word level, it is also possible to perform the operation of each of the states independently, in parallel.
- Finally, at the algorithm level, asymptotic behavior of the VA can be exploited to derive parallel processing architectures [11].

The bit-level parallelism can be incorporated into almost any arithmetic logic unit, regardless of the particular choice of parallel architecture. However, the utilization of word level parallelism is highly dependent on the computational fabric and resources. For a parallel processing

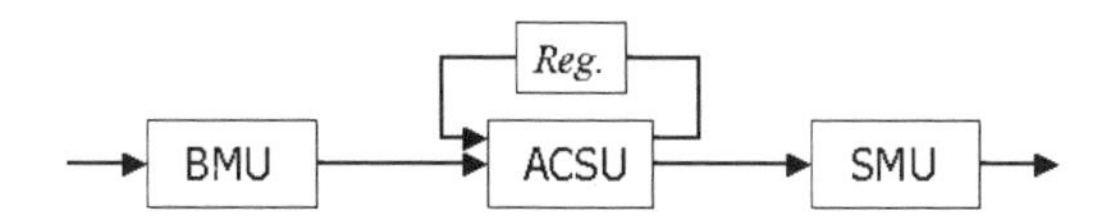

Figure 11.6 Viterbi decoder implementation architecture

engine, with a 'modest' number of ACS units at each node (e.g. less than 10), the feasible word level parallel execution is the concurrent state implementations, ideally with one butterfly operation at each node. It is also possible to assign a single state to each processing node, which might lead to excessive intermediate data movement between the nodes. Algorithm level parallelism offers the highest rewards in extracting the maximum parallelism. The only two drawbacks associated with algorithm level parallelism are, firstly, the increase in processing latency and, secondly, the overheads associated with the block overlap. There is usually a tradeoff between the two, where the block overlap overhead can be reduced at the expense of more throughput latency.

Although only the maximum a posteriori (MAP) decoders, used for decoding turbo codes [15, 16], provide the optimum decoder performance, other soft-output decoders, notably the soft-output Viterbi algorithm (SOVA), have also been proposed and successfully demonstrated for turbo decoding [16]. Further, it has been shown [16] that structurally, SOVA is similar to the log-MAP algorithm, where, the '*max*' operation of VA is further augmented by an additional correction factor $\ln(1+e^{|-x-y|})$. This correction factor can be implemented by an extra lookup table operation in the path matrix and L_k, the log-likelihood function, calculations. Therefore, the forward and backward recursion of the MAP algorithm for calculation of $\alpha_{k-1}(s')$ and $\beta_k(s)$ can be viewed as a 'generalized VA', and computation of L_k as a 'generalized dual-maximum' computation. For about 0.5 dB penalty in the *SNR* performance [17], it is also possible to exclude the correction factor in the calculations, resulting in an algorithm known as the max-log-MAP. Although turbo decoding requires more operations for the forward, backward, and log-likelihood calculations, and requires several iterations of both decoder blocks with interleaving operation in between, the parallel implementation of a turbo decoder can be based on similar spatial mapping previously discussed for the VA. While a turbo decoder has a smaller number of states, it does require the 'butterfly' operation on two directions, forward and backward, and requires log-likelihood calculation for every symbol, as the estimation is symbol-by-symbol, as opposed to a Viterbi decoder, which estimates the maximum likelihood for a sequence of symbols.

11.2.4. Interleaver/Deinterleaver

In current digital communications systems, at two different parts of the transceiver chain, data interleaving is employed for two fundamentally different reasons. The first location is in the turbo coder and decoder, where it is used to reduce the probability of generation of codes with low weights, at the output of the encoders. The interleaving operation reduces the probability of both encoders having the same inputs, which could cause low weight codes at the output of both parallel encoders. Of course, in the decoder, the same interleaving operation is required for the systematic data, between the two log-MAP decoders. The second location where interleaving is used is after the FEC encoding, for randomizing bursts of channel errors caused by multipath fading. This causes such errors to appear as random, as an AWGN channel, at the input to the FEC decoder. The random appearance of errors is essential for effective operation of the decoder. A complementary deinterleaving operation is required in the receiver to rearrange the data back into its original sequence.

While the channel interleaving, block or bit, is usually defined by mathematical or operational algorithms (cf. WCDMA first and second interleaving algorithms as defined in Ref. [18]), the turbo interleaving strategy is often designed, for maximum performance, for a particular channel arrangement, and as such may not conform to a definable standard algorithm. A parallel processing fabric, especially with reconfigurable interconnect, may be used for limited permutation changes and bit reordering operations. However, for long interleaving depths used in channel and turbo interleaving, little or no advantage can be gained from such platforms.

11.2.5. Modulation/Demodulation

Modulation is the process of conditioning a transmitted waveform in order to convey the information message to the receiver. Demodulation is the process of extracting this message from the conditioned signal. Modulation schemes can be categorized into two fundamental groups: analog, such as AM, FM and PM, and digital, such ASK, FSK and PSK [19]. All modern wireless systems use one or more types of digital modulation for voice and data communications. Digital modulation schemes differ greatly in performance and complexity. Modulation schemes are selected for their power and spectral efficiency, bit-error-rate (BER) performance and envelope behavior. While simple linear PSK modulation schemes such as QPSK and $\pi/4$-QPSK offer the best trade-off between power and bandwidth requirements, other more complex nonlinear modulations such as GMSK provide high spectral efficiency with the desirable constant envelope behavior, for low data-rate wireless systems. Recent requirements for high data-rate transmission have led to the inclusion of such schemes as 8-PSK, M-ary QAM and OFDM. There are other signal conditioning techniques, such as spread spectrum, used in CDMA cellular and GPS systems, that need to be included under the modulation heading. Although the modulation schemes vary in implementation complexity, it is the demodulation in the receiver that requires by far most of the processing. Linear modulations usually require mapping of signal to a particular constellation point, while the demodulation, after the signal detection, is based on a reverse operation. Nonlinear modulations, such as MSK and GMSK, require considerably more effort, both in transmitter and receiver end, compared to the linear counterparts. For example, GMSK modulation (with $BT=0.3$) used in GSM-based systems, requires a modified Viterbi algorithm, known as maximum-likelihood sequence estimator MLSE (or MLSD for MLS detector), for the demodulation and equalization of the received signal [20]. In MLSE, the branch metric calculations of the classic Viterbi algorithm are modified to include the inter-symbol interference (ISI) contribution of previous symbols. Therefore, in demodulation of GMSK modulated signals, a reconfigurable parallel processing fabric can be used in much the same way as it was discussed above for the Viterbi algorithm in the context of FEC.

Efficient implementation of OFDM modulation and demodulation is also possible on reconfigurable processing fabric platforms, since the inverse fast Fourier transform (IFFT) and fast Fourier transform (FFT) algorithms used for these operations exhibit a high degree of parallelism. The radix two FFT algorithm is performed using a butterfly operation, in much similar manner to Viterbi butterfly in terms of data arrangement and movement, where each butterfly operating on a set of input data can be performed on a single processing array node of the fabric. The nodes can then operate concurrently for the transform operation on the parallel data. Depending on the processing array dimensions it may be optimum to use larger radix structures for the FFT operation.

The spreading and despreading of data, performed for spread spectrum modulation, exhibit the most regular form of parallelism, where the same operation is carried out on all the data

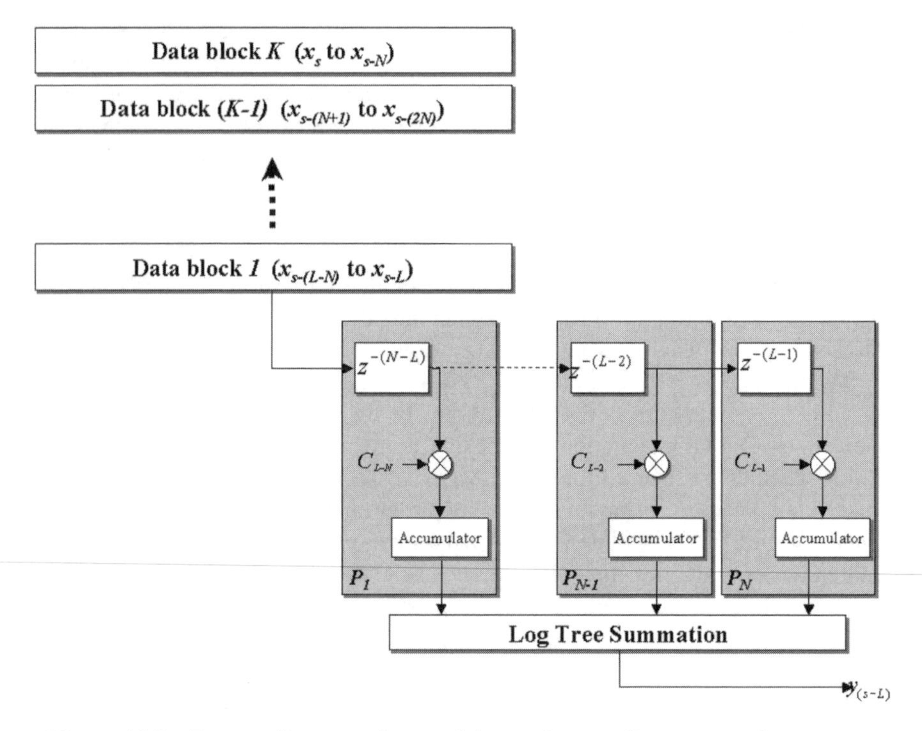

Figure 11.7 Despreading operation spatial mapping to a linear array of processors

samples, in parallel. For example, in a WCDMA terminal, the despreading operation of a symbol of an AMR channel, requires 128 multiplications of the received data with the appropriate complex code, and the accumulation of the multiplication results. This operation can be efficiently spatially mapped to a linear array of processors. Figure 11.7 shows such a spatial mapping example for a linear array of N processors, and a despreading of L chip samples.

In general, the incoming samples can be processed in K separate blocks, with $K=(L+1)/N$, where the samples in each block are processed in parallel inside the array. In the example in Figure 11.7, each processing element is made up of a complex code multiplier and an accumulator register. The complex chip samples of the block k are moved into the array, where they are multiplied by the despreading codes, and accumulated in each processing element. After the operation on the final block (K), a final summation is performed of all the accumulators of the processors. This operation can be carried out in $\log_2(N)$ steps with $N-1$ additions. Therefore, algorithms such as despreading, correlation and FIR filtering that are based on the similar structure on a reconfigurable processor array, should asymptotically be implemented N times faster as compared with a conventional sequential DSP engine.[†] The reconfigurable nature of the array plays a central role in the efficient execution of these operations whereby the array may operate in a SIMD fashion when partial sums are accumulated on individual elements, and switching to a MIMD mode when sums are accumulated across elements.

[†] Although a speedup factor of N is theoretically possible, a realistic speedup would be less that N due to I/O operations and data movement latencies that depend on the processor and memory architectures.

The above mapping example utilizes the word-level parallelism inherently present in the correlation algorithm. Thanks to the immense flexibility of the reconfigurable fabric, it is also possible to devise alternative spatial mappings that utilize the algorithm-level parallelism, where even higher computational gain should be possible.

11.2.6. Interpolation and Pulse-Shaping Filtering

Any wireless transmitter conforming to a standardized system definition, has to meet the spectral mask set by the particular standard. For example, the WCDMA spectral mask is defined by a root-raised cosine (α=0.22) pulse-shaping filter that is required to meet a specified adjacent channel leakage power ratio (ACLR) [21] of 45 dB @ 5 MHz. The pulse-shaping filter, which is often the last baseband operation in the transmitter chain, is also an interpolating filter, increasing the sample rate at the output of the filter. This increase in sample rate is desirable as it allows a wider filter transition band, reducing the required number of filter coefficients [22]. The higher sample rate also relaxes the requirements of the post digital-to-analogue converter lowpass filter [19]. Therefore, the interpolation and pulse shaping in the transmitter are performed by a single filter with real coefficients, placed on both in-phase and quadrature signal paths. Even though the inserted zero samples for the interpolation operation change the required filter structure from an FIR to a polyphase one, similar word and algorithm level parallelism still exists in this family of filters.

11.2.7. Inner Receiver

So far the discussion of the transceiver chain has included operations that are partitioned between the transmitter and the receiver. Some of these functions, such as source coding, require more computational effort at the encoder end, the transmitter, while the rest of the functions, such as FEC, place most computation demand at the receiver. What remains is a large number of baseband operations that are required only at the receiver. These functions are mostly required for frame, clock timing, and carrier synchronization, and also to mitigate and compensate for the signal distortions caused by the propagation channel, noise and interference. There are also several new functions such as digital down-conversion, channelization filtering, and sample-rate conversion, that have been introduced as SDR digital front-end baseband algorithms [22]. In the following sections, the parallel implementation of such functions are briefly discussed.

11.2.7.1. Digital IF (DIF) and Matched Filtering

Digital IF is a relatively recent addition to receiver baseband algorithms, which concept is discussed thoroughly in Ref. [22]. There is no common definition for DIF or digital front-end (DFE), which has also been referred to as zero IF (ZIF) in some publications. The major functionalities required from this block are:

- I/Q down-conversion;
- sample-rate conversion, and
- filtering.

Depending on the application and system specification, the complexity and the implementation of these functions could differ considerably. For example, whilst a handset may require filtering for a single channel of a single air interface, a basestation transceiver may require multisystem operation, leading to channelization and filtering of several channels of a number of air interfaces, concurrently. Therefore, whilst the channel filter for a handset may be a cascaded-integrated-comb (CIC), followed by one or several half-band filters, the base station DIF may require CIC filters for rate change, followed by a maximally decimated filter bank for each air interface [22].

Matched filtering is an essential part of the receiver front-end, as the optimum SNR performance is achieved at the output of the matched filter. This filter can be cascaded to the DIF unit or can even be part of the DIF, in both the handset or BTS receiver. In the handset, it can be the last stage, after the CIC filters, operating at twice the final sample rate, which helps to reduce the required number of filter coefficients. It is also possible to include the matched filter as part of the clock timing synchronization unit, where the samples are corrected by the timing error, by interpolation and resampling. If interpolation/resampling is to be performed independently, a simple structure such as a four-channel polyphase filter, with a two-tap FIR at each channel, can provide adequate performance for most wireless applications.

Almost all of the above operations are based on a transversal tap-delay structure that exhibits similar word and algorithm level parallelism, as was discussed for FIR filters in the modulation/demodulation section, resulting in nearly N times processing advantage (where N is the number of processors). Again, the reconfigurable nature of the fabric plays an important role here, whereby some groups of processors can perform the MAC functions while other groups can reassemble the input or output array in a way that optimizes memory I/O. However, if instead of an FIR structure, an IIR structure is used for any of the filtering operations, the algorithm level parallelism is lost, due to the inherent feed-back path in the IIR structure.

11.2.7.2. Channel Estimation

Radio signals that are subject to multipath propagation suffer attenuations from multipath fades, which are frequent and can be as deep as 30 dB. Multipath fading is sometimes referred to as 'fast' or 'Rayleigh' fading. The fading rate of the signal envelope varies depending on the speed, incident signal geometry, and channel coherence time. Similarly, depending on the propagation environment, the amplitude fading can have a *Ricean* or *Rayleigh* probability distribution function.

The first generation cellular systems, with relatively narrow band and analog modulations, used space diversity to combat the effects of multipath propagation, which was manifested as 'flat' fading over the full band of signal frequencies. Depending on the time dispersion of the propagation channel, a digitally modulated signal can also suffer from inter-symbol interference (ISI), which can only be removed by channel equalization or spatial filtering. Modern wireless systems such as IEEE802.11a have a signaling bandwidth larger than the channel coherence bandwidth, even in indoor environments, suffering from 'frequency-selective' fading, that can be corrected with frequency domain equalization. Therefore, irrespective of the transmitter modulation technique, a modern wireless digital receiver requires reliable knowledge of the channel behavior, known as channel estimation, on an on-going basis, for a nondeterministic time variant terrestrial or indoor channel.

Each channel estimate has three major time varying components [23] – the delay ($\tau_n(t)$), the attenuation ($\alpha_n(t)$) and, finally, the phase ($\theta_n(t)$) – associated with each significant resolvable propagation path. The terminology and the estimation method of these three components may differ for different air-interface technologies. In CDMA-based air interfaces, a 'path searcher' unit is used to estimate the delay component, whilst either the common pilot or dedicated pilot symbols are used for the estimation of the other two components. In an OFDM-based air-interface, such as IEEE802.11a, dedicated symbols such as the 'long' symbols in the preamble are used to estimate the frequency response of the channel. In such systems, the Fourier transform of the channel estimate (i.e. Fourier transform of the channel impulse response, $C(\tau,t)$) is used for channel equalization. Whilst the path searcher of the CDMA-based systems uses correlation algorithms based on a transversal filter, OFDM channel estimation is based on FFT algorithms. Both these algorithms enjoy a high degree of parallelism with their parallel implementation already discussed in the previous sections.

11.2.7.3. Equalization

Equalization is a technique mainly designed to deal with inter-symbol interference (ISI), which is usually observed in high data-rate communication systems. The motivation for equalizer receivers dates back to the 1960s. Early work on conventional receivers for various synchronous data transmission systems (QAM, PAM, etc.) suggest a structure with a linear receiver filter (or equalizer) followed by sampling and symbol-by-symbol decision [24, 25]. It has been shown [25] that under several performance criteria the receiver filter can be expressed by a cascade of a matched filter and a transversal (or lattice) filter with tap spacing equal to the symbol interval. The matched filter is designed to optimize the signal-to-noise ratio while producing a residual ISI at its output. The transversal filter, however, establishes a reduced ISI with a penalty in noise immunity. Although these structures are known in the literature as sub-optimal linear receivers, they can offer adequate performance in some applications. The noise enhancement problem of linear receiver filters can be avoided if symbol-by-symbol decision is replaced by a technique based on the detection of the entire sequence (maximum-likelihood sequence estimation or MLSE) or employment of some highly nonlinear processing methods such as a decision feedback equalizer (DFE) [23]. While the linear equalizer is based on the transversal filter structure which exhibits good word and algorithm level parallelism (as discussed for FIR), the nonlinear DFE, as with IIR filters, only benefits from word level parallelism. MLSE equalizer implementation is based on modified Virterbi algorithm [20], and is discussed in Section 11.2.5.

Spread spectrum systems such as WCDMA and cdma2000 do not require adaptive equalizers when operating channels with high spreading factor, with channel equalization provided by the rake receiver. On the other hand, when the spread spectrum processing gain (PG) does not provide sufficient gain for the wanted signal to overcome the channel noise, multi-user interference and self-interference caused by multipath propagation, there is again a need for ISI mitigation (at chip rate), which can be performed adequately by adaptive linear equalizers [26]. Similarly, wireless systems based on OFDM modulation do not require time domain equalization. Instead, a one-tap frequency equalizer (FEQ), together with a cyclic prefix, is used to combat multipath effects in such systems [27].

11.2.7.4. Recovery Loops

In a digital receiver, typically four recovery loops are employed: amplitude, phase, timing, and carrier. These recovery mechanisms are required because of imperfections in the transmitter and receiver hardware, and the distortions caused by the propagation channel. For example, the timing error exists because of the clock frequency error in the transmitter, clock frequency error in the receiver and the propagation path delay change of the received signal. Therefore, any digital receiver must reconstruct the timebase of the transmitter in order to convert the modulated waveform from continuous time into a sequence of discrete data symbols. The traditional solution for achieving clock synchronization is to adjust the A/D sampling clock so that output samples are provided with the required frequency and phase. However, in modern SDR receivers the sampling is performed using a fixed, free-running oscillator. In such arrangements, the samples passed to the baseband receiver are nonsynchronous with respect to the transmitted signal. Clock synchronization is performed digitally using multirate sampling techniques, to resynthesize the received signal with the desired sampling frequency and phase. Evidently, the performance and operation of each loop depend considerably on the physical property of the parameter under control, and as such require different algorithms with varying complexity [28]. However, they all require three fundamental operations:

- the generation of an error signal;
- conditioning of the error signal, and finally
- the application of the correction signals in a closed-loop manner.

These three components of the recovery loops are based on nonregular algorithms, that, at first glance, exhibit little or no parallelism at the 'innermost loop' level. However, on closer examination, considerable parallelism can be extracted for each loop both at word and, more importantly, at algorithm level. For example, a WCDMA receiver requires up to eight rake fingers to run in parallel. Each of these eight fingers requires independent recovery loops that can be executed in parallel on a reconfigurable processing fabric. Further, at the expense of some data latency, each algorithm can be applied to several symbols in parallel.

11.2.8. Summary

In this section, a brief look has been taken at a generic transceiver and a breakdown of the functionalities of transmitter and receiver baseband algorithms. By way of an example, it was shown that an arbitrary L-tap transversal filter structure can be supported efficiently on a linear array of processors. The parallel implementation of the Viterbi algorithm was also discussed, and possible word-level and algorithm-level parallelism was identified. Discussion of the other algorithms in the transceiver chain revealed that units such as correlators, FIR filters, and linear equalizers, are based on the transversal filter structure. Further, it was also discussed that several major algorithms in the receiver, such as convolutional and turbo decoders and MLSE equalizers (with minor modifications), are based on the Viterbi algorithm structure. There exists a number of irregular algorithms that initially appear unsuitable for platforms that have a fixed parallel structure. However, it was shown that even algorithms such as vocoders and receiver recovery loops exhibit word and algorithm level parallelism, and can thus benefit from a parallel implementation on a reconfigurable fabric.

11.3. The MS1 Reconfigurable DSP (*r*DSP)

11.3.1. The MS1 rDSP Architecture

Following the discussion in the previous section of the suitability of parallel computing architectures for digital signal processing algorithms used in current and future wireless systems, it is interesting to consider an example reconfigurable processor, currently available commercially. Morpho Technologies† provides a reconfigurable DSP (*r*DSP) solution, based on a reconfigurable array processing paradigm, known as MS1 *r*DSP, which is also used as the reconfigurable compute fabric (RCF) in the Motorola MRC6011 family of products. In this section, MS1 *r*DSP architecture is discussed further.

As in conventional reconfigurable systems [29], the MS1 core contains a reconfigurable block, called the *r*DSP fabric or RC array, and a 32-bit RISC processor, as the core controller, known as mRISC (Figure 11.8). The RC array, which is shown in more detail in Figure 11.9, consists of reconfigurable cells (RC) interconnected by a reconfigurable network. Both the functionality of the RCs and the network interconnections are determined by a configuration program, called context, residing in context memory. By writing the appropriate context, the developer can use the RC array to exploit the parallelism available in the application. The mRISC is a typical 32-bit RISC processor, implemented as a five-stage scalar pipeline. It contains sixteen 32-bit data registers, a scalable direct-mapped instruction and data caches, four programmable timers, and an interrupt controller. In addition to typical RISC instructions,

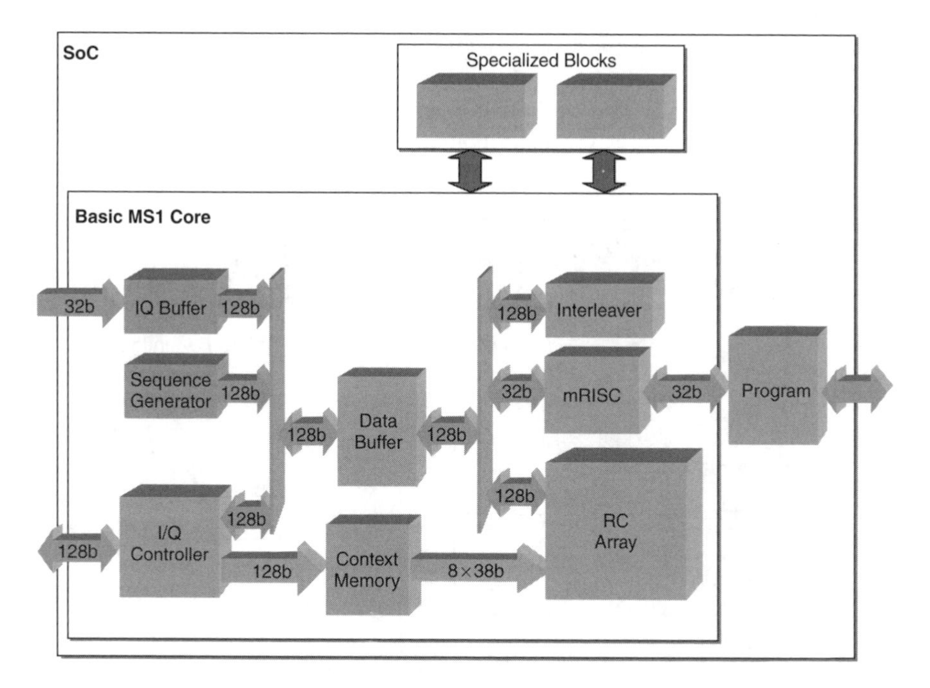

Figure 11.8 The MS1 *r*DSP core architecture

† Morpho Technologies is a privately held company located in Irvine, California.

the instruction set of the mRISC controller is augmented with special instructions to control the operation of the other core's components. The mRISC determines the application's control flow, executes the sequential tasks of the application, and starts transfers to/from the off-core memory. The instructions for the core controller are stored in the instruction memory.

The MS1 core also contains a context memory, a data buffer and an I/O controller.

- The context memory stores the context words for the RC array; the context words configure the operations performed by the RCs and their interconnectivity.
- The data buffer stores the input data to the RC array as well as the produced results; the data buffer allows data input/output to be overlapped with data processing.
- The I/O controller performs context loading from the off-core memory into the context memory and data transfers between the data buffer and the off-core memory; the I/O controller receives transfer requests from the core controller and generates an interrupt once the transfer is complete.

The reconfigurable cell is the programmable element in the RC array (Figure 11.9). Its 16-bit datapath performs arithmetic, logical, complex correlation and multiply-accumulate operations and includes 16 data registers. Three input multiplexers select the input data source. A context register stores the context word received from the context memory. The bits from the context register directly control the datapath components. The RC array works in a SIMD or MIMD fashion. Context is broadcast from the context memory along the rows (or columns) of the RC array on each clock cycle. The RCs along a given row (column) perform the same operation as indicated by the broadcast context. Depending on the selection of the input multiplexers, each RC processes data either from its internal registers, from another RC in the RC array or from the data buffer. In the latter case, data from the data buffer is broadcast row-(column-) wise if context is broadcast column-(row-) wise. Different rows or columns

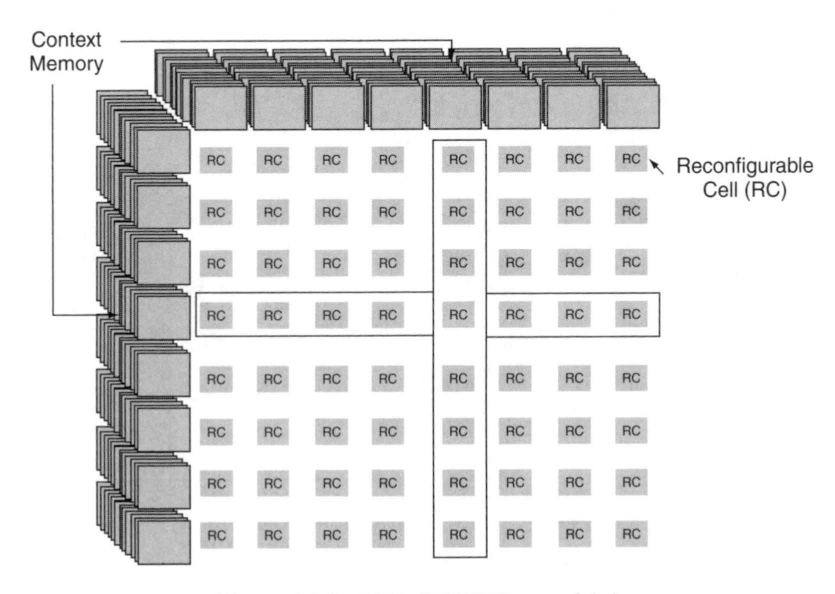

Figure 11.9 MS1 *r*DSP RC array fabric

can process operands from different sources. Furthermore, it is possible to change between row/column context broadcast mode on a clock cycle basis. The operation of the RC array is coordinated by the mRISC controller, which generates sequences of addresses to the context memory and data buffer. The context memory and data buffer respond to an address sequence from the mRISC controller by broadcasting context and data, respectively, to the RC array.

The MS1 family of cores can be easily integrated into a system-on-a-chip that may include special peripherals for a particular class of applications. For example, the MS1-16 chip incorporates the following peripherals for baseband processing in third generation wireless systems: a buffer to temporarily store I/Q samples, known as IQ buffer, a sequence generator to produce codes for spreading and despreading, and an interleaver engine to perform interleaving and deinterleaving.

11.3.2. Development Tools for the MS1 rDSP

Programs for the MS1 core consist of two parts: context code and core controller code. Both components are automatically generated from a single Morpho-C program, which is a C/C++ program (labeled 'User Code' in Figure 11.10) with special mStatements. mStatements are high-level statements that the programmer embeds into regular C/C++ code in order to control the operation of the MS1 core. In a Morpho-C program, the programmer can schedule tasks in order to overlap context loading, data input/output and processing, in many different ways. The Morpho-C code is preprocessed by the mTranslator. The output of mTranslator is a plain, pre-processed C program that is recognizable by the GNU gcc compiler. Both the mTranslator and the gcc compiler are part of the toolset supplied with the MS1 core. Therefore, the programming approach for MS1 is similar to the familiar paradigm employed by today's DSP processors, with additional instructions for the *r*DSP fabric.

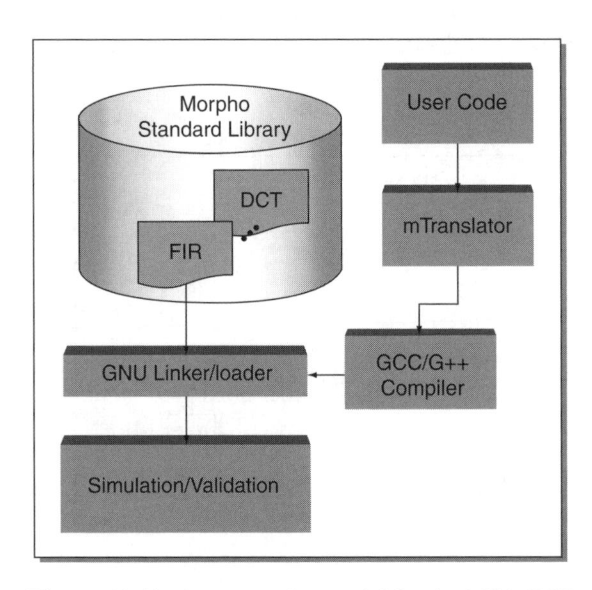

Figure 11.10 Programming model for the MS1 *r*DSP

11.4. Implementing a 3G Air Interface in SDR

In this section we describe a high-processing-load 3G wireless application that has been successfully implemented and benchmarked using the MS1 *r*DSP as a testbed example to demonstrate the real-time processing capability of the *r*DSP for a complex and realistic SDR requirement. An MS1 *r*DSP implementation of a forward-link WCDMA adaptive multirate (known as AMR) speech channel receiver has been developed and is described below.

The AMR channel is of particular interest, as it supports circuit-switched speech service, with a transmission timing interval (TTI) of 20 ms, tolerating low delay constraints, and supporting full macro-cell mobility. The short delay constraint demonstrates the real-time nature of the service, and the high mobility requirement implies a need for more complex physical layer algorithms to cope with the high Doppler spread, large time dispersion, and macro-diversity support associated with this type of cell environment.

Since the real-time algorithms implement the receiver side of the link, simulation models of the transmitter and the propagation channel were developed to provide the baseband signal at the input to the receiver. Therefore, as shown in Figure 11.11, the transmitter blocks, such as the enhanced full rate (EFR) encoder, AMR transport and physical channels, the common pilot channel (CPICH), and the root-raised cosine (RRC) pulse-shaping filter, are all implemented as simulation models. Further, impairments such as frequency offset and clock timing errors, multipath propagation and AWGN noise and system interference, are also generated by simulation and superimposed on the transmitter signal.

The receiver functions were supported on MS1 testbed and comprised a programmable eight-finger rake receiver, transport layer functions such as three Viterbi decoders, cyclic redundancy checks (CRC), and first and second interleavers. All the EFR decoder functions were also supported on the MS1 testbed. The interface between the simulation model and the testbed is based on file transfer. The functions on the MS1 testbed are executed in real time, as far as the receiver operation is concerned.

A detailed discussion of the WCDMA system is avoided here, as it is beyond the scope of this chapter. Only the points important for the discussion are mentioned, with references to

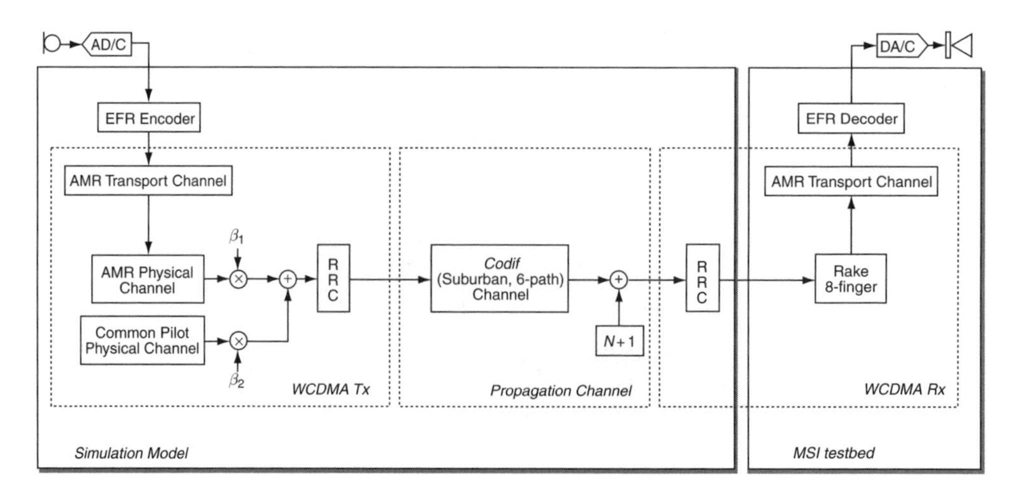

Figure 11.11 Overall system diagram

the relevant standards documents. Interested readers are encouraged to visit the 3GPP website (www.3gpp.org), for more information on the WCDMA system definition.

11.4.1. The WCDMA AMR Channel

The WCDMA AMR channel is defined in Ref. [30]. It is a channel that supports data rates from 4.75 to 12.2 kbits/s with three different BERs of 5×10^{-4}, 1×10^{-3} and 5×10^{-3}. In our implementation, for reasons of algorithm availability, the GSM EFR has been used as the speech codec; the GSM EFR vocoder has the same input and encoded parameters as the AMR-NB vocoder in mode MR122 [3].

At the transport layer, the 20 ms speech frames are divided into three classes of data, for mapping to three different radio access bearer (RAB) services, supporting differing levels of error protection. As shown in Figure 11.12, the RAB subflow No. 1, labeled 'A', has a 12-bit CRC parity, and a 1/3 rate convolutional encoder ($K=9$). RAB subflow No. 2, labeled 'B', has also a 1/3 rate convolutional encoder ($K=9$), without any CRC parity bits. Finally, RAB subflow No. 3, labeled 'C', has only a 1/2 rate convolutional encoder ($K=9$), and no CRC parity bits.

The first discontinuous transmission (DTX) insertion is used when a fixed number of bits are reserved for a traffic channel (TrCH). However, due to the scope of the demonstrated channel, where a dedicated control channel (DCCH) is not included, insertion of DTX is used to ensure that the number of bits at the input to the first interleaver is an integer multiple of the number of radio frames in the transmission time interval (TTI). First interleaving is performed on each RAB making the total 20 ms block, while the second interleaving is performed on 10 ms data segments. The AMR transport layer at the receiver performs the inverse operations to the transmitter, with CRC parity provided to EFR decoder, as a 'bad frame indicator'.

The AMR transport blocks are then mapped to the dedicated physical channel, as defined in Ref. [31] and shown in Figure 11.13. The above arrangement provides a spreading factor (SF) of 128. No TFCI is present in this slot format, as the channel supports a single service.

11.4.2. Transmitter and Propagation Channel

A top-level block diagram of the transmitter chain is shown in Figure 11.14. The voice signal is low-passed filtered and sampled by the A/D converter to 13-bits/sample, at 8000 samples/s.

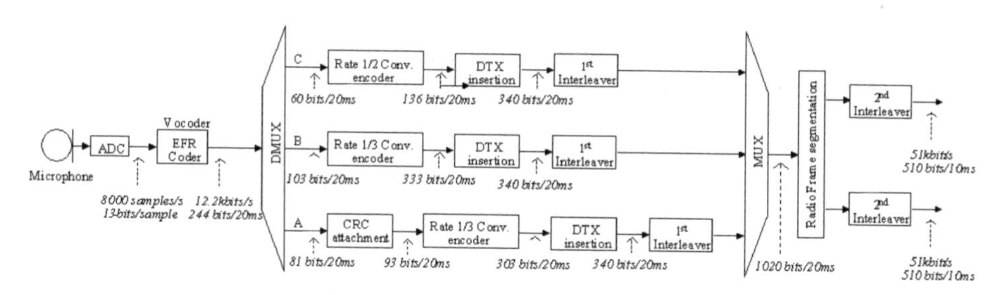

Figure 11.12 AMR transport channel

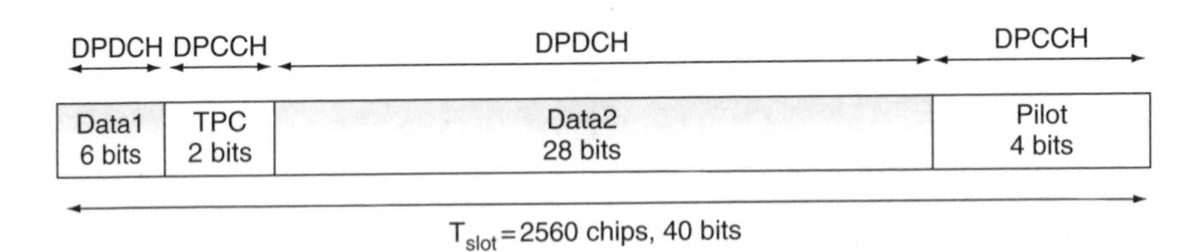

Figure 11.13 DPCH slot format used for the downlink 12.2 kbits/s speech channel

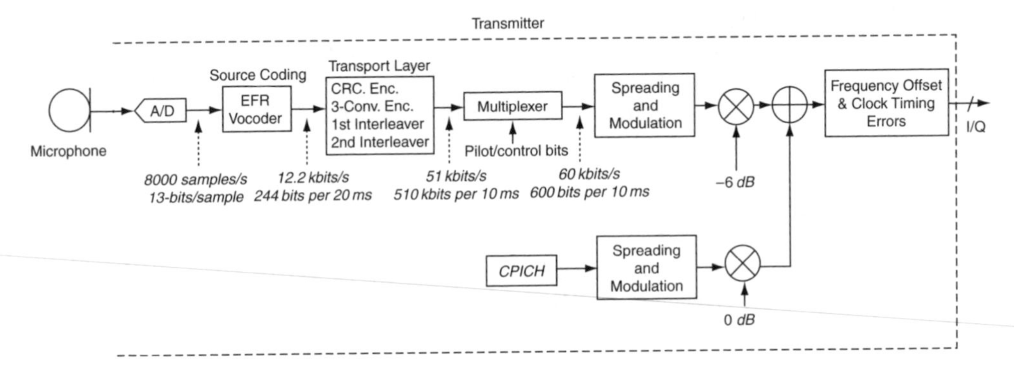

Figure 11.14 Transmitter block diagram

This provides a data stream of 104 kbits/s, which is then compressed by the EFR voice coder to 12.2 kbits/s. The encoded data is then processed by the transport layer block, which performs three convolutional codings (as shown in Figure 11.12), CRC coding, followed by first and second interleavers. Pilot and control bits are then multiplexed with the coded information bits, which are then spread by the channelization code and QPSK modulated, before spreading by the long complex scrambling code to the chipping rate of 3.84 Mchips/s. A CPICH channel is also generated and added to the AMR DPCH channel, with 6 dB more chip energy. The waveform is then filtered by a 49-tap RRC FIR pulse-shaping filter. The waveform is further modified, corrupted by frequency offset and clock-timing errors, before it is passed to the propagation channel model. The clock-timing error alters the output-sampling rate from its nominal rate, by resampling the output signal using linear interpolation. A numerically controlled oscillator (NCO) controls the sampling frequency at the output of the interpolator. A fractional-delay filter is used to modify the phase and frequency of the sampling clock. This function is needed to test the recovery loops at the receiver end. A perfect closed-loop power control algorithm is implemented for the AMR DPCH, while the CPICH signal power is subject to multipath fading.

The other computer simulated block is the propagation channel which is based on the CODIT channel model [32, 33]. The CODIT model is a measurement based wideband channel model, defined for different environments such as urban, suburban, rural, suburban hilly and rural hilly. In each environment, a number of scatterers is generated, with each scatterer characterized by an average power Ω_i, an excess time delay T_{bi}, a mean angle of arrival at the mobile α_i, and Nakagami-m fading with a specified m_i value. The CODIT model is valid for the local area behavior of the radio channel, where the channel is assumed to be wide-sense stationary uncorrelated scattering (WSSUS). This assumption is valid over approximately tens of wavelengths (well within the simulated speech duration). In the channel model, additive-white-Gaussian noise (AWGN) and system interference are also added to the transmitted waveform.

11.4.3. A Software-Defined Receiver Implementation

A top-level block diagram of the receiver chain is shown in Figure 11.15. The received signal is first filtered by a 49-tap RRC FIR filter. The baseband signal is then fed into the MS1 testbed where it is processed by the rake receiver, which performs the tasks of despreading and demodulation, along with the correction for channel and transmitter impairments. The output of the rake is then demultiplexed to remove the various physical layer control information such as pilot and power control bits, and passed to the second interleaver. After the collection of two radio frames, the first interleaving is performed on the received data, after which the DTX bits are removed. The received encoded bits are then processed by the three Viterbi decoders, for the 1/3 and 1/2 rate coders with a constraint length $K=9$ (256-state Viterbi algorithm), to correct errors caused by the channel. CRC parity decoding is performed on the RAB subflow No.1 decoded bits to detect any uncorrected errors on this class of data. If CRC parity indicates an error, the 20-ms speech frame is discarded, and the EFR vocoder is informed to repeat the last speech frame. If the CRC check does not detect any errors, the data stream is demultiplexed to its original state at 12.2 kbits/s. These bits are then processed by the GSM EFR decoder, to recover the original PCM speech frame at 104 kbits/s. A digital-to-analogue converter (DAC) processes the audio output signal, and the analog audio signal is played through a speaker. The above mentioned tasks are performed in a regular, time-multiplexed schedule within a CPICH symbol duration (256 chips) on the *r*DSP fabric.

From the perspective of benchmarking, in terms of performance and bit-exact operation and algorithms, all of the receiver blocks are well defined, with the exception of the rake receiver, where the computational load and algorithm complexity depend on the required performance, and can differ considerably for different implementations. Since more than 70% of the total receiver processing, shown in Figure 11.15, is consumed by the rake receiver block, it is important to discuss the operation and performance of this block further, so that a better appreciation of the computational load can be gained and thus a benchmark value of this block provided.

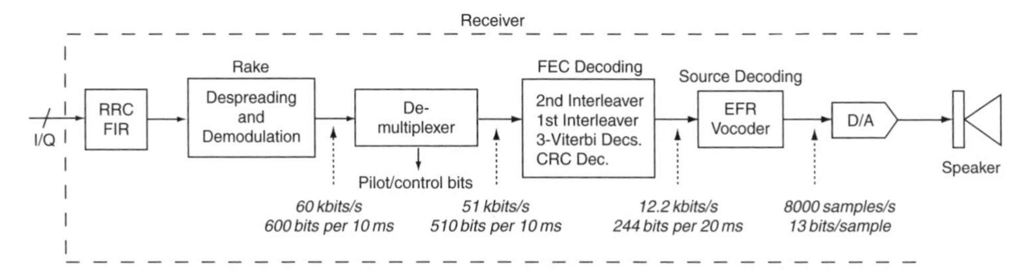

Figure 11.15 Receiver block diagram

11.4.3.1. Design of the Rake Receiver

As the chip-rate implementation of the rake receiver is all performed in software, a number of parameters that traditionally were 'hardwired', can be defined by the operating condition and the required overall performance of the system (i.e. they can be truly software defined). For example, as shown in Figure 11.16, at the expense of more processing, it is possible to reduce the memory requirement of the input sampled data, by operating at twice the chip rate (i.e. 2×OVS), and using an interpolation filter to achieve the desired time resolution for fine timing adjustments. It is also possible to set, and change at run-time, the desired number of rake fingers and the spreading factor of each finger. Finally, maximal ratio combining (MRC) is used for path combining. Alternatively, any other combining scheme can be integrated in the software, and selected 'on-the-fly'. The iteration rate of the rake receiver is a single CPICH symbol duration, which is 256 chips. The short iteration time ensures that TPC detection is fast, and only delayed by two to three CPICH symbols, depending on the maximum expected propagation channel excess delay.

The internal functions of each rake finger are shown in Figure 11.17. As can be seen, channel estimation, clock and carrier synchronizations are all performed independently for each finger. An external, path-searching algorithm provides the necessary timing information to select the optimum sampling instant prior to decimation. The path searcher output, although providing the initial timing epoch, is not sufficient for a circuit-switched channel, as the timing epoch changes during a long active session. Therefore, the synchronization system must be capable of correcting for an initial timing error, and then tracking the timing within a fraction of a chip, for optimum performance. With input samples at twice the chip rate, the delay-locked-loop (DLL), used for clock synchronization, is capable of correct synchronization, in the presence of worst-case expected noise and interference, with an initial acquisition error in excess of half a chip period. After timing correction by interpolation, the received samples are decimated by a factor of 8 for subsequent chip-rate processing by the correlation/despreader units. Soft output data are generated for the DPCH and CPICH channels, at the corresponding symbol rates. Early and late correlation values are also generated for the CPICH channel at ±½ chip periods away from the current (punctual) sampling instant. These are used for timing

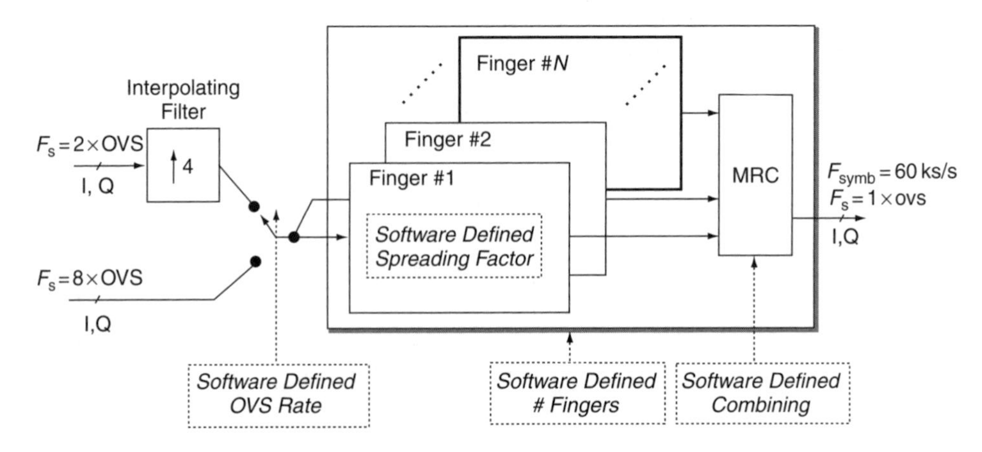

Figure 11.16 Rake receiver arrangement

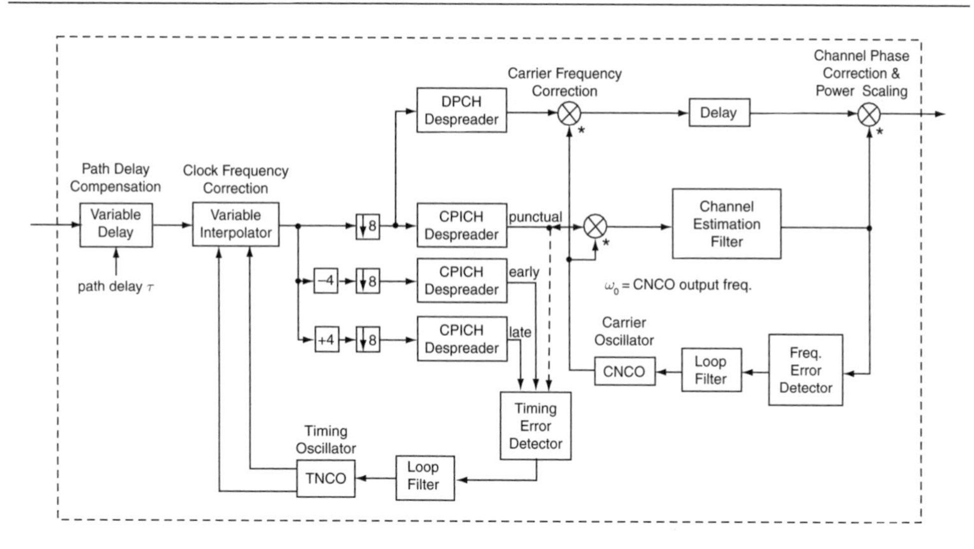

Figure 11.17 Rake finger block diagram

synchronization purposes. Although clock timing can be locked to the carrier frequency in the transmitter and the receiver, the clock and carrier synchronization loops of each finger are designed to operate independently, thus enabling the use of asynchronous clocks in the receiver, if desired. Timing adjustment is achieved using a noncoherent clock recovery loop, which is insensitive to the carrier phase of the received signal. The instantaneous timing error is updated at CPICH symbol intervals.

Channel estimation and compensation are based on estimates generated by an adaptive weighted multi-slot averaging (WMSA) algorithm [34]. Instantaneous channel estimates are computed at CPICH symbol intervals and low-pass filtered for improved SNR. The SNR improvement is directly proportional to the filter length, which depends ultimately on the coherence time of the propagation channel. As the channel conditions dictate, the filter length can be defined and changed on-the-fly. However, for most macrocell operations, a nine-tap filter provides reasonable channel estimation performance.

Since the local oscillator of the RF section is ‘free-running’, and is not locked to the baseband clock oscillator, an automatic frequency control (AFC) algorithm is provided to estimate and correct the residual frequency offset of the recovered symbols in baseband. This simplifies the analog front-end design of the receiver, at the expense of more baseband signal processing. This feature is provided for each finger separately, to provide maximum performance in severe multipath propagation environments, with large Doppler spread.

Table 11.1 summarizes the rake receiver specification used in the testbed demonstrator. These parameters are only for demonstration purposes, and as such can easily be changed or extended.

The BER performance of the rake receiver is shown in Figure 11.18, for a six-finger rake receiver, a mobile speed of 100 km/h, a carrier frequency offset of 0.15 ppm and a clock frequency offset of 10 ppm. A six-path ‘suburban’ CODIT channel is used for the simulations.

Table 11.1 Rake receiver specification

Function	Parameter value	Comments
Number of fingers	1–8	Software defined
Spreading factor (SF)	4–256	Software defined
Over sampling factor	2 and 8	Software defined
Path combining	MRC	Software defined
ADC clock frequency error	±10 ppm	'Free-running' oscillator and DLL
Carrier frequency error	±1 ppm	'Free-running' oscillator and AFC (0.15 ppm specified in Ref. [30]).
Channel estimation filter	Nine-tap	Based on CPICH and WMSA algorithm

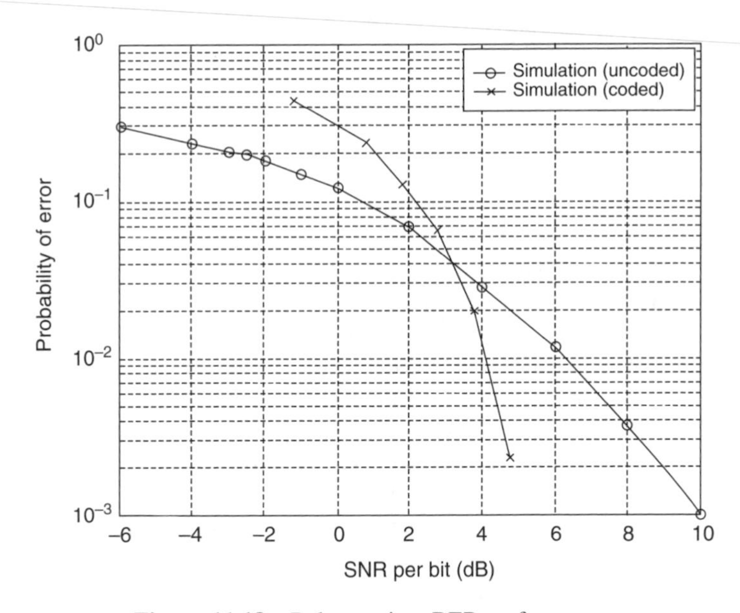

Figure 11.18 Rake receiver BER performance

11.4.4. Performance Capability

Recalling that the rake fingers operate at symbol-rate (66.7 μs), the second interleaver at the radio frame rate (10 ms), and all the other operations at speech frame rate (20 ms), the testbed implementation has proven that a unified time multiplexed approach of all the functions, based on the lowest time granularity, is a feasible processing approach for the implementation of real-time software defined systems. It also demonstrated that the *r*DSP reconfigurable platform is capable of providing the required processing power for the AMR receiver algorithms, for both the chip-rate and symbol-rate processing units. The processing requirement of the chip-rate algorithms (mainly the rake receiver) is around 70%, and the symbol rate (the transport layer) is 30% of the total processing load required by the above defined functions, with traditional DSPs only supporting a very small portion of this load, mainly the EFR decoding. In summary, it can be stated that an eight-finger rake receiver requires 20%; transport layer, including the three Viterbi decoders requires less than 2%; and the EFR decoder consumes less than 4% of the available processing time of a MS1-16 *r*DSP core.

The AMR receiver was fully implemented using the compilation tool-chain described in Section 11.3 and validated both on a cycle-accurate simulator as well as on a hardware development system as shown in Figure 11.19. In fact, the testbed results, and supporting cycle-accurate simulations indicate that all the transmit/receive baseband functions of a handset can be comfortably supported by a single *r*DSP core.

The discussion above has described a circuit-switched, forward-link AMR channel; however, further benchmarking and cycle-accurate simulations of a 384 kbits/s (SF=4) WCDMA channel, supporting packet transmission mode, have shown that the same core can support all the transmitter/receiver functions of such a channel. The 384-kbits/s channel requires a different rake receiver design, to work with the relatively shorter transmission duration of packet mode, and requires support of turbo encoding/decoding and MPEG codecs. Benchmarking simulation of a 384-kbits/s channel, based on the packet-switched rake receiver on a MS1-16 *r*DSP core showed that a four-finger rake receiver with two-branch diversity (eight fingers in total) requires 26% of the available processing time. Similarly, the turbo decoder (with five iterations) also consumes 26% of the total available processing time.

Figure 11.19 Hardware development platform

The above performance results were based on the initial MS1 *r*DSP core design and the prototype chip. By implementation of new features and through further optimization, the current commercial cores that are available from Morpho Technologies deliver significantly higher performance.

11.5. Conclusion

The aim of this chapter has been to introduce a new processing paradigm for applications such as wireless baseband and multimedia algorithms. This new paradigm is based on a carefully dimensioned reconfigurable array fabric, and a number of specialized blocks, all included in a system-on-chip (SoC). The new DSP architecture is supported by a new programming approach, based on spatial mapping of the signal processing algorithms to the computing fabric, using tools based around familiar C/C++ approaches.

A discussion of various baseband algorithms and mapping techniques to exploit their inherent parallelism was provided, which hopefully will pave the way for a better acceptance of this new processing paradigm and more in-depth discussions in the future by the SDR community.

A commercially available example of such a processor – the MS1 *r*DSP – was also described, and a real-time example application, supported by the *r*DSP, was presented. Based on numerous benchmarking and simulation activities, it has been shown that the new reconfigurable fabric can support chip-rate and symbol-rate processing requirements of cellular systems such as UMTS/WCDMA, cdma2000, GSM/EDGE/GPRS, wireless data systems such as IEEE802.11a, b and g, and location technologies such as 'classic' and 'assisted' GPS.

Such a novel approach to wireless baseband computing will hopefully pave the way for future totally software defined radios.

Acknowledgments

The authors would like to thank the team at Morpho Technologies for all their contributions, especially Mark Davies, Rafael Maestre, Saeid Safavi and Afshin Niktash.

References

[1] W.J. Bouknight, S.A. Denengerg, D.E. McIntyre, J.M. Randall, A.H. Sameh and D.L. Slotnick, 'The Illiac IV system,' *Proc. IEEE*, 1972, **60**, 369–88.

[2] K.E. Batcher, 'Design of massively parallel processor,' *IEEE Trans. on Computers*, 1980, **C-29**, 836–40.

[3] *'3GPP-AMR-NB with ETSI-EFR Implementation on the StarCore SC140 Core'*, (AN2280/D), Rev 0, 5/2002, Motorola application note.

[4] 3GPP TS 46.051, 'Enhanced full rate (EFR) speech processing functions: general description', *Release 4*, Technical Specification Group Services and System Aspects, Third Generation Partnership Project.

[5] GSM 06.53: *Digital Cellular Telecommunications System (Phase 2+)*; *ANSI-C Code for the GSM Enhanced Full Rate (EFR) Speech Codec*.

[6] J. Daemen and V. Rijmen, *AES Proposal: Rijndael*, AES Algorithm Submission, September 3, 1999, available at *http://www.nist.gov/CryptoToolkit*.

[7] Specification of the 3GPP Confidentiality and Integrity Algorithms; *Document 1: f8 and f9 Specification*.

[8] J.D. Gibson, *The Mobile Communications Handbook*, V. Bhargava and I. Fair (Eds), CRC Press, Inc., Boca Raton, 1996, Chapter 10.

[9] C. Berrou, A. Glavieux and P. Thitimajshima, 'Near Shannon limit error-correcting coding and decoding: turbo codes,' *Proc. 1993 Int. Conf. Comm.*, pp. 1064–70.

[10] A.J. Viterbi, 'Error bounds for convolutional coding and an asymptotically optimum decoding algorithm,' *IEEE Trans. Information Theory*, 1967, **IT-13**, 260–69.
[11] G. Fettweis and H. Meyr, 'Feedforward architectures for parallel Viterbi decoding,' *Kluwer J. VLSI Signal Proc.* 1991, No. 3, 105–19.
[12] L. Jia, Y. Gao, J. Isoaho and H. Tenhunen, 'Design of a super-pipelined Viterbi decoder,' *IEEE ISCAS'99*, Orlando, Florida, May, 1999.
[13] G. Fettweis and H. Meyr, 'A modular variable speed Viterbi decoding implementation for high data rates,' *North-Holland Signal Processing IV, Proc. EUSIPCO '88*, 1988, pp. 339–42.
[14] G. Fettweis and H. Meyr, 'High-speed parallel Viterbi decoding: algorithm and VLSI architecture,' *IEEE Communication Magazine*, 1991 (May), 46–55.
[15] W.E. Ryan, *A Turbo Code Tutorial*, New Mexico State University, Box 30001 Dept. 3-O, Las Cruces, NM 88003.
[16] A.J. Viterbi, 'An intuitive justification and a simplified implementation of the MAP decoder for convolutional codes,' *IEEE JSAC*, 1998 (February), 260–64.
[17] I. Verbauwhede and C. Nicol, 'DSP architecture for next-generation wireless communications,' *ISSCC 2000*, DSP Tutorial.
[18] 3GPP TS 25.212, *Multiplexing and Channel Coding (FDD)*, Release 4, Technical Specification Group Radio Access Network, Third Generation Partnership Project.
[19] M.E. Frerking, *Digital Signal Processing in Communication Systems*, Van Nostrand-Reinhold, 1994.
[20] G. Ungerboeck, 'Adaptive maximum-likelihood receiver for carrier-modulated data-transmission systems,' *IEEE Trans. Comms.*, 1974, **22**, 624–36.
[21] 3GPP TS 25.104, *BS Radio Transmission and Reception (FDD)*, Release 4, Technical Specification Group Radio Access Networks, Third Generation Partnership Project.
[22] T. Hentschel and G. Fettweis, *Software Defined Radio: Enabling Technology*, W.H.W. Tuttlebee (Ed.), John Wiley & Sons, Ltd, Chichester, 2002, Chapter 6.
[23] J.G. Proakis, *Digital Communications*, fifth edition, McGraw-Hill, New York, 2001.
[24] J.G. Proakis and J.H. Miller, 'An adaptive receiver for digital signaling through channels with intersymbol interference,' *IEEE Trans. Inform. Theory (Corresp.)*. 1969, **IT-15**, 484–97.
[25] T. Ericson, 'Structure of optimum receiving filters in data transmission systems,' *IEEE Trans. Inform. Theory (Corresp.)*, 1971, **IT-17**, 352–53.
[26] S. Werner and J. Lilleberg, 'Downlink channel decorrelation in CDMA systems with long codes,' in *Proc. IEEE Vehicular Tech. Conf.*, Houston, USA, 1999, Vol. 2, pp. 1614–17.
[27] S. Boumard and A. Mammela, 'Channel estimation versus equalization in an OFDM WLAN system', in *Proc. IEEE Vehicular Tech. Conf.*, 2001, pp. 653–57.
[28] H. Myre, M. Moeneclaey and S.A. Fechtel, *Digital Communications Receivers, Synchronization, Channel Estimation, and Signal Processing*, John Wiley & Sons, Inc., NY, 1998.
[29] W.H. Mangione-Smith *et al.*, 'Seeking solutions in configurable computing,' *IEEE Computer*, 1997 (December), 38–43.
[30] 3GPP TS 34.108, *Common Test Environments for User Equipment (UE) Conformance Testing*, Release 4, Technical Specification Group Radio Access Network, Third Generation Partnership Project.
[31] 3GPP TS 25.211, *Physical Channels and Mapping of Transport Channels onto Physical Channels (FDD)*, Release 4, Technical Specification Group Radio Access Network, Third Generation Partnership Project.
[32] W.R. Braun and U. Dersch, 'A physical mobile radio channel model', *IEEE Transactions on Vehicular Technology*, 1991, **40**(2).
[33] J. Jimenez, *et al.*, *Final Propagation Model*, Deliverable from the RACE CODIT Research Project, R2020/TDE/PS/DS/P/040/b1, June 1994.
[34] S. Abeta, M. Sawhashi and F. Adachi, 'Performance comparison between time-multiplexed pilot channel and parallel pilot channel for coherent RAKE combining in DS-CDMA mobile radio', *IEICE Trans. Comms.*, 1998, **E81-B**, 1417–25.

12

The picoArray: A Reconfigurable SDR Processor for Basestations

Rupert Baines

picoChip

Increasingly sophisticated algorithms are being used everywhere in communication systems to support higher data rates and richer services. This is most visible in mobile systems, where the move to third generation is driving significant changes in component design for telecoms equipment. In addition to basic voice and messaging, third generation UMTS technology paves the way for telecom operators to offer sophisticated data services that industry analysts predict are essential for revenue growth over the next decade.

The market for UMTS basestations is estimated to be around $10 billion a year for the next few years, as deployments are made around the world both for new networks and as capacity is added to existing networks to support the growth in customer numbers and calls.

However, the complexity of UMTS technology and the competitive pressures combine to make this a uniquely challenging market for manufacturers. With wireless carriers pushing OEMs to deliver lower cost infrastructure with higher performance and more traffic processing capacity, most traditional approaches simply cannot deliver the processing horsepower required at an acceptable cost and within an acceptable time-scale.

This chapter introduces a radically new processing architecture, based upon a massively parallel array of microprocessors, optimized for wireless applications, which can address these challenges. As well as delivering more than ten-times the performance of legacy architectures, the new approach can reduce development time, thus enabling the advantages of a software-defined system to be brought to market in an economic time-scale and cost.

Software Defined Radio: Baseband Technologies for 3G Handsets and Basestations. Edited by W. Tuttlebee
 ISBN: 0-470-86770-1

12.1. Introduction

12.1.1. Shannon, Moore and the Design Gap

The demand for increased capacity exists against a background of data throughput getting ever closer to the theoretical capacity of a channel – the Shannon limit – as people strive for higher data rates or longer reach over fixed channels.

The computational requirements of a system are rising ten- to a hundred-times faster than Moore's law can deliver increased performance in silicon. Estimation and detection algorithms in today's communication systems require the number of operations per second to grow by a factor of ten every 4 years [1] – a compound annual growth rate (CAGR) of 78%. That compares to the increase in processor speed from Moore's law of a factor of 10 every 6 years, or a compound growth of 58%. This demonstrates the reality of the design gap and quantifies it. If we compare the GSM and WCDMA systems, the CAGR figure is even greater (Table 12.1).

Improvements in error correction coding theory over the last 60 years have come ever closer to the Shannon limit, and the advent of turbo codes in 1993 almost totally closed the final '2-dB gap' between practical codes and the Shannon limit – but all of this requires significantly more processing.

While Moore's law holds well for general purpose processors, memories and some other devices, the difficulty of integrating ever bigger systems means that the growth curve for complex System-on-Chip (SoC) devices is significantly slower with a CAGR of just 22% There is therefore a further design gap which increases every year (Figure 12.1). Requirements in communications systems ('Shannon') are growing more complex and more quickly than Moore's law, but implementation capabilities such as SoC are falling further behind.

This indicates that current architectures (FPGA, DSP) are inadequate, while dedicated solutions (SoC) are not only uneconomic, but are also falling further behind. Clearly, a new approach is indicated. This is reflected by some commentators with a historical perspective of industry cycles [2].

12.1.2. The Drivers for Flexibility

Not only must equipment deliver the required performance, but design times are under constant pressure and budgets are stressed. With rapidly developing technology and evolving standards, systems also need to be flexible enough to accommodate changes. Such changes may be the emergence of a new standard or algorithm, system design improvements, or learning from real experience after deployment; all three factors get tougher with successive generations of technology.

Table 12.1 Complexity growth of mobile systems

Year	System	Processor requirement	Spectrum efficiency
1992	GSM	10 MIPS/channel	1 bits/s/Hz
2000	WCDMA	3000 MIPS/channel	3 bits/s/Hz

Processing requirement CAGR of 104% (cf. Moore's law of 58% CAGR)

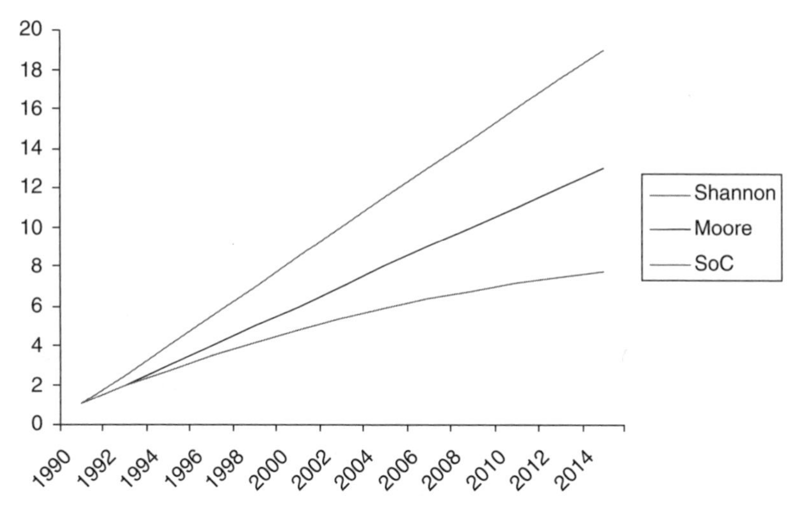

Figure 12.1 The design gap

New technologies are causing a fundamental change in the way equipment is designed, and there is a growing awareness of the attractiveness of reconfigurable DSP, flexible architectures or other software defined radio (SDR) approaches [3]. Several technologies are being developed to implement these flexible approaches efficiently.

12.2. The 3G Design Challenge

One of the key challenges is the standard 3G (WCDMA) basestation and its baseband signal processing requirements, which include chip rate, symbol rate and baseband control functionality. The core detection and estimation algorithms involved are essentially similar for the various CDMA standards: IS95B, cdma2000, 3GPP-FDD, with 3GPP-TDD and TD-SCDMA imposing some variation. However, this section will focus on WCDMA FDD, as it reflects both the most commercially significant system, at 85% of the 3G market, and highest level of complexity for any deployed technology today. Moreover, the techniques are equally applicable in other technology domains.

12.2.1. The Challenge of 3G WCDMA Basestations

With revisions to the core specifications still ongoing, the development of WCDMA Node-B basestations remains an expensive challenge for design teams (Figure 12.2). The demands of the complex baseband protocol stack push the limits of available silicon, which has a knock-on effect on development time. So, design teams are keen to find out about the capabilities of any new device that could ease this pain.

12.2.1.1. Technology Evaluation Platforms

To demonstrate that they have silicon available to ease the job of design, a number of vendors have prepared benchmark studies. Many of these Node-B evaluation platforms are focused

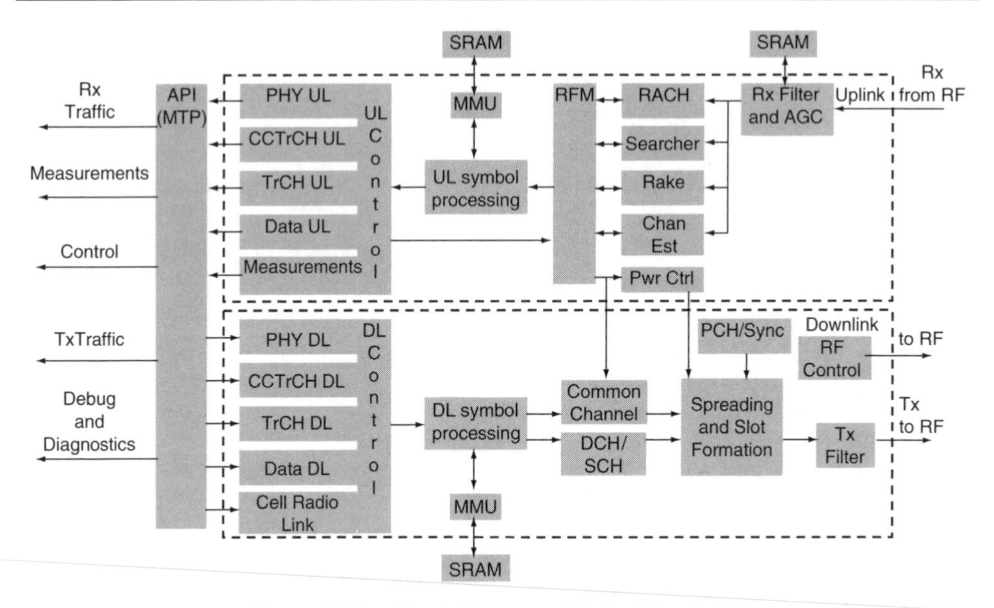

Figure 12.2 Block diagram of Node B baseband

on a narrow set of 'ideal' conditions, designed to suit a particular operator need. There is a benefit to this approach. It provides a reasonable target for evaluation and provides an indication of how much processing power will be needed for certain situations, such as 64 channels of voice traffic within a medium-sized cell. However, the indications from the operators who will buy the basestations are that they do not want to pick basestations for specific functions.

12.2.1.2. The Operator Need

The requests for information (RFI) that have been issued by operators recently have focused on flexibility as a primary concern. Keen to be able to derive revenue as quickly as possible from 3G services, the operators do not want to be in the position of being forced to miss out on a lucrative revenue stream because their chosen basestation does not support the required (but yet to be identified) mix of traffic.

A further crucial aspect to flexibility is the ability to incorporate new algorithms, enabling updates for increased performance. It is widely recognized that algorithms can be improved after a period of experience, learning and optimization – but this only has a value if the changes can actually be incorporated as updates to deployed systems. Given that an improvement in link budget of 1 dB may be worth £1million per basestation for a busy site (improved performance translates to capacity translates to revenue), this is clearly important.

12.2.1.3. The Conventional Design Approach

In principle, the conventional approach to basestation design should yield flexible systems. Many designers have embraced the combination of field-programmable gate array (FPGA), digital signal processor (DSP) and general-purpose processor to implement the baseband functions for WCDMA. The design offers apparent flexibility as FPGAs and DSPs are reprogrammable. However, each uses a different set of tools and development techniques. That means designs need to be implemented separately on each, generally by different teams of engineers. It is only during the integration phase that the development team can see the real interaction between each of these components. The situation is further complicated by the fact that many processors, DSPs and FPGAs need to be used in a basestation linecard to provide sufficient horsepower to handle the complex WCDMA baseband protocols.

In principle, the layered protocol structure of WCDMA as defined by the 3G Partnership Project eases the job of building a baseband card using a variety of components. There is an apparently clear distinction between the different parts of the protocol, which can be broadly separated into chip rate, symbol rate, transport layer and operations, administration and management (OAM) components.

Indeed, in the Global System for Mobile communications (GSM) protocol that acts as the antecedent for WCDMA, a clear separation between the physical, network-access, transport and management layers exists. This is not the case for WCDMA. There are complex interactions between the different layers that have to be taken into consideration during design. Many benchmark designs often ignore these interactions for the sake of simplicity and will pick a particular ('point') set of traffic and air-interface classes to demonstrate an operational system. Examples include inner-loop power control, or the impact of different traffic types on bearers and performance.

With the right architecture, however, it is possible to align the operators' need for flexibility with the complex interactions that the 3GPP standards demand of a complete WCDMA baseband design.

12.2.2. Increasing Complexity Demands

There are two primary degrees of freedom in Node-B basestation requirements. The first is the size of the basestation itself. Ideally, operators would like to be able to select from a menu of basestation options, ranging from picocells that can be concentrated on a building or shopping mall to full 50-km-range macrocells. The second concerns the traffic types that the basestation needs to be able to support.

Traffic on a WCDMA network can vary widely, from voice through a number of data types to the latest 'broadband Internet' addition to the protocol, high speed downlink packet data access (HSDPA). Voice calls are handled differently to data transfers not only in terms of the forward error correction used, but also in the treatment of the frame at the physical layer. Based on cell conditions and operator requirements, the data rate for data packets can vary widely due to changes in the spreading factor.

A high spreading factor, which increases the number of chips per symbol, will reduce the data rate but reduce the error rate if fading conditions are bad. If conditions are bad, however, an intelligent basestation design may also deploy more resources to augment the number of paths handled by the rake receiver. This has a knock-on effect on the chip- and symbol-rate processing section of the baseband as different levels of resources need to be deployed to process the incoming data. Under good conditions, the limiter may be the performance of the turbo-code section, which may be deployed on a DSP armed with specialized assistance engines or on an FPGA. Under poor conditions, the chip-rate section is likely to take on more of the burden.

12.2.3. Conventional Architectures

Most basestations today use a combination of DSP software and FPGA hardware in the baseband. DSP software might be the preferred option (as was the case in earlier generations), but with the processing requirement being some 100-times that of the previous generation, peripheral hardware accelerators for the DSP host are essential to make up for the lack of processing power in the DSP. These have been initially designed as ASICs.

12.2.3.1. Capability and Limitations of ASICs

The ASICs typically handle the chip-rate processing, taking sample data directly from the RF section, while the DSP handles the symbol-rate processing of the protocols and decodes the data in the stream. All this is typically managed by a powerful RISC processor such as a PowerPC.

While the trend was initially towards ASICs to provide dedicated processing power, the rapid evolution of 3G standards coupled with the broad range of implementation options is driving hardware development back towards flexibility. This is largely due to spiralling development costs, the need for differentiating features and risk of specification change.

To develop a complex ASIC for a basestation in contemporary silicon technology requires a large team of engineers; a development cost of $50M and a 36-month design gestation are not untypical. If there is a design problem, the re-spin of the device will add cost and verification time, increasing the time-to-market. Given the pace of standards evolution and the relative immaturity of the market, this is a severe restriction (Figure 12.3).

The long development time of a custom device may not matter in some environments, but it clearly is a concern in a fast-paced market such as 3G; indeed, now the development cycle of the ASIC may be longer than the release life of the product. This unfortunate state of affairs is made worse by the 'synchronous' deployment of UMTS around the world, compared to the slower development and staggered deployment of GSM.

12.2.3.2. The Increasing Capability of FPGAs

On the other hand, FPGA devices have used the increasing performance of silicon technology to reach million gate densities and gigabit-per-second interface speeds, so high-end FPGA devices can now be used for chip-rate processing. FPGAs are attractive in that they can be

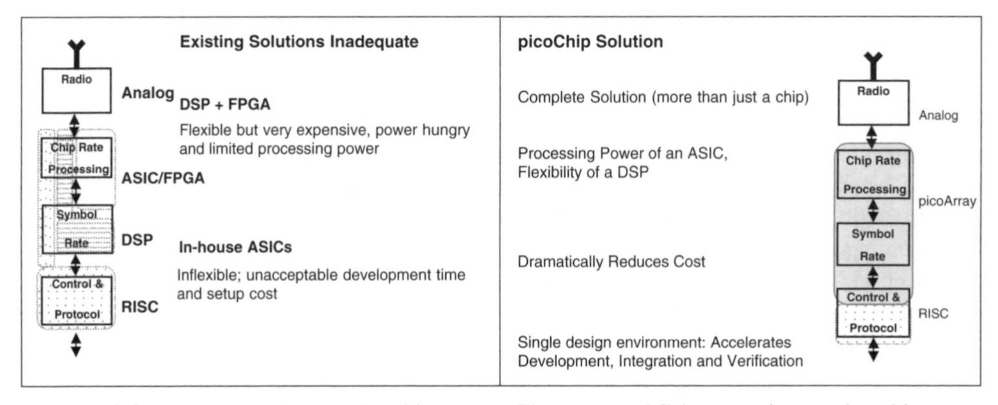

Figure 12.3 3G basestation card architectures. The current 3G base station card architecture, compared with the integrated development approach based around a picoArray

re-programmed at the engineer's workbench in a design flow that is similar to that of an ASIC, using a high-level language such as VHDL and synthesis tools. While FPGAs have improved in density more rapidly than ASICs, this is largely due to increasing use of dedicated blocks such as multipliers and processors – they still suffer from the same factors that limit ASICs to 22% per annum complexity growth. This is largely as a result of the verification problems and the challenge of developing and testing something that complex.

FPGAs have traditionally been extremely versatile and universal – the same device may be designed into a router for layer 3 MAC processing, or into a Node B for DSP PHY processing. Unfortunately this has a trade-off with optimization for a particular function; the gate level granularity they offer is ill suited to efficient and quick implementation of complex tasks. In an attempt to address this problem, FPGA vendors are now including DSP elements alongside the control and data-handling elements in their chips, but cost and power remain high.

12.2.3.3. Advances in General Purpose DSP

Another approach is to use general purpose digital signal processors. DSP software clearly delivers design flexibility, but traditionally has not been able to deliver the sheer process power required in these systems, with processing capability only rising with Moore's law, not with system requirements. Historically, this increase in performance came primarily through increased clock rate, and consequently increased power, but it is now being supplemented in high-end devices with exploitation of instruction level parallelism, implemented as very long instruction word (VLIW), in devices such as the TIC6xxx, the ADI TigerSHARC and the Broadcom Firepath.

In order to achieve such performance, complex pipelines and multiple instruction-execution units are used, requiring a sophisticated compiler to exploit them. While contemporary DSP compilers are more efficient than previous generations, and claim to allow a significant proportion of the programming task to be written in C, there is a performance impact. Some critical tasks will still have to be written in assembler, and producing code on such a complex device is hard and time consuming.

Some DSP parts incorporate hardware to accelerate computationally intensive functions specific to these algorithms, or special-purpose instructions (e.g. spread/despread), but even at 600-MHz clock rate, the performance is insufficient to replace the FPGA or ASIC, and they burn considerable power. The solution is to use multiple devices, and some processors are expressly optimized for such multiprocessor systems. The drawback (apart from the cost of using many expensive processors and the power dissipation) is in the difficulty of device coordination, programming and design verification. Partitioning algorithms across multiple devices and ensuring that they are in synchronization is a challenging task. Indeed, it is usually impossible to give deterministic performance metrics, and massive testing is required to give statistical assurance. This reflects the fact that using a serial processor obscures the inherently parallel nature of most DSP applications. A better approach would be to exploit this inherently parallel nature of the application, and to use an architecture that matches this.

Another concern about DSPs is that they typically introduce significant latency. Partly, this reflects the 'block based' algorithms they employ (as opposed to more 'stream based' structures common in hardware); partly it reflects the inherent need to time-slice and stagger between tasks or processes. This can introduce significant latency, which is a significant concern in some systems: for example, in inner-loop power control.

12.2.3.4. FPGA-DSP Partitioning

Conventional wisdom has it that the chip-rate sections will go predominantly into the FGPA part of the baseband card. The symbol-rate portion generally fits better on a DSP; it is easy to see why by comparing the estimated processing speeds for given parts of the WCDMA protocol stack. Taking the receive section, the chip-rate part of that, which includes equalization, digital filtering, despreading and the rake receiver, constitutes most of the MIPS needed – perhaps 75% of the total (say 200 MIPS out 3000 MIPS for a macrocell receiver per channel).

It makes sense then to put the chip-rate section into an FPGA, or so it seems, because that many MIPS demands a lot of DSPs (perhaps 1 per channel). However, many of the algorithms in WCDMA are control intensive and rely on large numbers of integer multiplications. Although FPGAs can handle this kind of workload, these workloads put a lot of strain on the designer. Control-intensive designs imply a large state space that may prove difficult to implement cost effectively in terms of logic density and design time on an FPGA. Similarly, multipliers on FPGAs, unless they have specially designed hard cores embedded for the purpose, need specialized bit-serial architectures, which are slow unless highly parallelized. This, again, slows down the design process because engineers have to spend time working on implementation hardware description language (HDL) code and verifying it, not on the algorithms that form the core intellectual property of the company.

12.2.3.5. Optimizations and Simplifications

Closer examination of the specification often reveals optimizations that can drastically reduce the computational load. For example, algorithms have been proposed for the path-selection part of the rake receiver that can cut the workload by almost two orders of magnitude compared with a 'brute force' data-path-oriented hardware design. However, they are algorithms that fit best on a processor with a reasonable amount of memory attached to it. While such an algorithm may improve the cost–performance ratio of the basestation, it complicates the design process because we now have key parts of the chip-rate subsystem sitting in both the FPGA and processor sections. The interactions between these stages become very complex; debugging and verification can be expensive and time consuming.

Furthermore, the existence of this interface becomes a significant bottleneck for data. One consequence is that once a system is defined and optimized, it is very hard to change. Although the individual components are flexible, the solution as a whole has effectively lost its flexibility in the process of system integration.

There are other portions of the chip-rate section that apparently fit better on a processor. These are typically the real-time measurement functions that look at channel behaviour. Both the rake receiver and the equalizer rely on channel impulse response (CIR) estimation. The least-squares calculations that may be used for CIR estimation tend to be matrix intensive, apparently lending themselves to a DSP implementation. However, the size of the matrix can lead to a slow update rate, incurring long error bursts if channel conditions change. That can force a rethink of the algorithm to make it fit a hardware substrate. So, the design may move back from the software to the hardware partition and into the hands of a new design-implementation team.

The situation gets more complex with the introduction of HSDPA. Under HSDPA, both the modulation and coding are adapted to channel, traffic and user requirements. The user equipment estimates the condition of the channel and translates it into a metric, which is transmitted to its current Node B host. There is clearly further potential for algorithms to move backwards and forwards across the software–hardware partition as more efficient algorithms are uncovered.

12.2.3.6. Other Issues

There are further interactions between the various parts of the baseband module. The 3GPP technical standard 25.215 demands that a number of measurements are taken at various points within the transmit and receive chains. These are functions that are often forgotten from the benchmarks but which can have a surprisingly large effect on the implementation of the baseband because of the need for units to report on their status at given times or at the request of the OAM code running in a general-purpose processor somewhere in the module.

12.3. Reconfigurable Architectures

12.3.1. A Variety of Approaches

Recognizing the above issues, a growing number of companies have developed specialized reconfigurable baseband devices, striking a different balance between wide applicability and optimization, to deliver efficient solutions to just this problem-set. To simplify wildly, there are three approaches:

The first, which might be called 'FPGA+' is to add a number of higher-level or higher-complexity functional blocks to a general purpose device in order to optimize it for a specific purpose such as wireless. Examples of this may be Chameleon, MathStar, Elixent, or the newer devices from FPGA suppliers. These may include some very rich function blocks or architectures. While supporting functionality at a higher level than conventional FPGA, and with a great degree of versatility, these devices still share the flexibility of that path and facilitate OEMs to include their own IP. However, as general purpose devices there is a trade-off of universality versus suitability for particular applications.

A second approach is to develop a reconfigurable system based around a 'programmable application specific standard product' (P-ASSP), which consists of a general purpose core supplemented with a number of optimized coprocessors or kernels (e.g. for the multipath searcher or equalizer). A good example is the 'wireless systems processor' of Morphics, where kernels are targeted to support a class of operations found in a set. Each kernel implements both data flow and control associated with a task, contains sufficient memory for that role and implements local communications through a configurable interconnect.

This has clear attractions in terms of power dissipation and cost, in that some large and complex blocks will have been optimized to perform specific operations. In addition, the use of multiple devices is well aligned to the parallel nature of the application.

However, an obvious concern is that the flexibility and reconfigurability of the system is totally limited to the functionality coded into these blocks. Given the pace of change of systems and architectures, it is hard to be confident that any hard-coded kernel will indeed be suitable (or even usable) when the requirement for an update arrives. Additionally, in many cases the core IP and expertise resides in the system manufacturer, which is then difficult to include in the device.

A third approach is based around a parallel array of processors is used by picoChip, Morpho or PACT. This is described in detail in the following sections.

12.3.2. Reconfigurability and 'Reconfigurability'

One final issue is the time-scale associated with changes and how reconfigurable a device is. Some applications require intermittent changes – for example, updating a major piece of network functionality, say, for a new release of a standard, or a significant algorithm improvement. This will be an infrequent occurrence, only undertaken after the new code has undergone significant testing. This type of change may be used more rapidly, but still in a managed or scheduled way – for example switching between different basestation modes to reallocate resources according to traffic patterns.

Other environments may seek more rapid changes – perhaps between different discrete applications (for example, Elixent discuss using a reconfigurable fabric embedded in a consumer device to change between an MP3 player and image compression for a camera). At an even more extreme case, QuickSilver have discussed changing functionality at a 60-kHz rate, allowing different functions within the same system to be switched in 'on the fly' as required.

Clearly, the type of 'reconfiguration' and the associated time-scales must align with the requirements of that system. However, the verification and test consequences are significant and should not be overlooked (Table 12.2).

The quasi-static or embedded case is well defined and understood: code is developed and tested in a controlled way, and updates or switches are done in a managed process. This is familiar from the development process of FPGAs or embedded processors. Indeed, one

Table 12.2 Timescales for reconfigurability

	Timescale of change	Example application	Example	Comment
Static	None	Hardcoded	ASIC	
Quasi-static/ embedded	Hours, days	Primary algorithm, updated occasionally	Typical basestation requirement. DSP, FPGA or Reconfigurable	New code release with strict regression test; changing between major modes
Application	Seconds, minutes	As needed, change in application, e.g. from MP3 to camera in a terminal	DSP or reconfigurable	Switch between a set of defined tasks as and when required
On the fly	Milliseconds	Switch between different modes or reusing capabilities very dynamically	General purpose CPU; some reconfigurable	Verification challenge

description of the slower changing systems is perhaps best regarded as 'field programmable' by analogy with an FPGA (the field programmable processor array, or FPPA).

In contrast, more dynamic switching that may happen at any time (essentially as an interrupt) is more analogous to general purpose code running on a microprocessor. This implies the need for an ultra-stable kernel to manage the transitions and arbitrate between them, and a need for memory management system that ensures that data consistency is maintained between context switches. Test and verification also become significantly more complex as not only must each state be tested individually, but so must any permutation in order to check that there are no unexpected interactions.

Given the cost of integration and verification in a system environment (or the even higher cost of a bug in the field), the economics of this level of versatility must be weighed against the cost of development.

12.3.3. The Parallel Processor Array

One way of addressing these issues is based around a parallel array of processors. Instead of a small cluster of very powerful discrete CPUs exploiting instruction parallelism, these have a very large number of 'appropriately sized' devices on a single die interconnected by a fast fabric, with on-chip bandwidth much greater than bus bandwidth.

The key to leveraging the power of such an array of processors lies in an interconnect that helps system designers to minimize the amount of local storage needed by optimizing communication between processors and a development environment that allows multiple algorithms running on different processors to be hooked together easily.

In such a computing fabric, tasks can be mapped directly onto CPUs almost as easily as drawing a block diagram. An attraction of this approach for signal processing (as opposed to more general computation) is that it matches the parallelism inherent in the DSP algorithms and across them for multiple data streams.

12.3.4. Problems with Parallelism

One of the biggest stumbling blocks to the use of massive parallelism is the difficulty of passing data between processing elements. The programming environments for most processors assume a small number of threads of control in action at any one time. This is in spite of the fact that the WCDMA is highly amenable to parallelization, especially in the chip-rate portions, just as long as a flexible control structure, allowing for both coarse- and fine-grained control, can be put in place.

The data flows between processes running on different cores can be predicted at compilation time, allowing the use of time-multiplexed interconnects, which helps reduce the amount of wiring needed on-chip. With a deterministic interprocessor fabric, it then becomes possible to borrow some useful features from the hardware world, which are designed to express parallelism, and bring them to a predominantly software-focused environment (Table 12.3).

12.3.5. The picoArray

The picoChip device (the 'picoArray') was explicitly architected and optimized for complex wireless infrastructure applications (Figure 12.4).

Table 12.3 Qualitative summary of different technology approaches

	ASIC	Traditional DSP	FPGA	Reconfig. 1 'FPGA+'	Reconfig. 2 'P-ASSP'	Reconfig. 3 parallel
Cost (see Figures 12.7 and 12.9 for computational density)	****	*	*	**	*** (a)	***
NRE	–	***	***	**	**	***
Power MOPS/ mW (see Figures 12.8 and 12.10)	**** >100	** 3	* <0.02	** 1	*** >10	*** 10
Time to develop (months)	* 32	** 23	** 28	** ?	*** ?	*** 15
Flexibility	–	***	***	***	? (a)	***
Ease of programming/ verification	***	* (b)	** (b)	** (b)	** (a)	***

Notes:
(a) Depends on how 'hard-coded' are the dedicated co-processors and how 'open' they are to configuration; also impacts verification
(b) Assumes a 'mixed' environment with different technologies that will require integration.
(*more stars* = better)

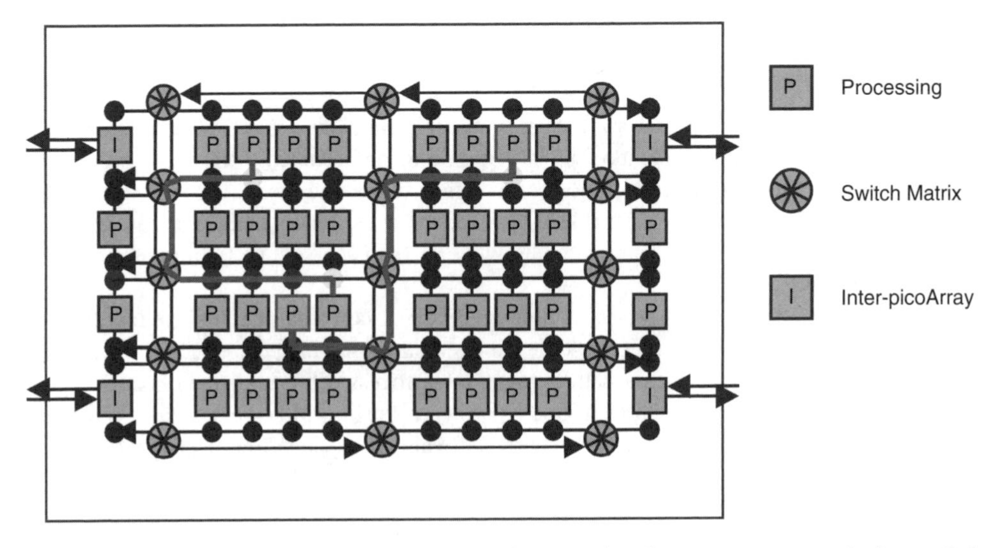

Figure 12.4 The picoArray architecture. The picoArray is a heterogeneous massively parallel array of processors, with distributed memory, and linked with a deterministic switch fabric

It features a heterogeneous array [5] of many hundreds of programmable 16-bit processing elements with a deterministic high-speed switching matrix that can handle the high-speed chip-rate functions, the lower-speed symbol-rate processing and some of the control functions. Each element has its own arithmetic units, processing elements and memory, and is programmed individually during device initialization.

The ability to use a small (optimal) processor design, and then 'clone' it multiple times allows incredibly high processing density (performance per mm^2, performance per $ or performance per W) to be delivered.

The PC102 (current implementation of picoArray, as at the time of writing) is a heterogeneous array of 333 programmable 16-bit RISC processors (three-way LIW, Harvard architecture with local memory), and 15 coprocessors (function accelerator units or FAU). The term array element (AE) is used to describe either processors or coprocessors (i.e. there are 348 AEs in the array). Both the structure of the array and the architecture of the processors was optimized for modern wireless systems (with specific acceleration built in for spread spectrum and OFDM).

There are three processor variants which all have the same basic structure. These three AE types are: standard AE (STAN), control AE (CTRL) and memory AE (MEM). Memory configuration and the numbers of communications ports vary between AE types. It is important to stress that these are 'genuine' microprocessors, with fully featured instruction set, 16-bit architecture and full capability, not merely an ALU. Indeed, there is an interesting 'Darwinian evolution' argument to suggest that this 16-bit, three-way LIW architecture is optimal for embedded processors.

A brief description of the four AE variants and a breakdown of the internal memory distribution is given in Table 12.4.

The application specific instructions within the core processors can deliver significantly higher performance. For example, the complex spread and despread instruction inside the

Table 12.4 picoArray processor variants and memory distribution

Type	Description	Number	Memory (bytes)
STAN	Standard: A standard AE type includes multiply accumulate peripheral as well as special instructions optimized for CDMA spread and despread.	260	768
MEM	Memory: An AE having multiply unit and additional memory.	69	8704
FAU	Function accelerator unit: A coprocessor optimized for specific signal processing tasks (FEC, preamble detect, FHT, etc.). Includes dedicated hardware for trellis operations, etc.	15	N/A
CTRL	Control: An AE type with a multiply unit and larger amounts of data and instruction memory optimized for the implementation of base station control functionality.	4	65 536
	Totals per PC102 device:	**348**	**1 062 400**

STAN processor (of which there are 260) can replace each 40 conventional DSP operations in a single cycle.

Additionally, the coprocessors include configurable hardware for correlation, selection and comparisons which can efficiently accelerate both detection and FEC.

- Preamble detection and multipath searching mode, as a scalable correlation engine for systems including W-CDMA, cdma2000 and TD-SCDMA. Each FAU provides two 128-tap, two samples per chip, matched filters to assist in acquisition and multipath searching tasks. When used for preamble detection, each FAU allows two preamble signatures to be correlated. Provides necessary resource to provide four-signature detection of a 30-km macrocell per PC102.
- Forward error correction (FEC) mode. Supports a wide variety of trellis-based FEC decoding techniques (Viterbi, trellis coded modulation, log MAP, fast Hadamard transform, etc.) for data rates in excess of 100 Mbit/s, including standard 3G specifications (WCDMA, including HSDPA; cdma2000; TD-SCMA) as well as other communications standards (802.11a/b/g WLAN, 802.16 fixed wireless, 802.20, etc.).

12.3.6. A Common Processing Environment

Because of the complexity of the interactions in the WCDMA stack, it is best to try to keep as much of the design as possible within the same environment, instead of being forced to split the development effort between teams with different implementation expertise. FPGA design is a specialized discipline that calls for knowledge not only of a HDL but experience in what works on an FPGA in terms of hardware implementation.

Just about all communications algorithms are modelled in C first and simulated on a processor. As we have seen, additional hardware support has gradually been added to DSPs to support certain pieces of the WCDMA protocol, such as Viterbi and turbo codes. However, even the long-instruction word architectures, which allow multiple instructions to be run in parallel, used by advanced DSPs can only go so far in being able to handle more than a small piece of the total protocol.

Although high-end DSPs have access to megabytes of on- and off-chip memory, many of the algorithms needed to implement core chip- and symbol-rate functions have kernels measured in bytes and do not need to store much in the way of data. Consequently, the majority of the processors within the picoArray have a fast local memory store, but it can be well optimized.

It is the high-level OAM functions that typically need much more code and data memory. A few of the processors have larger memory for control tasks, together with off-chip memories where required.

12.3.7. Control and Data

In a parallel device, it is important that the granularity of these elements is well aligned to the tasks within a communications system, striking a balance between the very fine granularity of a universal FPGA, or the ‘big chunks’ of a powerful DSP. In general, there are two distinct classes of operations:

Data flow – where operations will be regular and predictable (whether stream or block) and may be fast (e.g. chip-rate processing). This will typically require many elements to be 'clumped' together, and it is important that interconnect arrangements are both fast and deterministic. There is a large degree of parallelism (both within algorithms and across multiple instances).

Control – which is 'diffused' across the entire system and must interact with many individual blocks. Typically these tasks are individually quite simple, but can be aggregated. This code will be serial, and will need to support many different options or switches for specific cases or modes.

The picoArray is optimized for wireless communications tasks in two ways. At one level, the structure of the array and the arrangement of the different element types across the chip reflect the balance of requirements of a wireless system. Secondly, the characteristics and instruction set of elements include support for specialist operations such as spread/despread or compare-add-select, or larger memory complements for control type operations.

12.3.8. Programming and Verification

A typical concern is that the parallel approach is impossible to program, or will be unmanageable to debug.

In fact, the reverse is the case: the orthogonality of tasks between processors makes development, debug and test much easier than in a conventional architecture.

The ability easily to partition algorithms across elements, and then run them independently, allows many of the advantages of high-level object oriented languages (specifically, encapsulation and hierarchies) to be applied to complex embedded systems.

Each processor can only communicate with those it is explicitly instructed to do so, and only then with signals of strict type. As a result, processes are essentially 'orthogonal', with no real-time decisions (no run-time arbitration or scheduling) and processes have no 'side effects'. These features make development much easier.

In conventional architectures, the complexity and risk grows with the interactions between blocks (as operations can have side effects, affecting timings or bus allocations in a dynamic way) – meaning that verification and test become proportional to the permutation of all blocks (i.e. for N blocks, time is proportional to $N!$ or to a product function).

In contrast, the picoArray enforces strict typing and independence. This isolation or 'orthogonality' is very similar to object oriented programming in that it allows 'encapsulation' of independent blocks, which can only interact in predictable ways. Consequently, debugging, verification and test all become significantly easier (i.e. time is proportional to N or to a sum function).

An underappreciated problem is that the interaction between the DSP and ASIC is complex as processing resources are distributed across multiple channels, and the RISC typically handling task scheduling and arbitrating between functional blocks to ensure a smooth flow of data from system input to output. Arbitration is particularly difficult as it is important that the system does not stall or lose data when contention for processing resource arises. As a consequence, verifying that data integrity is maintained under every interrupt or contention scenario is extremely difficult, and typically consumes many months of exhaustive testing.

There are also several fundamental problems with the traditional approach. Baseband processing algorithms do not map well onto the typical heterogeneous combination of DSP, ASIC and FPGA. At the root of the problem lies an imbalance between the data processing and control schemes that each class of device supports, and restrictive communication between each device. Marrying the orthogonal requirements of device control, communication and data processing within the cluster is complex, and requires significant software overhead to manage the scheduling of processing resource and to arbitrate effectively when contention for resource arises.

The challenge for the system developer is to guarantee that when contention for resource arises as a result of, say, a new subscriber entering the cell and requiring bandwidth for a high-speed data call. The system has to ensure that information is not lost or corrupted, but given that the availability of resources at any given time cannot be guaranteed, the system verification process requires exhaustive testing of every loading scenario. Exhaustive testing is prohibitively complex and time consuming, but more practical performance testing cannot guarantee that the specified performance can be achieved for every scenario.

So, while the use of accelerators and DSP coprocessors improve the performance of the individual devices, the inherent problem of guaranteeing that processing resource will be available for every conceivable loading scenario still remains. With all the different design environments it is very difficult to verify that the system works until it is actually built, and this is a multi-million dollar bet. The design objective, therefore, is to check, as far as possible, the proper functioning of hardware and software before the system is built. This reduces the time in testing and trials, and increases the quality of the end system.

While requiring different environments, the conventional approach does allow use of familiar development environments: VHDL or Verilog for programming the ASIC or FPGA, and C or assembler for the DSP (although, as noted, the increasing complexity of high performance DSPs makes the use of assembler increasingly difficult).

A better approach is to use the inherently parallel nature of the task. The picoArray takes the straightforward approach of integrating these two into a single, unified toolset. Each element can be programmed either in C or a highly efficient assembly code, while a structural description is used to describe the inherently parallel, interprocessor interconnect and timing. This approach allows the algorithms to be efficiently partitioned and mapped onto specific processing elements at a relatively high level. It also allows the use of new or existing C code to add functions, optimizing code reuse and exploiting existing programming skills for rapid prototyping. It is worth mentioning this is a complete (and completely standard) implementation of C, not a sub-set or proprietary variant.

While reconfigurable devices have attractive hardware features, unless they can deliver a ‘comfortable’ programming environment for engineers and, critically, a powerful and trusted simulation and verification regime, they will not be used. Consequently, this is an area where reconfigurable devices need to make significant efforts. The picoArray addresses this in three ways:

Firstly, combining control and DSP functionality in one device and coding in one single environment dramatically simplifies the design implementation and verification time. While there are different ‘flavours’ of array elements, they are from the same family and the programming environment is entirely consistent.

The traditional system design approach employs totally distinct, and possibly incompatible, design processes for the ASIC, FPGA and DSP system components. In contrast, the picoArray approach uses the same development process for the data path and control functions within the system. Thus, the homogeneous picoArray development environment avoids significant product integration risks; the danger is that products developed using the older, heterogeneous,

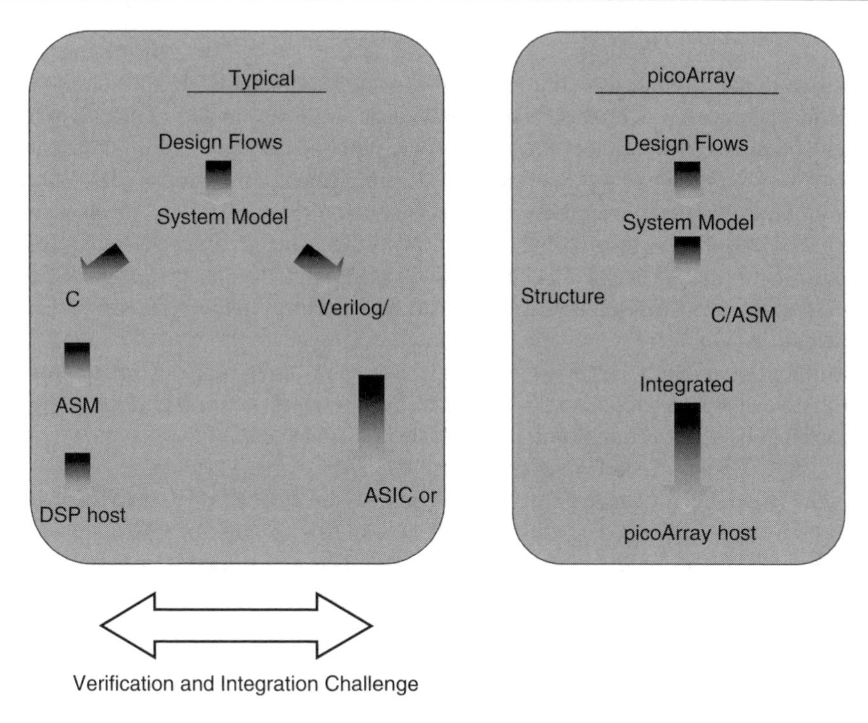

Figure 12.5 A comparison of code development flows

design methods require significant redesign during the integration of the ASIC, FPGA and DSP product elements. This concern intensifies when designing across technology generations. While migrating from one version of a product to the next is a manageable effort, the need to re-architect, reintegrate and codevelop using different architectures and devices, does not scale, increasing the effort and risk of moving to new generations. In contrast, a unified design environment like the picoArray means code and architectures will port from one generation to the next without requiring significant effort or repeated codevelopment (Figure 12.5).

Secondly, the granularity of the architecture means that tasks are decomposed into manageable 'chunks' which are statically mapped to discrete elements. Not only are these elements small enough to test and validate, but because they are static and only interact in controlled ways, that validation is trustworthy.

The level of abstraction is such that a *process* aligns well to a *processor*. The system hierarchy is then a close map to the code structure, making it easier to identify or develop tasks.

Finally, the deterministic architecture eliminates scheduling and arbitration in the underlying architecture, so system loading is entirely fixed and does not have to use statistical multiplexing to use the processing resources. This is true *within* processors as there are no interrupts or complex pipelines with interlocks or bubbles, and true *between them*, as it is a deterministic interconnect.

All these factors combine to simplify verification and testing before the system is built: performance is deterministic and fixed at compile time, unlike a conventionally complex DSP where performance is determined at run time. Consequently, a designer can accurately predict the final performance from cycle-accurate deterministic (rather than statistical) simulations.

12.3.9. Control and Management

It is possible to use signals, as expressed in a language such as VHDL, to describe the data flows that pass between processes. C code simply reads and writes to named signals to communicate with other processes. In this way, it is easy to pipeline data between processors, each of which may work on small parts of the data needed to implement, for example, a rake receiver.

By describing the interconnectivity of processes using these HDL-like mechanisms, it is possible to borrow other useful elements of hardware design, such as place-and-route algorithms. A placer can use broadly accepted techniques such as simulated annealing efficiently to place processes that need close communication near to each other, relieving the design team from the burden of this job.

It takes more than one IC, even armed with hundreds of on-chip processors, to implement a WCDMA Node-B baseband module. However, by extending the time-multiplexed communication system between ICs, it is possible to scale up the capacity of the system with relative ease. The same C-based design environment can be used.

A massively parallel architecture provides a good degree of flexibility when it comes to the interface between OAM and data-path code. It is tempting to think of OAM code as being best implemented on a single, high-speed processor armed with megabytes of local memory. However, a distributed architecture has advantages here too.

Centralized control and management can be easier to manage, since we are only required to develop code for a single processor. Provided the interfaces to the signal processing chain is well defined, and relatively small in number, developing centralized control can be very effective. However, shoe-horning too many control functions into one processor can lead to performance problems, particularly where fine-grained control is needed over functions such as rake-receiver and equalizer processing.

A fundamental problem with legacy architectures, such as high-end DSPs, is that it is not easy to combine coarse-grained and fine-grained control tasks. Although it is possible to handle most coarse-grained control activities, the tasking structure used by most kernels makes it difficult to get timely, fine-grained control.

With a parallel-processor array, it is possible to make use of a hierarchical control structure. A high-level OAM processor may take care of global operations and administration but hand off fine-grained functions to semi-autonomous controllers distributed through the array, each one specialized to a small number of functions relevant to the data-path processors around them. The underlying architecture can be the same. The OAM code may simply not make use of specialized instructions used by the data-path processes, such as Viterbi decoding or despreading. The logic needed to implement these functions is typically small relative to the processor core, meaning that there is little in the way of waste silicon.

12.4. Comparison of Different Approaches

12.4.1. Simple Performance Metrics

While comparing performance is always hard, especially across different architectures, there are some fundamentals which it is useful to share. The processing capacity figures given in Table 12.5 are the net processing calculations for the processors (i.e. coprocessor figures are not included). The device processing capacity figures were calculated as follows.

Table 12.5 Key performance metrics of the picoArray

Device characteristics	Parameter	Unit
LIW RISC processors	333	per device
Coprocessor	15	per device
Processor clock	160	MHz
Peak processing capacity	213.1	GIPS
Total on-chip data bandwidth	4.2	Tbit/s
Peak 16-bit MACs	42	G-MACs
Internal bus speed	5.12	Gbit/s
On-chip RAM	1062	kbytes
External SRAM External SDRAM	8 128	Mbytes Mbytes
SRAM/SDRAM access speed	40/80/160	MHz
DMA channels	4 * 2.24	Gbit/s
IPI (16-bit inter-device channels)	4 * 2.56	Gbit/s
ADI (16-bit I/O channels)	8 * 2.4	Gbit/s

- Peak processing capacity = (peak of four LIW instructions per operation) × (160 MHz clock) × (333 processors per device) = 213.1 GIPS. The four LIW instructions in the peak processing capacity calculation are based upon three execution units (three-way LIW) plus a left or right logical shift included in the operand of the first instruction.
- Average processing capacity = (average of two LIW instructions per operation) × (160 MHz clock) × (333 processors per device) = 106.6 GIPS (based on observed data from typical complex systems).
- Within the peak signal processing capacity of the device, the peak rate for 16-bit multiply accumulate instructions is = (260 STAN processors per device) × (one multiply-accumulate instruction per operation) × (160 MHz clock) = 42 giga multiply-accumulate (GMAC) instructions per second.

In addition, there are (69 MEM + four CTRL processors per device) × (one multiply instruction per operation) × (160 MHz clock) = 11.2 giga multiply instructions per second.

The application specific instructions within the core processors can deliver significantly higher performance. For example, the complex spread and despread instruction inside the STAN processor (of which there are 260) can replace each 40 conventional DSP operations in a single cycle.

Additionally, the coprocessors include configurable hardware for correlation, each FAU can execute:

- 320 Mbyte FHT butterflies/second = (two butterflies) × (160 MHz clock)
- 3.84 giga complex convolutions/second = (128 taps) × (two samples per chip) × (two signatures) × (3.84 Mchips) (word size = 8 bits).

The architecture of the picoArray emphasizes independence of each processor, and of bus resource; as a result, there is no arbitration, scheduling or contention. This 'orthogonality' significantly eases the task of predicting performance, and a bit-accurate/cycle-accurate simulator is available. This also significantly increases the sustained throughput (average compared to peak utilization), as any processor can operate at full speed independently of any other, without blocking or stalling. For example, while the STAN devices could be fully utilized to deliver 42 giga-MACs sustained performance, the remaining discrete processors could be operating independently on other tasks, to deliver an additional 50 giga instructions/s.

12.4.1.1. Inter-processor Communications

Within the picoArray core, AEs are organized in a two dimensional grid, and communicate over a network of 32 bit buses (the picoBus) and programmable bus switches. They are connected to the picoBus by *ports*. The ports act as nodes on the picoBus and provide a simple interface to the bus based on *put* and *get* instructions.

- The inter-processor communication protocol is based on a time division multiplexing (TDM) scheme, where data transfers between processor ports occur during time slots, scheduled in software, and controlled using the bus switches. The bus switch programming and the scheduling of data transfers is fixed at compile time.

  The total internal data bandwidth is 348 processors × two buses × 32 bits × (160 MHz clock) = 3.6 terabits per second.

Communication time slots throughout the picoBus architecture are allocated automatically by the toolchain, according to the bandwidth required. Faster signals are allocated time slots more frequently than are slower signals, ensuring system latency and bandwidth requirements can be defined and implemented.

This structure can scale seamlessly across multiple picoArrays, limited only by algorithmic constraints.

12.4.2. Benchmarks

A detailed benchmark comparison of several different approaches, analysing the computation density and power consumption for two classes of benchmark, a FFT and a Viterbi decoder is described in.

Figures 12.6 to 12.9 are taken from this comparison and include estimates for the picoChip PC101. To make a fair comparison, this analysis scales all the investigated devices to the

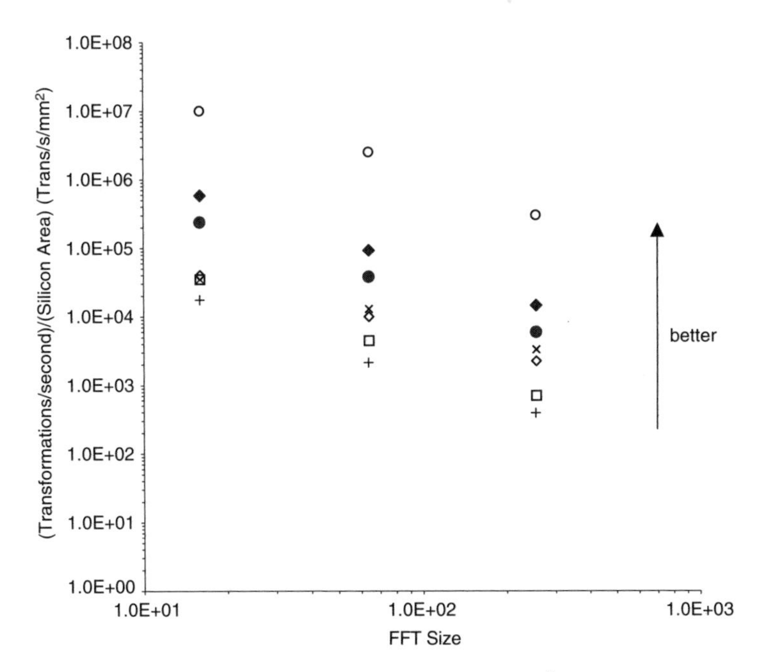

Figure 12.6 Computational density of FFT (transforms/s/mm^2). For key, see Figure 12.8

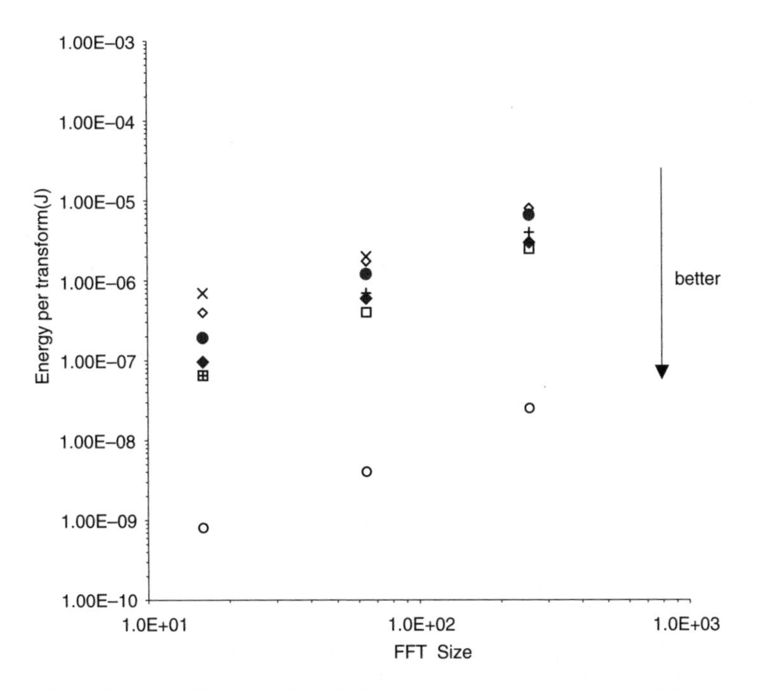

Figure 12.7 Energy efficiency of FFT (J per transform). For key, see Figure 12.8

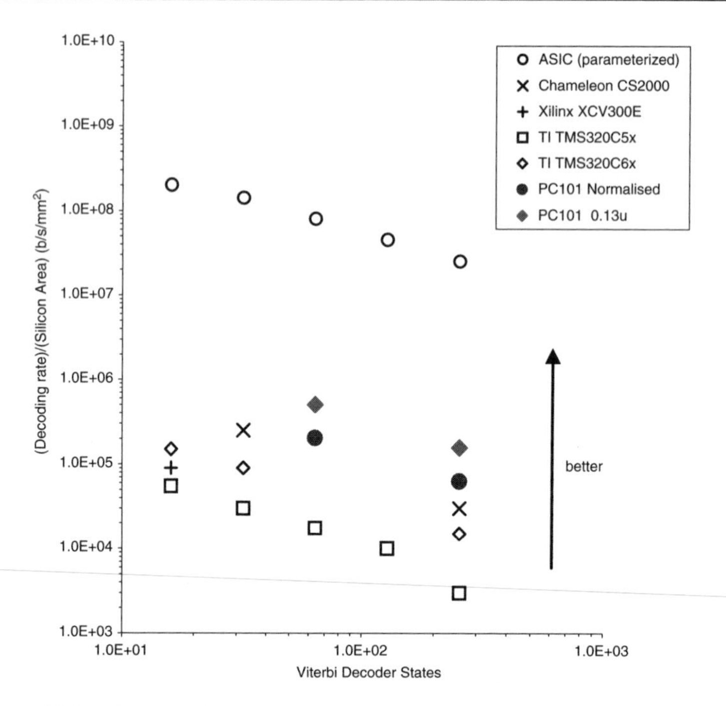

Figure 12.8 Computation density comparison of Viterbi decoder (bits/s/mm^2)

same feature size (0.18 μm). This means that the normalized results shown are an accurate reflection of the architectural efficiency.

12.4.2.1. Devices Compared

The original (un-scaled) feature sizes and core voltages of the devices studied: DSP: Texas Instruments C6x DSP: 0.18 μm, 1.8 V; FPGA: Xilinx Vertex-E XCV300E: 0.18 μm, 1.8 V; Reconfigurable: Chameleon CS2000: 0.25 μm, 2.5 V; ASIC (Synopsys data-path compiler) 0.25 μm, 1 V; Reconfigurable parallel: picoChip PC101: 0.13 μm, 1.2 V.

12.4.2.2. Computational Density – FFT

On a normalized basis, comparing the pure efficiency of the architecture, the picoArray achieves approximately ten-times better computational density than any other reconfigurable or programmable approach. Since cost is essentially proportional to area for a given geometry, this contributes to the performance and cost benefits of the technology. It is worth emphasizing that as a normalized result, it reflects a sustained or inherent advantage of the architecture beyond any implementation or engineering effects.

When this is translated to a specific implementation (at the 0.13-μm technology), the advantage increases to approximately 30 times, reflecting the incremental benefit of the device in the latest process and providing a comparison of realizable devices.

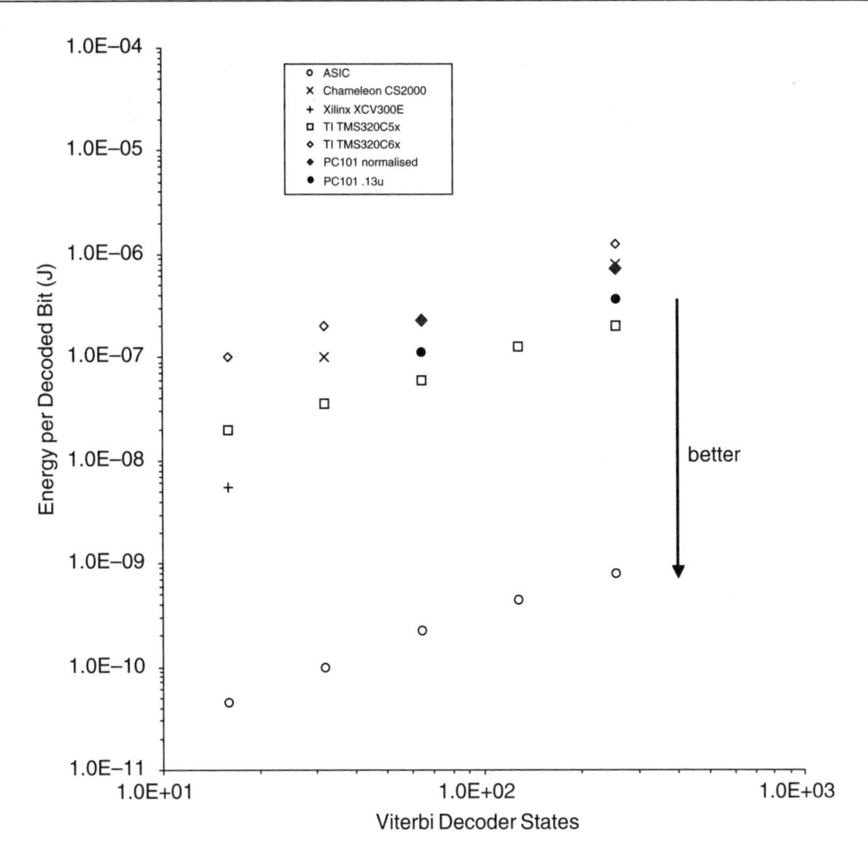

Figure 12.9 Energy efficiency of Viterbi decoder (J per decoded bit)

12.4.2.3. Energy Efficiency – FFT

The power efficiency shows that the picoArray is comparable or towards the best of the reprogrammable devices, but (not surprisingly) not quite as good as a DSP specifically optimized for handheld/low-power applications.

12.4.2.4. Computational Density – Viterbi

It is interesting to compare Figure 12.8 with Figure 12.6, and see how the relative ranking of the different devices has changed. Once more, the efficiency of both the architecture and the implementation of the picoArray are clear, with approximately 3 times and 10-times performance advantage over the next-best reconfigurable technology (and nearly two orders of magnitude better than a legacy DSP).

12.4.2.5. Energy Efficiency – Viterbi

Similarly to Figure 12.7, the results in Figure 12.9 are clear, and not surprising, with a dedicated ASIC being the most power-efficient, followed by a low-power DSP and with the reconfigurable devices being somewhat similar. The actual picoArray (dot) scores well, representing the power advantages of the latest fabrication geometry.

According to these estimates, the computational density, even on the normalized basis, of the PC101 is better than any of the reconfigurable alternatives, for both FFT and Viterbi decoder benchmarks, by a factor of between 2 and 20, illustrating relative architectural advantage. Comparing actual devices gives an advantage to products fabricated in newer processes, but is more useful for an actual evaluation, since it is not possible actually to use a nominal, normalized device. In this circumstance, the advantage of the PC101 increases to between 4 and 40 (a scaling factor of 2 is valid for both density and power between 0.18 μm and 0.13 μm).

The energy efficiency of all the reconfigurable devices is roughly comparable, with the picoArray doing better than any except for the optimized low-power DSP (which, however, suffers accordingly in performance).

12.5. A Commercial 3G Node B Implementation

12.5.1. System Implementation

picoChip have developed a complete software defined basestation (or rather, a software defined baseband: from RX filters through to NBPA interface, implementing all layer 1 processing and Control). This meets 3GPP compliance for a macrocell architecture.

It is worth mentioning that the implemented design is a fully featured production quality design, not merely a benchmark exercise, with three sectors, 64 channels per cell, 30-km cell radius, two antennas per sector and full diversity on both transmit and receive (every antenna can be connected to every channel). The primary blocks to be considered include multipath searcher, RACH preamble detect, rake (four fingers per antenna), TX and RX filters. This includes a high performance RACH detector, with all 16 signatures in every access slot, multi-bit resolution and support for high mobility (Figure 12.10).

The architecture supports all data modes (including CPCH, CPCH-CD and–most importantly–HSDPA. It is expandable for other modes too, as they emerge.

As well as data path and control functions, diagnostics and measurements to the TS 25.215 standard are supported. This is important, and can have a significant impact on a design. This means having measurement data in the signal path that impose an overhead on the system, which is sometimes overlooked in benchmarks, but the intermittent nature of these test packets can be complicated for some architectures to implement.

The reconfigurable approach currently supports all channels with just eight picoArray devices and a PowerPC. This required roughly half the development time of a conventional architecture, a significant reduction in bill-of-materials, and lower power. What is more, the same platform could support multiple standards, or be seamlessly upgraded.

12.5.2. Total Cost of Ownership

One way to compare the relative advantages of different approaches is to perform a comprehensive analysis of all the cost drivers, and assess the 'total cost' of an approach.

To consolidate all of these different items is difficult and very situation specific. However, Figure 12.11 shows the result of such an analysis.

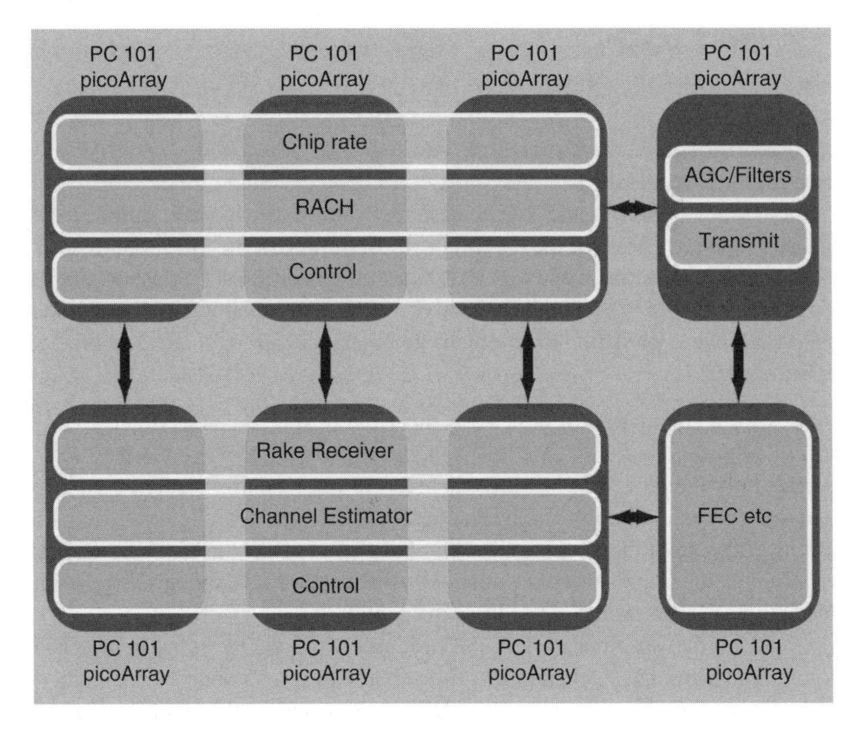

Figure 12.10 Mapping algorithms onto the picoArray

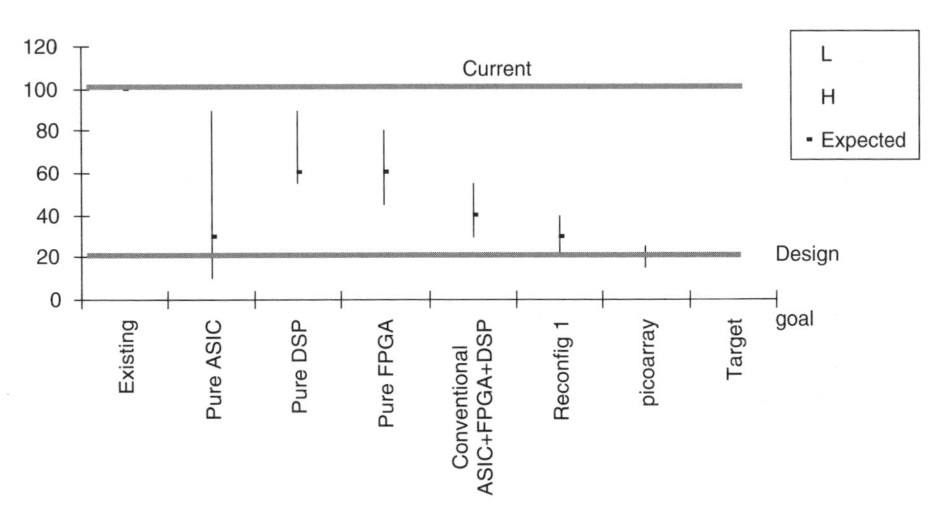

Figure 12.11 Total cost of ownership. A consolidation of the total cost of ownership of six different design approaches for a Node B. Existing design is 100; target for cost reduced version is 20

Factors considered included:

- unit cost;
- power;
- project NRE costs (mask making, tooling, etc.);
- development NRE costs (allowing for some cross-project economies);
- time: includes development and verification/test. However, it only includes direct cost elements and does *not* allow for the opportunity cost of time or the impact of time-to-market.
- Flexibility: an allowance was made for flexibility in terms of impact on debugging. However, no attempt was made to include the qualitative value of, for example, field upgrade.
- Volume: values were estimated to shipment volumes depending on a range of success and market life scenarios.

Three cases were calculated for best/worst/most likely spread and a Monte Carlo analysis performed.

In this case, only the pure ASIC (if there is high volume) and the picoArray systems could achieve the design goal, although another reconfigurable approach might do so in a best case analysis; none of the 'traditional' approaches (DSP, FPGA or DSP+FPGA hybrid could do so). This was influenced primarily by unit cost ($/MIPS required), although development time ($/engineer-month) and power ($/watt) were influential too.

As important as the absolute cost is its variability, driven by spreads in schedule and production volume forecasts. Development time estimates reflect uncertainties in both design and test durations (e.g. statistical testing of DSPs). It is noteworthy that the reconfigurable approaches ensure more controlled development times and hence more predictable costs. In information theory, 'surprise' is proportional to information value, but in business surprise is not viewed positively. Indeed, in financial theory, variation is explicitly proportional to risk, which then commands a risk premium.

The second element is the impact of production volumes, particularly for the ASIC approach with its high NRE and its longer design time. In the best case, with a long, stable production run across which to amortize costs, it is unmatched. However, in the worst case the combination high NRE from development and a limited volume (perhaps attributable to commercial problems following late launch) result in cripplingly expensive unit costs.

This analysis used only quantitative development data and excluded the 'qualitative' or strategic value of reconfigurability. To a degree that omission could be said to have penalized an SDR approach, but even with this omission the economic benefits of flexibility can be compelling.

12.6. Other Applications for the picoArray

Once such a flexible, software-based architecture has been embraced, it becomes easier and quicker for design teams to work on related radio communications systems. Although not based on the WCDMA protocol, mobile broadband systems, such as the recently launched IEEE 802.20 project, are embracing similar themes of flexible radio protocols that can respond quickly to changes in fading conditions and bursty traffic such as Internet protocol (IP) packets. As WCDMA evolves to become more IP focused, we can expect to see even more interaction between the control and data-path elements of the protocol. That will drive

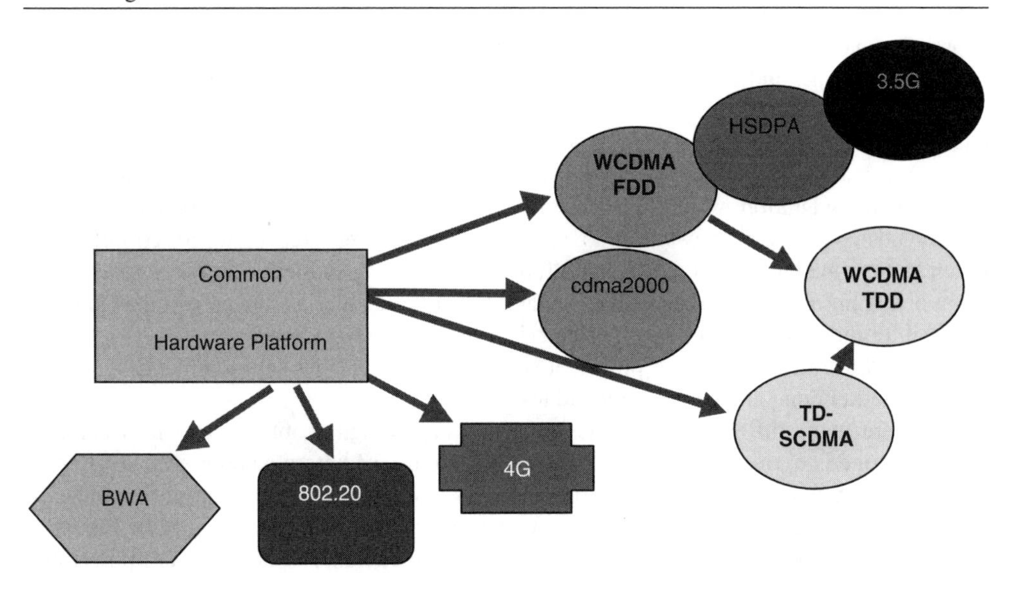

Figure 12.12 Software defined basestation supporting a variety of air interfaces

the emphasis towards development environments based on a unified view of the system that can provide the design flexibility and responsiveness needed.

Clearly, there are many other protocols or modes which can be supported (Figure 12.12). The picoArray was architected to address a range of air interfaces, with specific support for the complexities of spread spectrum and OFDM. As such, it is well suited the various flavours of 3G, but also the various other protocols (802.16, etc.) that exist , or the proposed new standards for beyond IMT-2000.

12.7. Summary

There is growing interest in the use of reconfigurable baseband processing, fuelled by the tension between ever-growing complexity of communications systems and implementation pressures.

The ability to use a small (optimal) processor design, and then ‘clone’ it multiple times allows incredibly high processing density (performance per mm^2, performance per $ or performance per W) to be delivered. The ability easily to partition algorithms across elements, and then run them independently, allows many of the advantages of high-level object oriented languages (specifically, encapsulation and hierarchies) to be applied to complex embedded systems.

The picoArray approach can dramatically accelerate development (leading to a faster time-to-market), reduce both direct (bill-of-materials) and total cost, and enable a fully flexible (software defined) basestation.

Benchmarks show the difference in computational density and energy efficiency in different architectures. These were extended to describe a detailed design of a full Node B WCDMA basestation. These designs face enormous execution challenges, rapidly evolving technical standards and the constant need for new features to differentiate products. Traditional design approaches based on a mix of DSP, ASIC and control processors have difficulties with power, cost, technical risk and development and verification time. Reconfigurable devices have the prospect of addressing some of these challenges. Additionally, a picoArray design approach enables field upgrade of basestations to new standards and with new features.

Once reconfigurable equipment is deployed in the field, systems tuning, functional changes and standards migration, which otherwise require expensive hardware modifications, can be accommodated purely in software. Also, scaleable processor resource allows for cost-effective redundancy to be built into the baseband function, providing scope for software fixes in the event of system component failure. This scalability, cost effectiveness and ease of development are vital for the successful rollout of 3G services around the world.

As well as direct factors (cost, power) it is important to consider subjects like design time, ease of programming and verification and test.

There are many different approaches to addressing design problems, and it is naïve to claim that one is optimum in all cases. However, the use of heterogeneous, massively parallel devices with a deterministic performance has many attractions. In particular, the ease of programming and verification across a variety of different functions has been shown to accelerate development, while the extremely high computational density reduces power and unit price.

References

[1] R. Subramanian, 'Shannon vs Moore: digital signal processing in the broadband age', *Proc 1999 IEEE Communications Theory Workshop*, May 1999, Aptos, CA.

[2] O. Makimoto and O. Manners, *Living with the Chip*, 1995.

[3] O. Baines, 'The DSP Bottleneck', *IEEE Communications*, 1995 (May).

[4] Lange *et al*, 'A software solution for chip rate processing in CDMA wireless infrastructure', *IEEE Communications*, 2002 (February).

[5] A. Abnous and J. Rabaey, 'Low power domain specific multimedia processors', *IEEE Workshop on VLSI Signal Processing*, 1996 (October), pp. 459–68.

[6] R. Subramanian *et al.*, 'Novel application specific signal processing architectures for wideband CDMA and TDMA applications' www.morphics.com/technology/whitepaper

[7] N. Zhang and R.W. Brodersen, 'Architectural evaluation of flexible digital signal processing for wireless receivers', http://www.es.lth.se/home/vikt/LowPower/articles/ning.pdf from Berkeley Wireless Research Centre, Dept. of EECS, University of California, Berkeley

Part IV

Epilogue: Strategic Impact

SDR TECHNOLOGY, SAY SOME, REPRESENTS A PARADIGM SHIFT. ARGUABLY, THE TECHNOLOGIES DESCRIBED IN EARLIER CHAPTERS WILL IMPACT BEYOND SIMPLY MOBILE COMMUNICATIONS. NEW BUSINESS MODELS, SHORTER MARKET WINDOWS, REDUCED PRODUCT COSTS, NEW APPLICATIONS – WHAT ELSE WILL RECONFIGURABLE BASEBAND PROCESSING ENABLE?

13

The Impact of Technological Change

Dr Walter Tuttlebee

Mobile VCE, UK

Since the advent of mass market wireless telephony in the 1980s, the commercial wireless industry has seen sustained rapid development. Even with the economic slowdown of the early '00's, market potential and growth in major emerging economies such as China and India, continues to drive investment and growth. The fast growth and proliferation of cellular markets as a global phenomenon, to the point where subscribers worldwide exceed 1 billion, has emerged as a potent driver of radio technology development.

The early years of the 21st century have already seen significant changes in the wireless industry, resulting from such technological progress, including the commercialisation of new, lower cost, short range wireless technology, embodied in Bluetooth-enabled mobile handsets and in Bluetooth- and IEEE 802.11 WLAN-enabled PCs. The changes now emerging in baseband processing look set to have an impact as profound as, and in the longer term perhaps more than, such earlier developments.

In this epilogue, we endeavour to draw together some of the messages emerging from the earlier chapters and to discuss briefly the potential impact of the technological changes that are underway upon the wireless industry.

13.1. New Technology

13.1.1. Perspectives

The contributors to this book are experienced industry practitioners in digital baseband processing semiconductor technology. They have approached their chapters each from their own unique perspective, thus reflecting a range of personal beliefs, developed over time, which may be considered either as 'strengths and experience' or as 'limitations to their field of view'. The reality is, of course, that often they can be both – our very strengths, without

Software Defined Radio: Baseband Technologies for 3G Handsets and Basestations. Edited by W. Tuttlebee
 ISBN: 0-470-86770-1

self-knowledge, constitute our weaknesses; they can represent opposite sides of the same coin. Such varied perspectives reflect the commercial approaches of the semiconductor industry to software defined radio. The beliefs and approach of a major existing supplier are, understandably, quite different from those of a small start-up company. For the latter, their new technology is often their *raison d'etre*, seen as offering major advantages to potential customers. For such a company, the radical new approach represents an, or even 'the', essential element in securing market entry and expanding the company. Such differences in perspective, however, stem from more than just marketing arguments. The conceptual approach to SDR of an engineer with a long history of traditional DSP evolution will quintessentially be quite different from that of a programmer, or of an embedded-processor architect, considering the application of his skills to new wireless technology. It is this transplantation of skills across application domains, at the same time recognising the step changes in wireless complexity and related challenges associated with 3G, that has given birth to some of the novel technologies described in the earlier chapters.[†]

13.1.2. The Technology S-curve

Many readers of this book will be familiar with the concept of the technology S-curve [1], which seeks to describe the pace of the innovation, development and maturity of a new technology and its potential to supplant existing solutions.

The essence of innovation is that a new approach to an established problem is identified, which, if it can mature and become *both* fully functional *and* more cost effective, has the potential to displace or supercede an existing technology solution – as, for example, the transistor displaced the vacuum tube. However, history teaches us that moving from early concept to 'full functionality' is non trivial and requires extensive research over a protracted period of time. And even when functionality is achieved, there remains the issue of cost competitiveness and the transition to mass-market acceptance. For such reasons, the rate of progress of technology development is usually slow in the early stages of the innovation cycle – the shallow gradient at the start of the S-curve.

As functionality is achieved, however, and the technology moves from being a theoretical concept to a proven implementation, its potential becomes increasingly recognized and investment increases, particularly if it has the possibility to be a mass-market displacement technology. At this stage, the rate of progress accelerates. As the technology moves on to early acceptance in the marketplace this process becomes self-reinforcing, with increasing market success motivating further investment. Increasing experience of, and familiarity with, the technology results in a higher and more rapid return, in terms of technology performance, on this investment. Thus the rate of the technology progress significantly accelerates. This process continues as market share grows and as new markets are enabled by the new technology, ones which the original (soon-to-be-displaced) technology was not capable of addressing, creating new product possibilities. This stage represents the rapid growth (or 'exponential') phase of the S-curve.

As time progresses, the technology and its new markets both become increasingly mature and the potential return on further development investment is reduced. The technology begins to run up against fundamental barriers that slow the rate of further progress advances – the technology matures. At this stage the hockey-stick curve of technology progress versus

[†] This can perhaps be best appreciated by reading the biographies of the respective chapter authors.

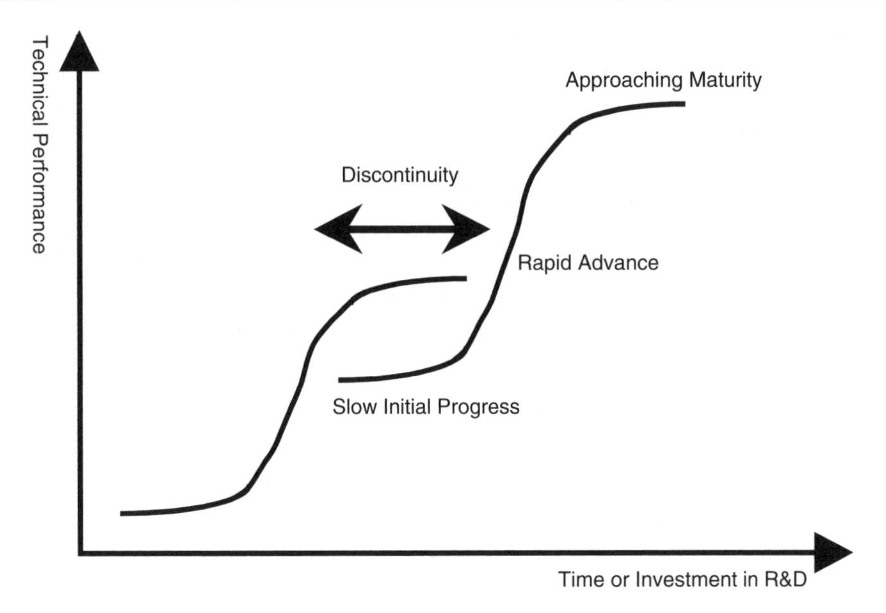

Figure 13.1 The Technology S-curve

time flattens off, to complete the S-curve. Figure 13.1 [1] illustrates (in a somewhat simplified textbook manner) the discontinuity that arises as a disruptive new technology emerges and the way that one S-curve takes over from that of its predecessor.

The concept of the S-curve has been applied at many different levels, from fundamental enabling technologies such as transistors and vacuum tubes, to application technologies such as cars and trains. At all levels, displacement is usually incomplete – there will always be some applications or markets for which the new technology is less well suited than the established one. The S-curve and its application to wireless technology is discussed by John Ralston, past steering committee chair of the SDR Forum, in an earlier volume in this series [2].

One of the key challenges when seeking to apply the concept of the S-curve, is the measurement and assessment of the relative position of the new and existing technologies on their respective, overlapping, curves. The major risk for the start-up company or for the investor lies in over estimating the maturity of an early technology, in which case the venture funding can run out before the product can be fully developed and become established in the market. Conversely, the major risk for the incumbent player with an established technology, is to underestimate the pace of advance of the new, and the resultant threat to the company's existing revenues.

Within the field of SDR baseband technologies, cases of the former are evident. We are still, arguably, at the stage where it is too soon to assess whether or not there will be significant instances of the latter. The presence in this book of contributions from major incumbent suppliers of semiconductor baseband products clearly shows that such companies have recognized the threats and are developing responses – generally as extensions of their existing solutions, rather than as radical innovations. Such an approach is clearly most appropriate for a strong incumbent, since it both builds upon their existing market presence and reinforces the current product base.

Of course, as described, the technology S-curve can be a somewhat simplistic model. Indeed, deciding whether an innovation should be considered as a new technology or as representing simply a technology extension, creating an inflexion in the curve and then a new period of rapid progress, can be and often is the subject of debate within companies when considering the merits of where to target new technology investment. Some would argue that the innovative SDR processor architectures represent new technology S-curves that will displace DSPs, whereas others would argue that they are enablers of a new period of rapid growth to existing DSP technology S-curves.

Whichever philosophical view is adopted, what is clear is that we are poised for a period of rapid advance in performance and capability of flexible baseband processing. Whilst the 3G wireless market has served as a key stimulus to this, as GSM did for DSP in the early 1990s, the impact of some of these advances will extend beyond wireless.

13.1.3. Technology Drivers and Trends

A common set of drivers and trends is evident from a reading of the earlier chapters. Most of the contributors have identified these, although with their emphasis varying, depending upon their technological heritage, industry position and their application focus (handset or basestation).

- *Processing requirements*: The acceleration of wireless-system complexity in the transition to 3G is very substantial and cannot be accommodated by simply relying on Moore's law. Recognition of this has been a key driver of innovation.
- *DSP-FPGA overlap*: One response to this need, from today's incumbent players, the DSP and FPGA suppliers, is that they are learning lessons and adopting approaches from each other. Thus, new DSP-based processors are increasingly incorporating co-processors, together with customized FPGA-like structures, designed to provide greater flexibility and step-increases in processing power. Likewise, DSP-like processor blocks are beginning to appear within the devices offered by the traditional FPGA suppliers.
- *New processor architectures*: If Moore's law is insufficient, an alternative approach must be sought to deliver the processing power required. Consequently, a range of innovative architectures is emerging, reflecting a departure from the longstanding Von Neumann/ Harvard DSP architectures. Foremost amongst these are parallel processing architectures, at a range of different scales; such solutions exploit the inherent parallelism of algorithms, and promise an alternative to the converged DSP/FPGA architecture.
- *New memory architectures*: High bandwidth memory architectures are increasingly being deployed, utilising large on- and off-chip memory and associated high speed buses, in all approaches, both the new architectures and in modern DSP/FPGA devices.
- *Processor functionality extension*: Processors are increasingly supporting the execution of control-oriented tasks, applications, and providing operating system support, as well as the signal processing functionality more traditionally associated with DSP devices.
- *Development flow*: The availability of an effective, efficient and easy to use development flow and programming toolsets will be a key determinant of marketplace success. The customer's prior investment in employee skills is substantial, and totally novel design flows would add unacceptably to product development timescales at a time when time-to-market is becoming more, not less, critical.
- *Soft silicon*: The drive to ever smaller silicon geometries is already changing the economics of ASICs. The costs of designing complex custom ASICs threaten to become prohibitive

unless economies of scale can address massive markets. This is a powerful argument for the creation of 'soft silicon' – reprogrammable, reconfigurable devices suited for multiple applications and markets – with the low power consumption and high performance normally associated with ASICs. The rewards for such technologies are perhaps the highest, as also are the challenges.

An implicit requirement for all the new and emerging semiconductor products is that the cost of the technology should represent a reduction, not an increase, as compared with the existing solutions. Whilst, in the early days of SDR, some premium may have been anticipated in return for the end-product flexibility offered by the technology, it is clear that the window for this premium is rapidly closing.

This does not mean that SDR baseband technology solutions will not mature. Quite the reverse – as can be seen from the earlier chapters, many technology solutions have now demonstrated functionality. The next stage will be to see which ones can achieve cost-effectiveness and secure viable market share, against both incumbents and other new entrants. Consider also that those VC-backed SDR baseband start-ups which have failed in the recent past, have subsequently been acquired by major semiconductor players – this has happened for good reason. Thus even those companies that fail to retain their independence may still be expected to find their technology influencing mainstream development in the coming years.

13.1.4. A Technology Taxonomy

Figure 13.2 offers one possible taxonomy of emerging SDR baseband technologies, together with a tentative and non-comprehensive mapping onto this taxonomy of the SDR processor solutions described this book, plus a few others.

This taxonomy cannot and should not be considered as a definitive classification, since in practice the boundaries between the categories are already blurred and continue to change, as technology developments in one area are adapted and incorporated into another. Further, this taxonomy does not differentiate between applications – infrastructure and terminals – and, as we have seen in earlier chapters, the requirements and proposed solutions for these two are quite different.

This book has focussed essentially on baseband semiconductor developments. For completeness however, and in view of its potential significance, an additional class is included within the taxonomy corresponding to the hybrid general purpose computing server running air interface software, with supporting radio I/O interface hardware, an approach espoused by Intel and Vanu. The I/O bandwidth on Gigabit Ethernet appears adequate to support basestation datastreams, new 64 bit CPUs would appear to offer adequate processing power and memory bandwidths are increasing – ie the technology limitations of this approach are receding. Further information on this approach may be found in [3].

The purpose of such a taxonomy is not to predict winners, but rather to classify, compare and contrast the various technologies. Indeed, no single winner will emerge. Rather, different solutions will be better suited to different applications – just as we see with handsets and basestations today – and arguably with room for multiple solutions. If, as we postulate, manufacturers increasingly learn and adapt from each other, so we may perhaps expect to see common architectural themes finding acceptance and incorporation in dominant designs, just as, over the past decade, has happened with conventional DSPs.

FPGAs	Altera Xilinx – adding DSP elements –
Intermediate Granularity Processor Chips	Elixent Mathstar Quicksilver Silicon Hive
Multiprocessor Chips	PACT picoChip Morpho
Hybrid Approach: DSP plus Co-/Multi-processor	Xilinx+PowerPC Infineon (In-house DSP+acquired MorphICs technology) Intrinsity LSI Logic Motorola (Starcore+Morpho)
DSP	ADI (optimised instruction set, plus 'smarts') Texas Instruments (hard macros) Sandbridge (optimised architecture) – adding FPGA elements –
Hybrid Approach: General Purpose Servers plus Interface Hardware	Intel/Vanu

Figure 13.2 A Possible Technology Taxonomy
(Technologies described in the book, and others)

13.2. Industry Impact

Having summarized above the key technology trends, we now extend our discussion to consider the potential impact of the new technology on wireless products and the value chain.

13.2.1. Multi-Standard Phones

Traditional ASIC/DSP processor approaches in terminals may be expected to give way increasingly to advanced, combination technology, i.e. architectures that support high processing power alongside low power consumption. Just as today most GSM phones are dual – or even tri-band – a complete shift from single-band in just 5 years – so it is possible to anticipate that by 2010 most phones could potentially embody a baseband platform capable of multiple standard support. Such a shift would most likely be driven not so much by the need for international roaming, but rather by the desire to improve economies of scale through the use of standard product designs for multiple geographical markets. The pace of such a shift could, however, be constrained by regulatory factors [4–6].

13.2.2. On-the-fly Terminal Reconfiguration

The market need for 'on-the-fly' reconfiguration – the ability of a phone to reconfigure dynamically in the field, or even whilst in-call – is perhaps much more open to question, at least from today's perspective. Should technological solutions emerge that offer such a capability with minimal cost penalty, then it is conceivable that these could find wide acceptance, since they could offer all the advantages of other approaches, yet with a potential for additional functionality.

Typical wireless applications today (eg for WAP, i-mode) are tailored to the air interface and/or impose specific infrastructure requirements (eg custom gateways). Recent research has demonstrated that middleware can be used to abstract the wireless terminal, enabling applications to be written independently of these two constraints [7]. The capability of communication between the lower protocol layers of network and terminal using, potentially downloadable, middleware promises increased flexibility, opening the market to third party application writers. Such developments could accelerate industry growth, just as a similar but more limited approach fuelled i-mode growth in Japan.

If 'on-the-fly' reconfiguration technology does find commercial success, this capability for dynamic reconfiguration may be expected also to drive the emergence of new applications beyond the wireless domain.

13.2.3. Basestations

The requirement for flexible basestation capability and the ability to upgrade functionality cost effectively, may be expected to continue to drive the development of basestation processing architectures over the next few years. The interaction between technology suppliers and the basestation manufacturers will be interesting to observe, as it could take different directions with different possible outcomes.

Some of the new technology start-ups are already well down the road of building relationships with 3G basestation manufacturers. Initially, such manufacturers have simply been evaluating their new products, but the picture could change significantly as first design-ins are secured. If such design-ins lead to successful basestation products with unique advantages, it is conceivable that some basestation manufacturers will wish to secure exclusivity of the source technology. Whether or not this would be possible without acquisition is, however, questionable – and arguably the costs involved in semiconductor design and manufacturing today make this look increasingly unlikely.

Acceptance of the new technologies by basestation manufacturers will of course increase the attractiveness of such companies to the established semiconductor suppliers, who might wish to acquire them as a means of consolidating their customer base and extending their technology capabilities. The acquisition in 2003 of two early pioneers which ran into difficulties – MorphICs by Infineon and BOPS by Altera [8] – demonstrates that the established incumbents are alert to such opportunities. The current industry trend is towards open architecture basestation products, as exemplified by the formation of the Open Base Station Architecture Initiative, OBSAI, in October 2002 [9]. Given this trend, and existing industry relationships, acquisition of start-ups by semiconductor players rather than by basestation manufacturers, would seem to be a more likely outcome. For the SDR baseband start-ups who can successfully demonstrate both functionality and cost effectiveness, the future is promising. However, they must do this in good time, whilst they have sufficient cash flow, and are ahead of their competitors, if they are to avoid the fate of MorphICs and BOPS.

The development of true programmable SDR processors suggests, longer term, the possible emergence of a basestation 'motherboard' industry, akin to the PC industry model. This would, arguably, reflect the present trends towards reduced cellular infrastructure costs and open architecture basestations, potentially commoditizing the basestation in the same way as happened with the computer. The emergence of a standard PC architecture and the resultant availability of cheap motherboards radically changed the shape of the computing industry, albeit in the end with enlarged markets. The potential impact of ready availability of low-cost air-interface-programmable basestation 'motherboards' is not easy to assess. Indeed, manufacturers have very different views on the desirability of such a development and their actions will, of course, influence the shape of any industry evolution. Established players might seek to maintain the basestation as a high-value product, whilst others might see competitive advantage in supplying commoditized low-cost basestation products, to change the ground rules of competition. Arguably, this latter development, if it materializes, could serve to accelerate the development and deployment of higher rate, short-range, picocellular systems, capable of 'intelligent relaying' and other possibilities. Two things are evident: if the PC-motherboard model is to happen, it will not happen overnight – if it does occur, however, it will have a profound and wide-ranging impact.

13.2.4. The Wider Electronics Industry

The inexorable drive to smaller geometry silicon (60 nm and less) would appear to be such that low-power, reconfigurable ICs must eventually emerge. The economic drivers demanding this, mentioned earlier, are arguably more profound than those that are driving the requirement for software radio. Indeed 'software radio is simply a subset of the wider set of reconfigurable electronic devices' [10] – such ICs will allow new reprogrammable capabilities to be incorporated into electronic devices of all types, not simply cellphones and basestations.

Creation of new approaches to exploit the capability of such a technology remains a challenge. For some years now, system upgrades for electronic consumer goods over the Internet has been possible – e.g. set top boxes for televisions. However, the reality is that most consumers never use such capabilities but prefer to take their device to a shop for someone else to upgrade for them. Unless very simple and effective usage models can be developed, this will continue to be the case.

Nonetheless, the commercial potential of the reconfigurable consumer electronic device market is huge, accompanied as it is by the associated opportunity for new business models and revenue streams. It is to be expected that such opportunities will motivate the development of user-friendly approaches, resulting in reconfigurable baseband processing finding its way into many products beyond radio in the coming decade. Just as the development of early DSP for GSM eventually resulted in a mature and powerful technology being increasingly used in other application fields, so we may expect to see the same for SDR baseband processing developed for 3G handsets and basestations.

13.3. Concluding Remarks

The contributors to this book have provided ample evidence of the advances in thinking and technology that are presently occurring as the semiconductor industry grapples with the challenges of SDR baseband processing for 3G wireless. These advances have not come without major investment and personal commitment from pioneering individuals; it is always easy to criticize a new idea and to wait for someone else to prove it, but somebody has to argue for a new approach and convince the budget holders to invest.

Today, however, we have reached the stage where the financial investment has been committed, where many of these new technologies have made the transition from concept to functionality, and in the 2004–2005 timeframe they will face the ultimate test of the marketplace. The development of mature, second generation, reconfigurable baseband technology over the next few years holds great promise. The full impact of the advances in SDR baseband processing that we have seen to date and which are described in this book will be profound, albeit perhaps still many years in the future.

Acknowledgements

The author wishes to acknowledge valuable conversations with all of the contributors to this book that have helped develop the thinking reflected in this chapter, as well as input from Shaul Berger. Discussions on the strategic impact of SDR technology with Stephen Blust and on Taxonomy options with Rupert Baines helped shape these aspects of the chapter.

References

[1] R. Foster, *Innovation: The Attackers Advantage*, Macmillan, London, 1986.

[2] 'A market perspective: software defined radio as the dominant design', J. Ralston, in *Software Defined Radio: Origins, Drivers and International Perspectives*, W.H.W Tuttlebee (Ed.), Chapter John Wiley & Sons, Ltd, 2002.

[3] 'Software Engineering for Software Radios', John Chapin, Chapter 10 of 'Software Defined Radio: Enabling Technologies', ed WHW Tuttlebee, pub Wiley, 2002.

[4] M. Grable, 'Regulation of software defined radio – United States', in *Software Defined Radio: Origins, Drivers and International Perspectives'*, W.H.W. Tuttlebee (Ed.), chapter 11, John Wiley & Sons, Ltd, 2002.

[5] P. Bender and S. O'Fee, 'European regulation of software radio', in *Software Defined Radio: Origins, Drivers and International Perspectives*, W.H.W Tuttlebee (Ed.), Chapter 10, John Wiley & Sons, Ltd, 2002.

[6] K. Moessner, N.J. Jefferies and W.H.W Tuttlebee, 'A solution for regulatory issues with SDR', URSI Conference, August 2002.

[7] 'Middleware Design Strategies for Future Wireless Services', Report from the Software Based Systems work area of the Mobile VCE Core 2 Research Programme, N Amanquah & J Irvine, 2003.

[8] MorphICs – http://www.morphics.com; BOPS – EE Times, 4th April 2003.

[9] Open Base Station Architecture Initiative – http://www.obsai.org

[10] W.H.W. Tuttlebee, 'Reconfigurable Radio', presentation given at 6th Framework preparatory meeting to DG INFOSOC, Brussels, May 2001.

Index

Software Defined Radio: Baseband Technologies for 3G Handsets and Basestations. Edited by W. Tuttlebee
 ISBN: 0-470-86770-1